SPECIAL FUNCTIONS OF
FRACTIONAL CALCULUS

Applications to Diffusion
and Random Search Processes

SPECIAL FUNCTIONS OF
FRACTIONAL CALCULUS

Applications to Diffusion
and Random Search Processes

Trifce Sandev

Macedonian Academy of Sciences and Arts, Macedonia

Alexander Iomin

Technion-Israel Institute of Technology, Israel

W World Scientific

NEW JERSEY · LONDON · SINGAPORE · BEIJING · SHANGHAI · HONG KONG · TAIPEI · CHENNAI · TOKYO

Published by

World Scientific Publishing Co. Pte. Ltd.

5 Toh Tuck Link, Singapore 596224

USA office: 27 Warren Street, Suite 401-402, Hackensack, NJ 07601

UK office: 57 Shelton Street, Covent Garden, London WC2H 9HE

Library of Congress Control Number: 2022945045

British Library Cataloguing-in-Publication Data
A catalogue record for this book is available from the British Library.

SPECIAL FUNCTIONS OF FRACTIONAL CALCULUS
Applications to Diffusion and Random Search Processes

ISBN 978-981-125-294-5 (hardcover)
ISBN 978-981-125-295-2 (ebook for institutions)
ISBN 978-981-125-296-9 (ebook for individuals)

For any available supplementary material, please visit
https://www.worldscientific.com/worldscibooks/10.1142/12743#t=suppl

Desk Editor: Nur Syarfeena Binte Mohd Fauzi

To my wife Irina.
TS

To the memory of my father Miron-Meer Iomin,
son of Herchen and Bluma, may his memory be blessed.
AI

Preface

Мы все учились понемногу,
Чему-нибудь и как-нибудь[1]
А. С. Пушкин
"Евгений Онегин"

Mathematics is the art of giving things misleading names.
The beautiful – and at first look mysterious – name the
fractional calculus is just one of the those misnomers
which are the essence of mathematics.[2]
I. Podlubny

Investigation of stochastic processes in complex media has been attracted much attention for years. Theoretical modeling of diffusion in heterogeneous and disordered media is a considerable part of these studies. Heterogeneous and disordered materials include various materials with defects, multi-scale amorphous composites, fractal and sparse structures, weighted graphs and networks, and so on. Diffusion in such complicated media with geometric constraint and random forces is often anomalous. Further developing of the theory for understanding and description of these random processes in a variety of realizations in physics, biology, social life and finance is a part of modern studies, what we called "complex systems". These complex phenomena entailing complex descriptions by fractional kinetics and fractional calculus.

The new mathematical approach sheds light on many questions are still exist for studying and opens new ones. Such an example is a random search

---

[1] We all learned a little, Anything and somehow (A. S. Pushkin, "Eugene Onegin").
[2] Ref. [13], Chapter 1.

process whose systematic research stems from projects involving hunting for submarines during World War II, while the modern study of first-passage, or hitting times covers a large area of search problems from animal food foraging to molecular reactions and gene regulation. In many cases, an introduction of a stochastic resetting in complex systems, has significant effects on the first-passage properties. Moreover, the random search processes in complex networks are important for various reasons, such as understanding animal food search strategies and improving web search engines, or for prolonging or speeding up the survival time in first-encounter tasks.

Many of the aforementioned processes can be described by various form of random walk models, including fractional Fokker-Planck equations and generalised Langevin equations, which in their turn describe a variety of completely different problems shared common features. In particular, a class of diffusion in heterogeneous environment is closely related to turbulent diffusion represented by an inhomogeneous advection-diffusion equation, and it also relates to generalized geometric Brownian motion, used to model stock prices.

The suggested book addresses technical issues surrounding the concept of special functions of anomalous transport, or fractional kinetics, described in the framework of fractional calculus. Much attention is paid to technical details in using the Fox H-functions and the Mittag-Leffler functions as well as functions related to them in description of fractional transport in random composite media. The latter defines an extremely large class of problems in life science and physics, medical physics and technology. We attempt to present the material in a self-contained way, such that the book can be a convenient reading not only for experts in the field but also for newcomers and undergraduate students interesting in the field of fractional calculus and fractional kinetics.

Skopje, Macedonia   *Trifce Sandev,*
Haifa, Israel   *Alexander Iomin*
                 April, 2022

Acknowledgments

It is our great pleasure to thank our colleagues for collaboration, numerous helpful and stimulating discussions. Their constant interest, support and encourage cannot be overestimated. Our thanks to Emad Awad (Alexandria), Lasko Basnarkov (Skopje), Aleksei Chechkin (Potsdam & Kharkov), Weihua Deng (Lanzhou), Viktor Domazetoski (Skopje), Sergei Fedotov (Manchester), Katarzyna Górska (Kraków), Andrzej Horzela (Kraków), Holger Kantz (Dresden), Ljupco Kocarev (Skopje), Ervin Kaminski Lenzi (Ponta Grossa), Vicenç Méndez (Barcelona), Ralf Metzler (Potsdam), Alexander Milovanov (Rome), Arnab Pal (Chennai), Irina Petreska (Skopje), Haroldo Valentin Ribeiro (Maringa), R. K. Singh (Bar-Ilan), Igor M. Sokolov (Berlin), Viktor Stojkoski (Skopje), Živorad Tomovski (Ostrava).

TS sincerely thanks Academician Ljupco Kocarev, President of the Macedonian Academy of Sciences and Arts, for the given trust, great collaboration and continuous support in my research all these years, and for providing a very professional and positive working atmosphere at the Academy.

TS expresses his special thanks to Prof. Dr. Ralf Metzler for the unforgettable moments in doing research in his group at the University of Potsdam, where a part of this book has been written, and for the great collaboration and support.

TS acknowledges DFG (Deutsche Forschungsgemeinschaft) research grant (DFG, Grant number ME 1535/12-1) for the project "Diffusion and random search in heterogeneous media: theory and applications" between Macedonian Academy of Sciences and Arts and University of Potsdam.

TS acknowledges financial support by the Alexander von Humboldt (AvH) Foundation of the project "Statistical Mechanics of Diffusion Processes in Disordered Media" (2020–2022), within the AvH fellowship programme for experienced researchers at the Institute of Physics & Astronomy, University of Potsdam.

TS also acknowledges support from the bilateral Macedonian-Chinese research project 20-6333, between the Macedonian Academy of Sciences and Arts and Lanzhou University, funded under the intergovernmental Macedonian-Chinese agreement.

AI sincerely thanks Distinguished Professor Mordechai (Moti) Segev for his continuous support.

AI also acknowledges financial support by the Israel KAMEA Program.

The authors thank the Editorial Consultant S. C. Lim and the Editor Mrs. Nur Syarfeena Binte Mohd Fauzi for the collaboration and their management of the publication process of this book.

Last, but not least, we are thankful to our families for all the patience and support in our work.

Contents

Acronyms and Symbols

CTRW continuous time random walk
FATD first arrival time distribution
FPE Fokker-Planck equation
FFPE fractional Fokker-Planck equation
GBM geometric Brownian motion
M-L Mittag-Leffler
MSD mean squared displacement
PDF probability density function
R-L Riemann-Liouville
lhs left-hand side
rhs right-hand side
$\mathcal{T}$ integral transform
$\mathcal{T}^{-1}$ inverse integral transform
$\mathcal{F}$ Fourier transform
$\mathcal{F}^{-1}$ inverse Fourier transform
$\mathcal{L}$ Laplace transform
$\mathcal{L}^{-1}$ inverse Laplace transform
$\mathcal{M}$ Mellin transform
$\mathcal{M}^{-1}$ inverse Mellin transform
$\mathcal{D}$ diffusion coefficient
$\Gamma(z)$ gamma function
$B(a, b)$ beta function
$\delta(z)$ Dirac delta function
$\Theta(z)$ Heaviside step function
$\Gamma(\nu, z)$ upper incomplete gamma function
$\gamma(\nu, z)$ lower incomplete gamma function
$(\gamma)_k$ Pochhammer symbol

$\mathrm{erf}(z)$ error function

$\mathrm{erfc}(z)$ complementary error function

$\bar{E}_n(z)$ generalized exponential integral function

$E_\alpha(z)$ one parameter Mittag-Leffler function

$E_{\alpha,\beta}(z)$ two parameter Mittag-Leffler function

$E_{\alpha,\beta}^\gamma(z)$ three parameter Mittag-Leffler function

$\mathcal{E}_\alpha(t;\lambda)$ associated one parameter Mittag-Leffler function

$\mathcal{E}_{\alpha,\beta}(t;\lambda)$ associated two parameter Mittag-Leffler function

$\mathcal{E}_{\alpha,\beta}^\gamma(t;\lambda)$ associated three parameter Mittag-Leffler function

$H_{p,q}^{m,n}(z)$ Fox H-function

I_{a+}^μ Riemann-Liouville fractional integral

$_{\mathrm{RL}}D_{a+}^\mu, D_{a+}^\mu$ Riemann-Liouville fractional derivative

$_{\mathrm{TRL}}D_{a+}^\mu$ tempered Riemann-Liouville fractional derivative

$_{\mathrm{C}}D_{a+}^\mu$ Caputo fractional derivative

$_{\mathrm{TC}}D_{a+}^\mu$ tempered Caputo fractional derivative

$_{\mathrm{W}}D_{-\infty}^\mu$ Weyl fractional derivative

$_{\mathrm{RF}}D_\theta^\mu$ asymmetric Riesz-Feller derivative

$_{\mathrm{RF}}D_0^\mu$ symmetric Riesz-Feller derivative

$\frac{\partial^\alpha}{\partial|x|^\alpha}, \partial_{|x|}^\alpha$ Riesz fractional derivative

$\mathbf{E}_{\rho,\mu,\omega,a+}^\gamma$ Prabhakar integral

$\mathbf{D}_{\rho,\mu,\omega,a+}^\gamma, {}_{\mathrm{RL}}\mathcal{D}_{\rho,\omega,a+}^{\gamma,\mu}$ Prabhakar derivative

$_{\mathrm{C}}\mathbf{D}_{\rho,\mu,\omega,a+}^\gamma, {}_{\mathrm{C}}\mathcal{D}_{\rho,\omega,a+}^{\gamma,\mu}$ regularized Prabhakar derivative

$_{\mathrm{TRL}}\mathcal{D}_{\rho,\omega,a+}^{\gamma,\mu}$ tempered Prabhakar derivative

$_{\mathrm{TC}}\mathcal{D}_{\rho,\omega,a+}^{\gamma,\mu}$ tempered regularized Prabhakar derivative

$_{\mathrm{RL}}\mathbf{G}_{\eta,t}$ generalized derivative operator in the Riemann-Liouville form

$_{\mathrm{C}}\mathbf{G}_{\gamma,t}$ generalized derivative operator in the Caputo form

$\mathcal{E}_{a+;\alpha,\beta}^{\omega;\gamma,\kappa}$ generalized integral operator

$L_\alpha(z)$ one-sided Lévy stable probability density

$\varphi(\alpha,\beta;z)$ Wright function

$_p\Psi_q(z)$ Fox-Wright function

$\langle\cdot\rangle$ ensemble average

$\lceil z\rceil$ ceiling function

$\lfloor z\rfloor$ floor function

$\Upsilon(x,t)$ jump probability density function

$\psi(t)$ waiting time probability density function

$w(x)$ jump length probability density function

$\wp_{\mathrm{fa}}(t)$ first arrival time distribution

$\mathcal{P}$ search reliability

$\mathcal{E}$ search efficiency

$B(t)$ standard Brownian motion

$h(u,t)$ subordination function

$\Re(z)$ real part of a complex number z

$\Im(z)$ imaginary part of a complex number z

$\imath = \sqrt{-1}$ imaginary unit

$\square$ Q.E.D. – *quod erat demonstrandum* – thus it has been demonstrated

$R = \mathbb{R}$ real values

$R_+ = \mathbb{R}_+$ non-negative real values

$C = \mathbb{C}$ complex values

$Z = \mathbb{Z}$ integer numbers

Chapter 1

Mathematical background

1.1 Integral transforms

Many linear differential and integral equations can be easily solved by employing integral transforms. This procedure maps a problem in task into the form of algebraic equations, which can be easily solved. A generic form of the integral transform reads

$$\mathcal{T}[f(z)](\xi) = \int_a^b K(\xi, z)\, f(z)\, dz = F(\xi), \tag{1.1}$$

where $K(\xi, z)$ is the kernel function, and the limits of the integration, (a, b), are determined by the problem in task. The kernel function establishes a relation between function $f(z)$, defined in z space, and its image $F(\xi)$ in its ξ space, and together with limits of the integration these determine the type of the transform. The inverse integral transformation reads

$$\mathcal{T}^{-1}[F(\xi)](z) = \int_u^v K^{-1}(z, \xi)\, F(\xi)\, d\xi, \tag{1.2}$$

where $K^{-1}(\xi, z)$ is the inverse kernel function, and u and v are the limits of integration. An appropriate form of the integral transform is determined by equations in task and corresponding boundary and initial conditions.

The basic property of the integral transforms is their linearity,

$$\mathcal{T}[c_1\, f_1(z) + c_2\, f_2(z)](\xi) = \int_a^b K(\xi, z)\, [c_1\, f_1(z) + c_2\, f_2(z)]\, dz$$

$$= c_1 \int_a^b K(\xi, z)\, f_1(z)\, dz + c_2 \int_a^b K(\xi, z)\, c_2 f_2(z)\, dz$$

$$= c_1\, F_1(\xi) + c_2\, F_2(\xi), \tag{1.3}$$

where $c_{1,2}$ are given constants.

In this book we will employ three forms of the integral transform: Fourier, Laplace and Mellin transforms, which are determined by the corresponding kernels of the transformation and the boundaries, as shown in Table 1.1.

Table 1.1   Table of integral transforms.

Transform	$K(\xi, z)$	a	b	$K^{-1}(z, \xi)$	u	v
$\mathcal{F}$	$e^{-\imath \xi z}$	$-\infty$	∞	$\frac{1}{2\pi} e^{\imath t \xi}$	$-\infty$	∞
$\mathcal{L}$	$e^{-\xi z}$	0	∞	$\frac{1}{2\pi \imath} e^{\xi z}$	$c - \imath \infty$	$c + \imath \infty$
$\mathcal{M}$	$z^{\xi - 1}$	0	∞	$\frac{1}{2\pi \imath} z^{-\xi}$	$c - \imath \infty$	$c + \imath \infty$

1.1.1   *Fourier transform*

As shown in Table 1.1, the kernel of the Fourier transform is exponential $K(k, x) = e^{-\imath kx}$, and taking the limits at infinity, $a = -\infty$ and $b = \infty$, we have

$$\mathcal{F}[f(x)](k) = F(k) = \int_{-\infty}^{\infty} f(z) \, e^{-\imath kx} \, dx, \qquad (1.4)$$

while the inverse Fourier transform reads

$$\mathcal{F}^{-1}[F(k)](x) = f(x) = \frac{1}{2\pi} \int_{-\infty}^{\infty} F(k) \, e^{\imath kx} \, dk. \qquad (1.5)$$

For applications, we also use "tilde" sign for the Fourier image, $\tilde{f}(k) \equiv F(k)$.

In addition to the basic property (1.3), the Fourier transform has the following properties:

(1) Translation:

$$\boxed{\mathcal{F}[f(x - x_0)] = e^{-\imath k x_0} F(k)} \qquad (1.6)$$

This can be easily shown from the definition of the Fourier transform,

$$\mathcal{F}[f(x-x_0)] = \int_{-\infty}^{\infty} f(x-x_0)e^{-\imath kx}\, dx \underset{(z=x-x_0)}{=} \int_{-\infty}^{\infty} f(z)e^{-\imath k(z+x_0)}\, dz$$

$$= e^{-\imath kx_0} \int_{-\infty}^{\infty} f(z)e^{-\imath kz}\, dz = e^{-\imath kx_0}\mathcal{F}[f(x)] = e^{-\imath kx_0}F(k).$$

(2) Modulation:

$$\boxed{\mathcal{F}\left[e^{\imath xk_0}f(x)\right] = F(k-k_0)} \tag{1.7}$$

Namely,

$$\mathcal{F}\left[e^{\imath xk_0}f(x)\right] = \int_{-\infty}^{\infty} e^{-\imath(k-k_0)x}f(x) = F(k-k_0).$$

(3) Scaling $(a \neq 0)$:

$$\boxed{\mathcal{F}[f(ax)] = \frac{1}{|a|}F(k/a)} \tag{1.8}$$

It directly follows from the definition of the Fourier transform,

$$\mathcal{F}[f(ax)] = \int_{-\infty}^{\infty} e^{-\imath kx}f(ax)\, dx$$

$$\underset{(z=ax)}{=} \frac{1}{|a|}\int_{-\infty}^{\infty} e^{-\imath(k/a)z}f(z)\, dz = \frac{1}{|a|}F(k/a).$$

(4) Convolution theorem:

$$\boxed{\mathcal{F}[f \star g] = \mathcal{F}\left[\int_{-\infty}^{\infty} f(\zeta)g(x-\zeta)\, d\zeta\right] = F(k)G(k)} \tag{1.9}$$

This property results from the change of the order of integration,

$$\mathcal{F}\left[\int_{-\infty}^{\infty} f(\zeta)g(x-\zeta)\, d\zeta\right] = \int_{-\infty}^{\infty} e^{-\imath kx}\left[\int_{-\infty}^{\infty} f(\zeta)g(x-\zeta)\, d\zeta\right] dx$$

$$= \int_{-\infty}^{\infty} f(\zeta)\left[\int_{-\infty}^{\infty} g(x-\zeta)e^{-\imath k\zeta}\, dx\right] d\zeta$$

$$= \int_{-\infty}^{\infty} e^{-\imath k\zeta}f(\zeta)\left[\int_{-\infty}^{\infty} g(x-\zeta)e^{-\imath k(x-\zeta)}\, dx\right] d\zeta$$

$$\underset{(x-\zeta=z)}{=} \int_{-\infty}^{\infty} e^{-\imath k\zeta}f(\zeta)\, d\zeta \int_{-\infty}^{\infty} e^{-\imath kz}g(z)\, dz$$

$$= \mathcal{F}[f(x)] \cdot \mathcal{F}[g(x)] = F(k) \cdot G(k).$$

Here we note that $\int_{-\infty}^{\infty} f(\zeta)\, g(x-\zeta)\, d\zeta = \int_{-\infty}^{\infty} g(\zeta)\, f(x-\zeta)\, d\zeta$, i.e., $f \star g = g \star f$.

(5) Transform of derivative of $f(x)$:

$$\boxed{\mathcal{F}\left[\frac{d^n}{dx^n}f(x)\right] = (\imath k)^n F(k)}$$

(1.10)

This can be shown by mathematical induction knowing that the Fourier transform of the first derivative is

$$\mathcal{F}\left[\frac{d}{dx}f(x)\right] = \int_{-\infty}^{\infty} e^{-\imath kx}\frac{df(x)}{dx}\,dx$$

$$= f(x)e^{-\imath kx}\Big|_{-\infty}^{\infty} + (\imath k)\int_{-\infty}^{\infty} e^{-\imath kx}f(x)\,dx$$

$$= (\imath k)\,F(k),$$

(1.11)

where we use integration by parts, and of the second derivative is

$$\mathcal{F}\left[\frac{d^2}{dx^2}f(x)\right] = \int_{-\infty}^{\infty} e^{-\imath kx}\frac{d^2 f(x)}{dx^2}\,dx$$

$$= \frac{df(x)}{dx}e^{-\imath kx}\Big|_{-\infty}^{\infty} + (\imath k)\int_{-\infty}^{\infty} e^{-\imath kx}\frac{df(x)}{dx}\,dx$$

$$= (\imath k)^2\,F(k) = -k^2\,F(k).$$

(1.12)

From the definition (1.4) follows the Fourier transform of the Dirac δ-function

$$\mathcal{F}\left[\delta(x)\right] = \int_{-\infty}^{\infty}\delta(x)\,e^{-\imath kx}\,dx = 1,$$

(1.13)

and exponential function

$$\mathcal{F}\left[e^{-a|x|}\right] = \int_{-\infty}^{\infty} e^{-(\imath kx+a|x|)}\,dx = \frac{2a}{a^2+k^2}, \quad \Re(a) > 0,$$

(1.14)

which means that the Fourier transform of a two-sided decaying exponential function is a Cauchy (or Lorentzian) function.

Remark 1.1. From the definition of the Fourier transform (1.4), it follows directly that

$$\boxed{\int_{-\infty}^{\infty} f(z)\,dz = \lim_{k\to 0}\int_{-\infty}^{\infty} f(z)\,e^{-\imath kx}\,dx = F(0)}$$

(1.15)

The Fourier transform and the inverse Fourier transform are implemented in Wolfram Language as `FourierTransform[f, x, k]` and `InverseFourierTransform[F, k, x]`, respectively. Here we note that the coefficients in front of the integrals in Wolfram Mathematica are symmetrical and taken to be $\frac{1}{\sqrt{2\pi}}$ in contrast to the integrals (1.4) and (1.5), where the coefficients are taken to be 1 and $\frac{1}{2\pi}$, respectively. In both cases the product of the coefficients in front of the integrals in the definition of the Fourier and the inverse Fourier transforms should be equal to $\frac{1}{2\pi}$.

1.1.2 Laplace transform

The Laplace transform is described by the kernel function $K(s,t) = e^{-st}$, and the limits $a = 0$ and $b = \infty$, see Table 1.1, i.e.,

$$\mathcal{L}[f(t)](s) = F(s) \equiv \hat{f}(s) = \int_0^\infty f(t)\, e^{-st}\, dt. \tag{1.16}$$

When dealing with the Laplace transform, we suppose that the function $f(t)$ equals zero for $t < 0$, i.e., the function is always multiplied by the Heaviside function $\Theta(t)$, which is equal to zero for $t < 0$. The inverse Laplace transform reads

$$\mathcal{L}^{-1}[F(s)](t) = f(t) = \frac{1}{2\pi i} \int_{c-i\infty}^{c+i\infty} F(s)\, e^{st}\, ds. \tag{1.17}$$

Besides the linearity basic property, the Laplace transform has also the following properties:

(1) Time shifting:

$$\boxed{\mathcal{L}[f(t-t_0)] = e^{-st_0} F(s)} \tag{1.18}$$

Here, we again use that the function $f(t-t_0) = 0$ for $t < t_0$, i.e., the function $f(t-t_0)$ is multiplied by the Heaviside function $\Theta(t-t_0)$. It can be presented as follows

$$\mathcal{L}[f(t-t_0)] = \int_0^\infty f(t-t_0)e^{-st}\, dt = \int_{t_0}^\infty f(t-t_0)e^{-st}\, dt$$

$$\underset{z=t-t_0}{=} e^{-st_0} \int_0^\infty f(z)e^{-sz}\, dz = e^{-st_0} \mathcal{L}[f(t)] = e^{-st_0} F(s).$$

A different result is obtained if one takes the Laplace transform of $f(t+t_0)$ since

$$\mathcal{L}[f(t+t_0)] = \int_0^\infty f(t+t_0)e^{-st}\, dt = e^{st_0} \int_0^\infty f(t+t_0)e^{-s(t+t_0)}\, dt$$

$$\underset{z=t+t_0}{=} e^{st_0} \int_{t_0}^\infty f(z)e^{-sz}\, dz$$

$$= e^{st_0} \left(\int_0^\infty f(z)e^{-sz}\, dz - \int_0^{t_0} f(z)e^{-sz}\, dz \right)$$

$$= e^{st_0} \left(\mathcal{L}[f(t)] - \int_0^{t_0} f(z)e^{-sz}\, dz \right).$$

Therefore, the following property holds true:

$$\mathcal{L}[f(t+t_0)] = e^{st_0}\left(\mathcal{L}[f(t)] - \int_0^{t_0} f(z)e^{-sz}\,dz\right) \tag{1.19}$$

(2) Frequency shifting:

$$\mathcal{L}\left[e^{s_0 t} f(t)\right] = F(s - s_0) \tag{1.20}$$

This property immediately follows from the integration

$$\mathcal{L}\left[e^{s_0 t} f(t)\right] = \int_0^\infty e^{-(s-s_0)t} f(t)\,dt = F(s - s_0).$$

(3) Time scaling $(a > 0)$:

$$\mathcal{L}[f(at)] = \frac{1}{a}F(s/a) \tag{1.21}$$

From the definition of the Laplace transform one finds

$$\mathcal{L}[f(at)] = \int_0^\infty e^{-st} f(at)\,dt$$

$$\underset{(z=ax)}{=} \frac{1}{a}\int_0^\infty e^{-(s/a)z} f(z)\,dz = \frac{1}{a}F(s/a).$$

(4) Transform of convolution:

$$\mathcal{L}[f \star g] = \mathcal{L}\left[\int_0^t f(\tau)g(t-\tau)\,d\tau\right] = F(s)G(s) \tag{1.22}$$

This property follows from Dirichlet's formula

$$\int_a^b dx \int_a^x f(x,y)\,dy = \int_a^b dy \int_y^b f(x,y)\,dx, \tag{1.23}$$

which yields

$$\mathcal{L}\left[\int_0^t f(\tau)g(t-\tau)\,d\tau\right] = \int_0^\infty e^{-st}\left[\int_0^t f(\tau)g(t-\tau)\,d\tau\right]dt$$

$$= \int_0^\infty f(\tau)\left[\int_\tau^\infty g(t-\tau)e^{-st}\,dt\right]d\tau$$

$$= \int_0^\infty e^{-s\tau} f(\tau)\left[\int_\tau^\infty g(t-\tau)e^{-s(t-\tau)}\,dt\right]d\tau$$

$$\underset{(t-\tau=z)}{=} \int_0^\infty e^{-s\tau} f(\tau)\,d\tau \int_0^\infty e^{-sz}g(z)\,dz$$

$$= \mathcal{L}[f(t)] \cdot \mathcal{L}[g(t)] = F(s) \cdot G(s).$$

We also note that $\int_0^t f(\tau)\,g(t-\tau)\,d\tau = \int_0^t g(\tau)\,f(t-\tau)\,d\tau$, i.e., $f \star g = g \star f$.

(5) Transform of derivative of $f(t)$:

$$\mathcal{L}\left[\frac{d^n}{dt^n}f(t)\right] = s^n F(s) - \sum_{k=1}^{n} s^{n-k} f^{(k-1)}(0) \tag{1.24}$$

From the Laplace transform of the first derivative

$$\mathcal{L}\left[\frac{d}{dt}f(t)\right] = \int_0^\infty e^{-st} \frac{df(t)}{dt}\, dt$$

$$= f(t)\, e^{-st}\Big|_0^\infty + s\int_0^\infty e^{-st} f(t)\, dt$$

$$= sF(s) - f(0), \tag{1.25}$$

and the second derivative

$$\mathcal{L}\left[\frac{d^2}{dt^2}f(t)\right] = \int_0^\infty e^{-st} \frac{d^2 f(t)}{dt^2}\, dt$$

$$= \frac{df(t)}{dt}\, e^{-st}\Big|_0^\infty + s\int_0^\infty e^{-st} \frac{df(t)}{dt}\, dt$$

$$= s^2 F(s) - s f(0) - f'(0), \tag{1.26}$$

by mathematical induction one obtains Eq. (1.24).

(6) Transformation of integral:

$$\mathcal{L}\left[\int_0^t f(\tau)\, d\tau\right] = \frac{1}{s} F(s) \tag{1.27}$$

This property follows from Dirichlet's formula, Eq. (1.23), which yields

$$\mathcal{L}\left[\int_0^t f(\tau)\, d\tau\right] = \int_0^\infty e^{-st} \left(\int_0^t f(\tau)\, d\tau\right) dt$$

$$= \int_0^\infty \left(f(\tau) \int_\tau^\infty e^{-st}\, dt\right) d\tau$$

$$= \frac{1}{s}\int_0^\infty e^{-s\tau} f(\tau)\, d\tau = \frac{1}{s} F(s). \tag{1.28}$$

(7) Integration in frequency domain:

$$\int_s^\infty F(\zeta)\, d\zeta = \mathcal{L}\left[\frac{f(t)}{t}\right] \tag{1.29}$$

This property can also be shown by exchanging the order of integration,

$$\int_s^\infty F(\zeta)\, d\zeta = \int_s^\infty \left(\int_0^\infty e^{-\zeta t} f(t)\, dt\right) d\zeta = \mathcal{L}\left[\frac{f(t)}{t}\right]. \tag{1.30}$$

(8) Transformation of periodic function $f(t + T) = f(t)$, $T > 0$:

$$\mathcal{L}\left[f(t)\right] = \frac{1}{1 - e^{-sT}} \int_0^T e^{-st} f(t)\, dt \qquad (1.31)$$

From the straightforward calculations, we have

$$
\begin{aligned}
\mathcal{L}\left[f(t)\right] &= \int_0^\infty e^{-st} f(t)\, dt = \int_0^T e^{-st} f(t)\, dt + \int_T^\infty e^{-st} f(t)\, dt \\
&= \int_0^T e^{-st} f(t)\, dt + \int_0^\infty e^{-s(\tau+T)} f(\tau + T)\, d\tau \\
&= \int_0^T e^{-st} f(t)\, dt + e^{-sT} \int_0^\infty e^{-s\tau} f(\tau)\, d\tau \\
&= \int_0^T e^{-st} f(t)\, dt + e^{-sT} \mathcal{L}\left[f(t)\right], \qquad (1.32)
\end{aligned}
$$

which yields Eq. (1.31).

One can easily find the following formulas for the Laplace transform of a Dirac δ-function, a power function and an exponential function, respectively,

$$\mathcal{L}\left[\delta(t)\right] = \int_0^\infty \delta(t)\, e^{-st}\, dt = 1, \qquad (1.33)$$

$$\mathcal{L}\left[t^\nu\right] = \int_0^\infty t^\nu e^{-st}\, dt = \Gamma(\nu + 1)\, s^{-\nu-1}, \quad \Re(\nu) > -1, \qquad (1.34)$$

$$\mathcal{L}\left[e^{-at}\right] = \int_0^\infty e^{-(s+a)t}\, dt = \frac{1}{s + a}, \quad \Re(s + a) > 0, \qquad (1.35)$$

where $\Gamma(z)$ is a gamma function (see Sec. 1.3.1).

Remark 1.2 (Limit results). There are useful limit relations between the function $f(t)$ and its Laplace transform pair $F(s)$. For example, from the result related to the Laplace transform of the first derivative (1.25) one finds

$$sF(s) = f(0) + \mathcal{L}\left[\frac{d}{dt} f(t)\right] = f(0) + \int_0^\infty e^{-st} \frac{df(t)}{dt}\, dt. \qquad (1.36)$$

Assuming that $f(t)$ is an analytic function (infinitely differentiable) and integrating by parts ($u = \frac{df(t)}{dt}$, $dv = e^{-st} dt$), one obtains

$$sF(s) = f(0) + \frac{1}{s} f'(0) + \frac{1}{s^2} f''(0) + \ldots . \qquad (1.37)$$

Then using the limit $s \to \infty$ one obtains the following useful relation

$$\lim_{s \to \infty} sF(s) = f(0+) = \lim_{t \to 0} f(t) \tag{1.38}$$

which is known as the initial value theorem. Continuing the procedure, one obtains

$$\lim_{s \to \infty} \left[s^2 F(s) - sf(0) \right] = f'(0). \tag{1.39}$$

In contrary, in the limit $s \to 0$, Eq. (1.36) yields

$$\lim_{s \to 0} sF(s) = f(0) + \int_0^\infty \frac{df(t)}{dt} dt = f(0) + f(\infty) - f(0). \tag{1.40}$$

Therefore,

$$\lim_{s \to 0} sF(s) = f(\infty) = \lim_{t \to \infty} f(t) \tag{1.41}$$

which is known as the final value theorem. It is valid if all poles of $sF(s)$ are in the left half-plane.

From the definition of the Laplace transform, one can find the following useful relation

$$\int_0^\infty f(t) \, dt = \lim_{s \to 0} \int_0^\infty e^{-st} f(t) \, dt = F(0) \tag{1.42}$$

The Laplace transform and the inverse Laplace transform are implemented in Wolfram Language as `LaplaceTransform[f, t, s]` and `InverseLaplaceTransform[F, s, t]`, respectively.

Example 1.1. Show that

$$\mathcal{L}\left[\frac{1}{\sqrt{t}} e^{-\frac{a^2}{4t}} \right] = \sqrt{\pi} s^{-1/2} e^{-|a|\sqrt{s}}. \tag{1.43}$$

From the definition of the Laplace transform, one has

$$\mathcal{L}\left[\frac{1}{\sqrt{t}} e^{-\frac{a^2}{4t}} \right] = \int_0^\infty \frac{1}{\sqrt{t}} e^{-\frac{a^2}{4t} - st} dt \underset{t=u^2}{=} 2 \int_0^\infty e^{-\frac{a^2}{4u^2} - su^2} du. \tag{1.44}$$

Then, we consider the integral

$$I(q, r) = \int_0^\infty e^{-q^2 u^2 - \frac{r^2}{u^2}} du. \tag{1.45}$$

Differentiating with respect to the parameter r, one obtains

$$\frac{\partial}{\partial r} I(q, r) = -2r \int_0^\infty e^{-q^2 u^2 - \frac{r^2}{u^2}} \frac{du}{u^2}. \tag{1.46}$$

Using the variable change $v = \frac{r}{q}u$, one finds

$$\frac{\partial}{\partial r}I(q,r) = -2q\int_0^\infty e^{-q^2 v^2 - \frac{r^2}{v^2}}\,dv = -2qI(q,r). \qquad (1.47)$$

The solution of this differential equation is

$$I(q,r) = C\,e^{-2qr}, \qquad (1.48)$$

where the constant C is obtained from the condition $r = 0$. Thus,

$$I(q,0) = \int_0^\infty e^{-q^2 u^2}\,du = \frac{\sqrt{\pi}}{2q}, \qquad (1.49)$$

from where it follows that $C = \frac{\sqrt{\pi}}{2q}$. Therefore, the integral becomes

$$I(q,r) = \frac{\sqrt{\pi}}{2q}e^{-2qr}. \qquad (1.50)$$

Using that $q = \sqrt{s}$ and $r = \frac{a}{2}$ we arrive at the result (1.43).

$\square$

Example 1.2. Show that

$$\mathcal{L}^{-1}\left[e^{-b\sqrt{s}}\right] = \frac{b}{\sqrt{4\pi t^3}}e^{-\frac{b^2}{4t}}, \qquad (1.51)$$

where $b > 0$.

Differentiating with respect to the parameter b and using the result obtained in the previous example, we have

$$\mathcal{L}^{-1}\left[e^{-b\sqrt{s}}\right] = \mathcal{L}^{-1}\left[s^{-1/2}\sqrt{s}\,e^{-b\sqrt{s}}\right]$$

$$= -\frac{\partial}{\partial b}\mathcal{L}^{-1}\left[s^{-1/2}e^{-b\sqrt{s}}\right]$$

$$= -\frac{\partial}{\partial b}\left(\frac{1}{\sqrt{\pi t}}e^{-\frac{b^2}{4t}}\right) = \frac{b}{\sqrt{4\pi t^3}}e^{-\frac{b^2}{4t}}.$$

$\square$

1.1.3 *Mellin transform*

The Mellin transform plays an important role in the definition of the Fox H-function. It is also a convenient technique for treating dilation operators like $[x\frac{d}{dx}]^n$. The transformation is described by the kernel function

$K(q, x) = x^{q-1}$, while the limits are set to $a = 0$ and $b = \infty$, see Table 1.1,

$$\mathcal{M}[f(x)](q) = F(q) = \bar{f}(q) = \int_0^\infty f(x) \, x^{q-1} \, dx. \tag{1.52}$$

The Mellin transform function (the Mellin image) $F(q)$ is defined on a complex plane with complex values $q = q_1 + \imath q_2$ such that the real part, q_1 is defined by $f(x)$. For example, if $f(x) = e^{-rx}$, Eq. (1.52) defines a gamma function

$$\Gamma(q) = \int_0^\infty x^{q-1} e^{-x} dx, \tag{1.53}$$

where $q_1 > 0$ (see Sec. 1.3.1 for details). In other example, for $f(x) = (x - x_0)_+^z \equiv \Theta(x)(x - x_0)^z$, the Mellin transform

$$\mathcal{M}\left[\Theta(x)(x - x_0)^z\right](q) = x_0^{z+q}/(z + q) \tag{1.54}$$

exists for $q_1 < -\Re(z)$. In general case, a region of all valid values of $\Re(q)$ is known as *a strip of definition* of the Mellin transform, see *e.g.*, Ref. [1]. Another important example for $f(x) = (1 + x)^{-1}$ defines a beta function:

$$F(q) = \int_0^\infty \frac{x^{q-1}}{1 + x} = B(q, 1 - q), \tag{1.55}$$

see Examples 1.5 and 1.6 of Sec. 1.3.1 for details.

A relation to the Laplace and Fourier transforms can be established by change of the variables $x = e^{-y}$ and $dx = -e^{-y}dy$. The integral (1.52) reads

$$F(q) = \int_{-\infty}^\infty g(y)e^{-qy} dy, \tag{1.56}$$

where $g(y) \equiv f(e^{-y})$. This expression corresponds to the two-sided Laplace transform. To obtain the Fourier transform, one uses $q = q_1 + \imath q_2 \equiv a - \imath k$ in Eq. (1.56), which according to definition (1.4) yields

$$F(k) = \mathcal{F}\left[g(y)e^{-ay}\right](k) \equiv \int_{-\infty}^\infty f\left(e^{-y}\right) e^{-ay} e^{-\imath ky} dy. \tag{1.57}$$

Expression (1.57) is used to define the inverse Mellin transform in the form of the inverse Fourier transform. That is

$$f\left(e^{-y}\right) e^{-ay} = \frac{1}{2\pi} \int_{-\infty}^\infty F(k)e^{\imath ky} dk. \tag{1.58}$$

Returning to the x variable, we have from Eq. (1.58) that the inverse Mellin transform is given by

$$\mathcal{M}^{-1}[F(q)](x) = f(x) = \frac{1}{2\pi\imath} \int_{c-\imath\infty}^{c+\imath\infty} F(q) \, x^{-q} \, dq. \tag{1.59}$$

It also follows from the inverse transform in Eq. (1.59), see [1], that if in the strip of definition, $F(q)$ is an analytic function and satisfies the inequality $|F(q)| < A|q|^{-2}$ for some constant A, then the function $f(x)$ is a continuous function of $x \in [0, \infty)$ and its Mellin transform is $F(q)$.

Additionally to the basic property of linearity see e.g., [1, 2], the Mellin transform has the following properties:

(1) Multiplication by x^a:

$$\boxed{\mathcal{M}\left[x^a f(x)\right] = F(q + a)} \tag{1.60}$$

We easily show that

$$\mathcal{M}\left[x^a f(x)\right] = \int_0^\infty x^{a+q-1} f(x)\, dx = F(x + a).$$

(2) Transform of $f(x^a)$:

$$\boxed{\mathcal{M}\left[f(x^a)\right] = \frac{1}{a} F(q/a)} \tag{1.61}$$

This property can be easily shown,

$$\mathcal{M}\left[f(x^a)\right] = \int_0^\infty x^{q-1} f(x^a)\, dx$$

$$\underset{(z=x^a)}{=} \int_0^\infty z^{(q-1)/a} f(z) \frac{1}{a} z^{(1-a)/a}\, dz$$

$$= \frac{1}{a} \int_0^\infty z^{q/a-1} f(z)\, dz = \frac{1}{a} F(q/a).$$

(3) Scaling property $(a > 0)$:

$$\boxed{\mathcal{M}[f(ax)] = a^{-q} F(q)} \tag{1.62}$$

From the definition of the Mellin transform, we have

$$\mathcal{M}[f(ax)] = \int_0^\infty x^{q-1} f(ax)\, dx$$

$$\underset{(z=ax)}{=} \frac{1}{a^q} \int_0^\infty z^{q-1} f(z)\, dz = a^{-q} F(q).$$

(4) Transform of convolution:

$$\boxed{\mathcal{M}\left[\int_0^\infty f(r)\, g\,(x/r)\, \frac{dr}{r}\right] = F(q)G(q)} \tag{1.63}$$

(5) Transform of derivative of $f(x)$:

$$\boxed{\mathcal{M}\left[\frac{d^n}{dx^n}f(x)\right] = (-1)^n \frac{\Gamma(q)}{\Gamma(q-n)}F(q-n)} \qquad (1.64)$$

(6) Transform of $\log^n x\, f(x)$:

$$\boxed{\mathcal{M}\left[\log^n x\, f(x)\right] = \frac{d^n}{dq^n}F(q)} \qquad (1.65)$$

For the latter expression, we have

$$\mathcal{M}\left[\log^n x\, f(x)\right] = \int_0^\infty x^{q-1}\log^n x\, f(x)\, dx$$

$$\underset{(z=\log x)}{=} \int_{-\infty}^\infty z^n e^{qz} f(e^z)\, dz$$

$$= \frac{\partial^n}{\partial q^n}\int_{-\infty}^\infty e^{qz} f(e^z)\, dz$$

$$\underset{(x=e^z)}{=} \frac{\partial^n}{\partial q^n}\int_0^\infty x^{q-1} f(x)\, dx = \frac{d^n}{dq^n}F(q)$$

The Mellin transform of the Dirac δ-function yields

$$\mathcal{M}\left[\delta(x-x_0)\right] = \int_0^\infty \delta(x-x_0)\, x^{q-1}\, dx = x_0^{q-1}. \qquad (1.66)$$

Another useful formula reads

$$\mathcal{M}\left[\frac{1}{\sqrt{4\pi\alpha}}e^{-\frac{\log^2 x}{4\alpha}}\right] = \frac{1}{\sqrt{4\pi\alpha}}\int_0^\infty e^{-\frac{\log^2 x}{4\alpha}}x^{q-1}\, dx$$

$$\underset{z=\log x}{=} \frac{1}{2\sqrt{4\pi\alpha}}\int_{-\infty}^\infty e^{-\left[\frac{z^2}{4\alpha}-qz\right]}\, dz$$

$$= \frac{e^{\alpha q^2}}{2\sqrt{4\pi\alpha}}\int_{-\infty}^\infty e^{-\left[\frac{z}{2\sqrt{\alpha}}-\sqrt{\alpha}\,q\right]^2}\, dz = e^{\alpha q^2}. \qquad (1.67)$$

In Wolfram Language, Mellin and inverse Mellin transforms are implemented as `MellinTransform[f, x, q]` and `InverseMellinTransform[F, q, x]`, respectively.

1.2 Asymptotic expansions

1.2.1 Tauberian theorems

The asymptotic behavior of a given function $r(t)$ can be analyzed by means of the Tauberian theorems [3]. One of the theorems states that if the asymptotic behavior of $r(t)$ for $t \to \infty$ is given by

$$r(t) \simeq t^{-\alpha}, \quad t \to \infty, \quad \alpha > 0, \qquad (1.68)$$

then, the corresponding Laplace pair $\hat{r}(s) = \mathcal{L}[r(t)]$ has the following behavior for $s \to 0$

$$\hat{r}(s) \simeq \Gamma(1 - \alpha)s^{\alpha - 1}, \quad s \to 0. \tag{1.69}$$

The theorem also works in the opposite direction, ensuring that $r(t)$ is the non-negative and monotone function at infinity.

This theorem can be formulated in the form of the so-called Hardy-Littlewood theorem. The theorem states that, if the Laplace-Stieltjes transform of a given non-decreasing function F with $F(0) = 0$, defined by Stieltjes integral

$$\omega(s) = \int_0^\infty e^{-st}\, dF(t), \tag{1.70}$$

has asymptotic behavior

$$\omega(s) \simeq Cs^{-\nu}, \quad s \to \infty \quad (s \to 0), \tag{1.71}$$

where $\nu \geq 0$ and C are real numbers, then the function F has asymptotic behavior

$$F(t) \simeq \frac{C}{\Gamma(\nu + 1)} t^\nu, \quad t \to 0 \quad (t \to \infty). \tag{1.72}$$

These Tauberian theorems are widely used in the theory of anomalous diffusion and in the theory of non-exponential relaxation processes.

The Tauberian theorem for *slowly varying functions* has also many applications in the theory of ultraslow diffusive processes and for analysis of strong anomaly. The theorem states that if a function $r(t)$, $t \geq 0$, has the Laplace transform $\hat{r}(s)$ whose asymptotics behaves as follows

$$\hat{r}(s) \simeq s^{-\rho} L\left(\frac{1}{s}\right), \quad s \to 0, \quad \rho > 0, \tag{1.73}$$

then

$$r(t) = \mathcal{L}^{-1}[\hat{r}(s)] \simeq \frac{1}{\Gamma(\rho)} t^{\rho - 1} L(t), \quad t \to \infty. \tag{1.74}$$

Here $L(t)$ is a slowly varying function at infinity, i.e.,

$$\lim_{t \to \infty} \frac{L(at)}{L(t)} = 1,$$

for any $a > 0$. The theorem is also valid if s and t are interchanged, that is $s \to \infty$ and $t \to 0$.

1.2.2   *Generating function formalism*

The generating function formalism or z-transform method[1] is often used for Markov chains, see *e.g.* Ref. [5]. By means of the generating function formalism, a probability function $f(t) = f_n$, defined at discrete times $t = n$, is transformed into its continuous counterpart, known as a generating function,

$$F(z) = \sum_{n=0}^{\infty} f_n z^n. \tag{1.75}$$

Here z is a complex variable, such that $|z| < R$, where R is the radius of convergence. In our considerations, $R = 1$. The advantage of the z-transform (1.75) lies in the reduction of convolutions, which involve the sequence $\{f_n\}$, as for example in Markov chains, to algebraic expressions for $F(z)$. The latter can be easily solved and explicit expressions for the generating function $F(z)$ can be obtained. Then the sequence $\{f_n\}$ can be obtained by the inverse z-transform,

$$f_n = \frac{1}{2\pi i} \oint_c F(z) z^{-(n+1)} dz. \tag{1.76}$$

Here C is a counterclockwise closed path encircling the origin and entirely in the region of convergence. Equation (1.76) can be obtained from Eq. (1.75) by multiplying both sides by $z^{-(n+1)}$ and integrating with respect to z. The lhs of Eq. (1.75) yields the rhs of Eq. (1.76), while for the rhs of Eq. (1.75) we obtain

$$\frac{1}{2\pi i} \sum_{m=0}^{\infty} \oint_c f_m z^{-(n-m+1)} dz = \frac{1}{2\pi} \sum_{m=0}^{\infty} f_m \frac{1}{2\pi} \int_0^{2\pi} e^{i(m-n)\theta} d\theta$$

$$= \sum_{m=0}^{\infty} f_m \delta_{m,n} = f_n. \tag{1.77}$$

Here we take the contour as a circle with the radius $R = 1$ and make the change of the variable $z = e^{i\theta}$.

In many cases, the knowledge of the explicit form of $F(z)$ does not help to calculate the integral, which can be very complicated. In that case, the discrete Tauberian theorem for power series [3] is useful, as it establishes a relation between f_n and $F(z)$.

Theorem 1.1 (The discrete Tauberian theorem [3]). *Let $f_n \geq 0$ and suppose that the power series*

$$F(z) = \sum_{n=0}^{\infty} f_n z^n \tag{1.78}$$

---

[1]We follow Sec. 3.2 of Ref. [4] and Sec. XIII.5 of Ref. [3].

converges for $0 \le z < 1$ and that

$$F(z) \sim \frac{1}{(1-z)^\rho} L\left(\frac{1}{1-z}\right), \quad z \to 1^-, \quad (1.79)$$

where L is a slowly varying function near infinity and $0 \le \rho < \infty$. Then Eq. (1.79) is equivalent to the following two relations,

$$f_0 + f_1 + \cdots + f_{n-1} \sim \frac{1}{\Gamma(\rho+1)} n^\rho L(n), \quad n \to \infty, \quad (1.80)$$

and

$$f_n \sim \frac{1}{\Gamma(\rho)} n^{\rho-1} L(n), \quad n \to \infty. \quad (1.81)$$

The latter expression also supposes that $\{f_n\}$ is a monotonic sequence.

1.2.3   *Laplace method*

The Laplace method is used to estimate integrals of the form

$$I = \int_a^b e^{-tf(z)} g(z) dz \quad (1.82)$$

for large t. It can be approximated by[2]

$$I \approx e^{-t\,f(z_0)} g(z_0) \sqrt{\frac{2\pi}{t|f''(z_0)|}}, \quad (1.83)$$

where z_0 is the extremum point of the function $f(z)$, i.e., $f'(z_0) = 0$, and if the extremum point is within the integration limits $(a < z_0 < b)$. If the extremum point is outside the integration limits $(z_0 > b)$, then the approximation result is calculated at $z_0 = b$.

Example 1.3. Let us evaluate the following integral

$$\int_0^t \frac{ab}{\sqrt{\pi t'}} \exp\left(-at' - \frac{b^2 y^2}{t'}\right) dt'$$

for large t and $a > 0$, $b > 0$.

---

[2]There is a vast library on the issue. We just mention Refs. [6, 7].

We introduce $t' = tz$, i.e., $dt' = t\, dz$ in the integral to obtain

$$I = \int_0^1 \frac{ab}{\sqrt{\pi t z}} \exp\left(-atz - \frac{b^2 y^2}{tz}\right) t\, dz$$

$$= \frac{ab\sqrt{t}}{\pi} \int_0^1 \exp\left(-t\left[az + \frac{b^2 y^2}{t^2 z}\right]\right) \frac{dz}{\sqrt{z}}$$

$$= \frac{ab\sqrt{t}}{\sqrt{\pi}} \int_0^1 e^{-tf(z)} g(z) dz, \tag{1.84}$$

where

$$f(z) = az + \frac{b^2 y^2}{t^2 z} \quad \text{and} \quad g(z) = \frac{1}{\sqrt{z}}. \tag{1.85}$$

For the extremum point we find

$$f'(z)\big|_{z=z_0} = a - \frac{b^2 y^2}{t^2 z_0^2} = 0 \quad \rightarrow \quad z_0 = \frac{b|y|}{t\sqrt{a}}. \tag{1.86}$$

This yields

$$f(z_0) = a\frac{b|y|}{t\sqrt{a}} + \frac{b^2 y^2}{t^2} \frac{t\sqrt{a}}{b|y|} = \frac{2b\sqrt{a}|y|}{t}. \tag{1.87}$$

$$f''(z)\big|_{z=z_0} = \frac{2b^2 y^2}{t^2 z_0^3} = \frac{2b^2 y^2}{t^2} \frac{t^3 a^{3/2}}{b^3 |y| y^2} = \frac{2a^{3/2} t}{b|y|} \tag{1.88}$$

and

$$g(z_0) = a^{1/4} \sqrt{\frac{t}{b|y|}}. \tag{1.89}$$

If $z_0 < 1$, then we apply the approximation formula (1.83) to obtain

$$I \approx \frac{ab\sqrt{t}a^{1/4}}{\sqrt{\pi}} \sqrt{\frac{t}{b|y|}} \sqrt{\frac{2\pi}{t\left|\frac{2a^{3/2}t}{b|y|}\right|}} e^{-t\frac{2b\sqrt{a}|y|}{t}}, \tag{1.90}$$

which gives

$$I \approx b\sqrt{a}\, e^{-2b\sqrt{a}|y|}. \tag{1.91}$$

This result will be used later in the analysis of the transition to the stationary state of a diffusing particle under stochastic resetting.

□

1.3   Special functions

1.3.1   *Gamma function*

The gamma function is defined for complex numbers with a positive real part by the integral

$$\Gamma(\alpha) = \int_0^\infty x^{\alpha-1} e^{-x} dx, \quad \alpha > 0. \tag{1.92}$$

It is extended by analytical continuation to all complex numbers, except the non-positive integers, where it has simple poles, see below. Integrating by part, we obtain a recursion formula,

$$\Gamma(\alpha + 1) = \alpha \Gamma(\alpha). \tag{1.93}$$

An alternative representation of the gamma function is via Euler's limit,

$$\Gamma(\alpha) = \lim_{j \to \infty} \left[\frac{j! j^\alpha}{\alpha(\alpha+1)\dots(\alpha+j)} \right], \tag{1.94}$$

where $\Gamma(1) = 1$. For the integer $\alpha = n$, the recursion (1.93) yields $\Gamma(n+1) = n\Gamma(n) = n!$. Recursion (1.93) also defines the gamma function for negative arguments as an analytical continuation $\Gamma(z-1) = \Gamma(z)/(z-1)$, where $\Gamma(0)$ and $\Gamma(-n)$ diverge, while the ratio is finite $\Gamma(-n)/\Gamma(-m) = m!/n!$.

The Gamma function is implemented in Wolfram Language by Gamma[z].

Example 1.4. Let us calculate $\Gamma(1/2) = (-1/2)!$:

$$\Gamma(1/2) = \int_0^\infty x^{-1/2} e^{-x} dx = 2 \int_0^\infty e^{-y^2} dy = \sqrt{\pi}. \tag{1.95}$$

$\square$

Another useful function is the beta function with the integral representation for $p, q > 0$,

$$B(p, q) = \int_0^1 x^{p-1}(1-x)^{q-1} dx, \tag{1.96}$$

while for arbitrary p and q, it is defined as a composition of gamma functions,

$$B(p, q) = \frac{\Gamma(p)\Gamma(q)}{\Gamma(p+q)}. \tag{1.97}$$

The beta function is implemented in Wolfram Language by Beta[a, b].

Using the beta function, some additional properties of the gamma function can be established as well, as shown in the ensuing examples.

Example 1.5. To prove formula (1.97) [8], we substitute $x = y/(1+y)$ in the definition (1.96) and obtain another integral representation of the beta function,

$$B(p,q) = \int_0^\infty y^{p-1}(1+y)^{-p-q}\,dy. \tag{1.98}$$

Next we establish the following identity

$$\int_0^\infty x^{p+q-1} e^{-(1+y)x}\,dx = \frac{\Gamma(p+q)}{(1+y)^{p+q}}. \tag{1.99}$$

Carrying out the variable substitution $z = (1+y)x$, we find

$$\int_0^\infty x^{p+q-1} e^{-(1+y)x}\,dx = \int_0^\infty \left(\frac{z}{1+y}\right)^{p+q-1} e^{-z}\,\frac{dz}{1+y}$$

$$= \frac{1}{(1+y)^{p+q}} \int_0^\infty z^{p+q-1} e^{-z}\,dz, \tag{1.100}$$

which leads to the relation (1.99), using the definition (1.92). We multiply Eq. (1.99) by y^{p-1} and integrate with respect to y. Changing the order of integration, we find

$$\int_0^\infty dx\, x^{p+q-1} e^{-x} \int_0^\infty y^{p-1} e^{-xy}\,dy = \int_0^\infty dx\, x^{p+q-1} e^{-x} \frac{\Gamma(p)}{x^p}$$

$$= \Gamma(p) \int_0^\infty x^{q-1} e^{-x}\,dx$$

$$= \Gamma(p+q) \int_0^\infty y^{p-1}(1+y)^{-p-q}\,dy, \tag{1.101}$$

where we have used the identity (1.99) to evaluate the second integral. Taking into account definitions (1.92) and (1.98), we obtain Eq. (1.97).

□

Example 1.6. Consider both expressions (1.96) and (1.97) for $q = 1-p$:

$$\Gamma(p)\Gamma(1-p) = B(p, 1-p) = \int_0^1 \left(\frac{x}{1-x}\right)^{p-1} \frac{dx}{1-x}. \tag{1.102}$$

The integral converges for $0 < p < 1$. Making the change of the variable $y = x/(1-x)$, which yields $dx/(1-x) = dy/(1+y)$, we obtain

$$\Gamma(p)\Gamma(1-p) = \int_0^\infty \frac{y^{p-1}\,dy}{1+y}. \tag{1.103}$$

We perform the analytical continuation in the complex plane with $y \to z = re^{\iota\phi}$ and with a branch cut along the real axis $(0,\infty)$, resulting in the contour integral

$$I = \oint_C \frac{z^{p-1}\,dz}{1+z}. \tag{1.104}$$

The contour of the integration C consists of four parts: (1) the integration over a circle with large radius $r = R \to \infty$, (2) the integration over a circle around $z = 0$ with small radius $r = \epsilon \to 0$, and (3) two integrals, one over the upper edge of the cut along the real axis (ϵ, R) and the second over the lower edge of the cut in the interval (R, ϵ). There is only a simple pole at $z = e^{\iota\pi}$, which yields $I = -2\pi\iota e^{\iota\pi p}$, according to the residue theorem. We have

$$I = I_R + I_\epsilon + I_{(\epsilon,R)}(\phi = 0) + I_{(R,\epsilon)}(\phi = 2\pi) = 2\pi\iota e^{\pi\iota p}, \tag{1.105}$$

and for the individual integrals,

$$I_{R=\infty} = I_{\epsilon=0} = 0, \tag{1.106a}$$

$$I_{0,\infty} = B(p, 1-p), \tag{1.106b}$$

$$I_{\infty,0} = -e^{2\pi\iota p}B(p, 1-p). \tag{1.106c}$$

Taking all these results into account, we finally obtain Euler's reflection formula,

$$\Gamma(p)\Gamma(1-p) = \frac{\pi}{\sin(p\pi)}. \tag{1.107}$$

$\square$

Equation (1.107) implies that $\Gamma(1/2) = \sqrt{\pi}$, see also Example 1.4.

Example 1.7. Another property that follows immediately from the beta function (1.96) is the Legendre duplication formula. From the integral (1.96), we have

$$B(p,p) = \int_0^1 [\tau(1-\tau)]^{p-1}d\tau. \tag{1.108}$$

Performing the change of the variable $s = 4\tau(1-\tau)$, which is symmetric with respect to $\tau = 1/2$, we find

$$B(p,p) = 2\int_0^{\frac{1}{2}} [\tau(1-\tau)]^{p-1}\,d\tau = \frac{1}{2^{2p-1}}\int_0^1 s^{p-1}(1-s)^{-1/2}\,ds$$

$$= \frac{1}{2^{2p-1}}B(p,1/2). \qquad (1.109)$$

Using the definition (1.97) and the fact that $\Gamma(1/2) = \sqrt{\pi}$, we obtain the Legendre duplication formula,

$$\Gamma(p)\Gamma(p+1/2) = 2^{1-2p}\sqrt{\pi}\,\Gamma(2p), \quad 2p \neq -1. \qquad (1.110)$$

$\square$

1.3.2 *Contour integral representation of* $\Gamma(p)$ *and* $1/\Gamma(p)$

We perform the analytical continuation in the complex plane $z = x - iy$ in the integration (1.92) to represent it as a contour integral

$$\int_C z^{p-1}e^{-z}\,dz = \int_C e^{(p-1)[\ln(z)]-z}\,dz, \qquad (1.111)$$

where the contour C starts at $+\infty$, runs around the point $z = 0$, and ends again at $+\infty$. Since $z = 0$ is a branch point, we take a branch cut along the non-negative real axis $(0, \infty)$. In this case, the contour consists of three parts: (1) the upper edge of the branch cut (∞, ϵ), (2) the circle C_ϵ of radius $\epsilon \to 0$ with the center at $z = 0$, and (3) the lower edge of the branch cut (ϵ, ∞). On the upper edge, $\ln z = \ln x$ is real, while on the lower edge $\ln z = \ln x + 2\pi i$. Therefore,

$$\int_C z^{p-1}e^{-z}\,dz = \int_\infty^\epsilon x^{p-1}e^{-x}\,dx$$

$$+ \int_{C_\epsilon} z^{p-1}e^{-z}\,dz + e^{2(p-1)\pi i}\int_\epsilon^\infty x^{p-1}e^{-x}\,dx. \qquad (1.112)$$

In the limit $\epsilon \to 0$, integration over the contour C_ϵ vanishes, and we obtain

$$\Gamma(p) = \int_0^\infty x^{p-1}e^{-x}\,dx = \frac{1}{e^{2(p-1)\pi i} - 1}\int_C z^{p-1}e^{-z}\,dz. \qquad (1.113)$$

To obtain a formula for the contour integral representation of $1/\Gamma(p)$, we replace p by $1 - p$ in Eq. (1.113),

$$\int_C z^{-p}e^{-z}\,dz = (e^{-2p\pi i} - 1)\Gamma(1-p). \qquad (1.114)$$

With the further substitution $z = \tau e^{\pi \iota} = -\tau$, we invert the contour C with respect to the y axis. In this case, the contour is known as a Hankel contour $\{Ha\}$. The contour integral (1.114) reads

$$\int_C z^{-p} e^{-z}\, dz = -e^{-p\pi \iota} \int_{\{Ha\}} \tau^{-p} e^{\tau}\, d\tau. \qquad (1.115)$$

Combining Eqs. (1.114) and (1.107), we find

$$\int_{\{Ha\}} \tau^{-p} e^{\tau}\, d\tau = 2\iota \sin(p\pi)\Gamma(1-p) = \frac{2\pi \iota}{\Gamma(p)}. \qquad (1.116)$$

This implies the following integral representation of the reciprocal gamma function,[3]

$$\frac{1}{\Gamma(p)} = \frac{1}{2\pi \iota} \int_{\{Ha\}} \tau^{-p} e^{\tau} d\tau. \qquad (1.117)$$

1.3.3 *Mittag-Leffler functions*

1.3.3.1 *One parameter Mittag-Leffler function*

The one parameter Mittag-Leffler (M-L) function is defined by [9]

$$E_\alpha(z) = \sum_{k=0}^{\infty} \frac{z^k}{\Gamma(\alpha k + 1)}, \qquad (1.118)$$

where $z \in \mathbb{C}$, $\Re(\alpha) > 0$) and $\Gamma(z)$ is the gamma function. It is an entire function of order $\rho = 1/\Re(\alpha)$ and type 1.

It is a generalization of the exponential function

$$E_1(\pm z) = \sum_{k=0}^{\infty} \frac{(\pm z)^k}{\Gamma(k+1)} = \sum_{k=0}^{\infty} \frac{(\pm z)^k}{k!} = e^{\pm z}. \qquad (1.119)$$

as well as trigonometric and hyperbolic functions

$$E_2(-z^2) = \sum_{k=0}^{\infty} \frac{(-1)^k z^{2k}}{\Gamma(2k+1)} = \sum_{k=0}^{\infty} \frac{(-1)^k z^k}{(2k)!} = \cos(z), \qquad (1.120)$$

$$E_2(z^2) = \sum_{k=0}^{\infty} \frac{z^{2k}}{\Gamma(2k+1)} = \sum_{k=0}^{\infty} \frac{z^{2k}}{(2k)!} = \cosh(z). \qquad (1.121)$$

---

[3]By deformation of the contour, this expression is the inverse Laplace transform.

Problem 1.1. Show that

$$E_3(z) = \frac{1}{3} \left[e^{z^{1/3}} + 2e^{-z^{1/3}/2} \cos \left(\frac{\sqrt{3}}{2} z^{1/3} \right) \right] \tag{1.122}$$

and

$$E_4(z) = \frac{1}{2} \left[\cos \left(z^{1/4} \right) + \cosh \left(z^{1/4} \right) \right]. \tag{1.123}$$

For $\alpha = 1/2$ one has

$$E_{1/2}(\pm\sqrt{z}) = \sum_{k=0}^{\infty} \frac{(\pm 1)^k z^{k/2}}{\Gamma(k/2 + 1)} = e^z \left[1 + \mathrm{erf}(\pm\sqrt{z}) \right], \tag{1.124}$$

where $\mathrm{erf}(z)$ is the *error function* defined by the series

$$\mathrm{erf}(z) = \frac{2}{\sqrt{\pi}} \sum_{k=0}^{\infty} \frac{(-1)^k z^{2k+1}}{k!(2k+1)}, \tag{1.125}$$

and by the integral

$$\mathrm{erf}(z) = \frac{2}{\sqrt{\pi}} \int_0^z e^{-x^2} dx. \tag{1.126}$$

Here we note that the result can be presented in terms of the *complementary error function* which is defined by

$$\mathrm{erfc}(z) = 1 - \mathrm{erf}(z) = \frac{2}{\sqrt{\pi}} \int_z^{\infty} e^{-x^2} dx, \tag{1.127}$$

which satisfies

$$\mathrm{erfc}(-z) = 2 - \mathrm{erfc}(z). \tag{1.128}$$

Therefore, one has

$$E_{1/2}(\pm\sqrt{z}) = e^z \left[2 - \mathrm{erfc}(\pm\sqrt{z}) \right] = e^z \mathrm{erfc}(\mp\sqrt{z}). \tag{1.129}$$

The error function and the complementary error function are implemented in Wolfram Language as `Erf[z]` and `Erfc[z]`, respectively.

Example 1.8. Show that for $n \in \mathbb{N}$ the following relation holds true:

$$\frac{d^n}{dz^n} E_n \left(\pm z^n \right) = \pm E_n \left(\pm z^n \right). \tag{1.130}$$

By using the definition (1.118), one finds

$$\frac{d^n}{dz^n} E_n\left(\pm z^n\right) = \frac{d^n}{dz^n} \sum_{k=0}^{\infty} \frac{(\pm 1)^k z^{nk}}{\Gamma(nk+1)} = \sum_{k=0}^{\infty} \frac{(\pm 1)^k \frac{d^n}{dz^n} z^{nk}}{\Gamma(nk+1)}$$

$$= \sum_{k=1}^{\infty} \frac{nk(nk-1)\ldots(nk-n)(\pm 1)^k z^{nk-n}}{\Gamma(nk+1)}$$

$$= \sum_{k=1}^{\infty} \frac{(\pm 1)^k z^{n(k-1)}}{\Gamma(n(k-1)+1)}$$

$$= \pm \sum_{k=0}^{\infty} \frac{(\pm 1)^k z^{nk}}{\Gamma(nk+1)} = \pm E_n\left(\pm z^n\right).$$

$$\square$$

The integral representation of the one parameter M-L function is

$$E_\alpha(z) = \frac{1}{2\pi\imath} \int_C \frac{s^{\alpha-1} e^s}{s^\alpha - z}\, ds. \tag{1.131}$$

The path C is a loop starting and ending at $-\infty$ and encircles $|s| \le |z|^{1/\alpha}$ in the positive sense $-\pi \le \arg s \le \pi$ on C. From the integral representation one can analyze the asymptotic behaviors of the M-L function.

For $0 < \alpha < 2$, the one parameter M-L function has the following asymptotics [10, 11]

$$E_\alpha(z) = \frac{1}{\alpha} e^{z^{1/\alpha}} - \sum_{k=1}^{n} \frac{z^{-k}}{\Gamma(1-\alpha k)} + O\left(|z|^{-1-n}\right), \quad |z| \to \infty,\ |\arg z| \le \theta, \tag{1.132}$$

$$E_\alpha(z) = -\sum_{k=1}^{n} \frac{z^{-k}}{\Gamma(1-\alpha k)} + O\left(|z|^{-1-n}\right), \quad |z| \to \infty,\ \theta \le |\arg z| \le \pi, \tag{1.133}$$

where $\pi\alpha/2 < \theta < \min\{\pi, \alpha\pi\}$.

Therefore, we can use the following asymptotics of the one parameter M-L function

$$E_\alpha(z) \sim \frac{1}{\alpha} e^{z^{1/\alpha}}, \quad z \gg 1, \tag{1.134}$$

for $0 < \alpha < 2$. Moreover, we will use the formula [10, 11, 12],

$$E_\alpha(-z) = -\sum_{n=0}^{\infty} \frac{(-z)^{-n-1}}{\Gamma(1-\alpha(1+n))} = -\sum_{n=1}^{\infty} \frac{(-z)^{-n}}{\Gamma(1-\alpha n)}, \quad z > 1. \tag{1.135}$$

for $0 < \alpha < 2$, to find the asymptotic behavior of the one parameter M-L function $E_\alpha(-z)$. For large z we have

$$E_\alpha(-z) \sim \frac{z^{-1}}{\Gamma(1-\alpha)} - \frac{z^{-2}}{\Gamma(1-2\alpha)}, \quad z \gg 1. \tag{1.136}$$

The most important one parameter M-L function in the theory of fractional differential equations is the associated one parameter M-L function

$$\mathcal{E}_\alpha(t; \lambda) = E_\alpha(-\lambda t^\alpha) \quad (\alpha > 0; \lambda \in \mathbb{C}). \tag{1.137}$$

Its Laplace transform reads [13]

$$\mathcal{L}\left[\mathcal{E}_\alpha(t; \mp\lambda)\right] = \frac{s^{\alpha-1}}{s^\alpha \mp \lambda}, \tag{1.138}$$

where $\Re(s) > |\lambda|^{1/\alpha}$.

From the asymptotic behaviors of the one parameter M-L function, one has

$$\mathcal{E}_\alpha(t; -\lambda) = E_\alpha(\lambda t^\alpha) \sim \frac{1}{\alpha} \exp\left([\lambda t^\alpha]^{1/\alpha}\right) = \frac{1}{\alpha} \exp\left(\lambda^{1/\alpha} t\right), \quad t \gg 1, \tag{1.139}$$

and

$$\mathcal{E}_\alpha(t; \lambda) = E_\alpha(-\lambda t^\alpha) \sim \frac{t^{-\alpha}}{\lambda\Gamma(1-\alpha)} - \frac{t^{-2\alpha}}{\lambda^2\Gamma(1-2\alpha)}, \quad t \gg 1, \tag{1.140}$$

for large t.

For $t \ll 1$, the one parameter M-L function behaves as follows

$$\mathcal{E}_\alpha(t; \mp\lambda) = E_\alpha(\pm\lambda t^\alpha) \sim 1 \pm \frac{\lambda t^\alpha}{\Gamma(\alpha+1)} \simeq \exp\left(\pm\frac{\lambda t^\alpha}{\Gamma(\alpha+1)}\right), \quad t \ll 1. \tag{1.141}$$

For $0 < \alpha < 1$ it is a stretched exponential, while for $1 < \alpha < 2$ it is a compressed exponential.

Example 1.9. Find the first term of the asymptotic behavior of the associated one parameter M-L function for $t \gg 1$, given by $\mathcal{E}_\alpha(t; \lambda) = E_\alpha(-\lambda t^\alpha) \sim \frac{t^{-\alpha}}{\lambda\Gamma(1-\alpha)}$ ($\lambda > 0$), by using the Tauberian theorem.

Let us use the Laplace transform formula (1.138) for the associated one parameter M-L function $\mathcal{L}\left[\mathcal{E}_\alpha(t; \lambda)\right](s) = \frac{s^{\alpha-1}}{s^\alpha + \lambda}$. By its series expansion for $s \to 0$ we have

$$\mathcal{L}\left[\mathcal{E}_\alpha(t; \lambda)\right] = \frac{1}{\lambda} \frac{s^{\alpha-1}}{1 + \frac{s^\alpha}{\lambda}} \underset{s \to 0}{\sim} \frac{1}{\lambda} s^{\alpha-1}\left(1 - \frac{s^\alpha}{\lambda}\right) \sim \frac{s^{\alpha-1}}{\lambda}.$$

By using the Laplace transform of power function (1.34) and applying the Tauberian theorem, we find the asymptotic behavior for $t \to \infty$,

$$\mathcal{E}_\alpha(t; \lambda) \underset{t \to \infty}{\sim} \mathcal{L}^{-1} \left[\frac{s^{\alpha-1}}{\lambda} \right] = \frac{1}{\lambda} \frac{t^{-\alpha}}{\Gamma(1 - \alpha)}, \quad \alpha < 1.$$

□

Example 1.10. Find the second term in the asymptotic expansion of the associated one parameter M-L function for $t \gg 1$, by using the Tauberian theorem.

We use the same approach as in Example 1.9. Therefore, we have

$$\mathcal{L}\left[\mathcal{E}_\alpha(t; \lambda)\right] = \frac{1}{\lambda} \frac{s^{\alpha-1}}{1 + \frac{s^\alpha}{\lambda}} \underset{s \to 0}{\sim} \frac{1}{\lambda} s^{\alpha-1} \left(1 - \frac{s^\alpha}{\lambda} \right) \sim \frac{s^{\alpha-1}}{\lambda} - \frac{s^{2\alpha-1}}{\lambda^2}.$$

Since $0 < \alpha < 1$ we note that one can not directly use the Laplace transform formula (1.34), $\mathcal{L}\left[t^\beta\right] = \frac{\Gamma(\beta+1)}{s^{\beta+1}}$, $\beta > -1$, for the second term. For this reason, we first divide the both sides of the equation by s, to have

$$\frac{1}{s}\mathcal{L}\left[\mathcal{E}_\alpha(t; \lambda)\right] \underset{s \to 0}{\sim} \frac{s^{\alpha-2}}{\lambda} - \frac{s^{2\alpha-2}}{\lambda^2},$$

and then we find the inverse Laplace transform. Thus, we find

$$\int_0^t \mathcal{E}_\alpha(t'; \lambda)\, dt' \underset{t \to \infty}{\sim} \mathcal{L}^{-1}\left[\frac{s^{\alpha-2}}{\lambda} - \frac{s^{2\alpha-2}}{\lambda^2} \right] = \frac{t^{1-\alpha}}{\lambda\Gamma(2 - \alpha)} - \frac{t^{1-2\alpha}}{\lambda^2\Gamma(2 - 2\alpha)}.$$

By differentiating both sides we finally find

$$\mathcal{E}_\alpha(t; \lambda) \underset{t \to \infty}{\sim} \frac{t^{-\alpha}}{\lambda\Gamma(1 - \alpha)} - \frac{t^{-2\alpha}}{\lambda^2\Gamma(1 - 2\alpha)}.$$

□

The one parameter M-L function (1.118) is implemented in the Wolfram Language as `MittagLefflerE[α, z]`.

1.3.3.2 *Two parameter Mittag-Leffler function*

The two parameter M-L function is defined by [14, 15, 16, 17]

$$E_{\alpha,\beta}(z) = \sum_{k=0}^{\infty} \frac{z^k}{\Gamma(\alpha k + \beta)}, \tag{1.142}$$

with $z, \beta \in \mathbb{C}$, $\Re(\alpha) > 0$. It is an entire functions of order $\rho = 1/\Re(\alpha)$ and type 1. The basic properties and relations of the one and two parameter

M–L functions appeared in the third volume of the Bateman project [8], Chapter XVIII Miscellaneous Functions.

It is a generalization of the one parameter M-L function, and for $\beta = 1$ we have

$$E_{\alpha,1}(z) = \sum_{k=0}^{\infty} \frac{z^k}{\Gamma(\alpha k + 1)} = E_\alpha(z). \tag{1.143}$$

Other well-known functions are special cases of the two parameter M-L function. For example,

$$E_{2,1}(z^2) = \sum_{k=0}^{\infty} \frac{z^{2k}}{\Gamma(2k+1)} = \sum_{k=0}^{\infty} \frac{z^{2k}}{(2k)!} = \cosh(z), \tag{1.144}$$

$$E_{2,1}(-z^2) = \sum_{k=0}^{\infty} \frac{(-1)^k z^{2k}}{\Gamma(2k+1)} = \sum_{k=0}^{\infty} \frac{(-1)^k z^k}{(2k)!} = \cos(z), \tag{1.145}$$

$$E_{2,2}(-z^2) = \sum_{k=0}^{\infty} \frac{(-1)^k z^{2k}}{\Gamma(2k+2)} = \sum_{k=0}^{\infty} \frac{(-1)^k z^{2k}}{(2k+1)!}$$
$$= \frac{1}{z} \sum_{k=0}^{\infty} \frac{(-1)^k z^{2k}}{(2k+1)!} = \frac{\sin(z)}{z}. \tag{1.146}$$

Example 1.11. Find the real and imaginary part of the M-L function $E_\alpha\left(\pm \imath t^\alpha\right)$.

Let us use the series expansion of the one parameter M-L function (1.118). Thus, we have

$$E_\alpha\left(\pm \imath t^\alpha\right) = \sum_{k=0}^{\infty} \frac{(\pm \imath)^k t^{\alpha k}}{\Gamma(\alpha k + 1)} = \sum_{k=0}^{\infty} \frac{(\pm \imath)^{2k} t^{2\alpha k}}{\Gamma(2\alpha k + 1)} + \sum_{k=0}^{\infty} \frac{(\pm \imath)^{2k+1} t^{(2k+1)\alpha}}{\Gamma((2k+1)\alpha + 1)}$$
$$= \sum_{k=0}^{\infty} \frac{(-1)^k t^{2\alpha k}}{\Gamma(2\alpha k + 1)} \pm \imath t^\alpha \sum_{k=0}^{\infty} \frac{(-1)^k t^{2\alpha k}}{\Gamma(2\alpha k + \alpha + 1)}$$
$$= E_{2\alpha}\left(-t^{2\alpha}\right) \pm \imath t^\alpha E_{2\alpha,\alpha+1}\left(-t^{2\alpha}\right). \tag{1.147}$$

Therefore, we find the real and imaginary part, which are given by $\Re\left(E_\alpha\left(\pm \imath t^\alpha\right)\right) = E_{2\alpha}\left(-t^{2\alpha}\right)$ and $\Im\left(E_\alpha\left(\pm \imath t^\alpha\right)\right) = \pm t^\alpha E_{2\alpha,\alpha+1}\left(-t^{2\alpha}\right)$, respectively. □

The integral representation of the two parameter M-L function is

$$E_{\alpha,\beta}(z) = \frac{1}{2\pi i} \int_{Ha} \frac{s^{\alpha-\beta} e^s}{s^\alpha - z} \, ds, \tag{1.148}$$

where the contour of integration Ha is the Hankel path, a loop starting and ending at 1, and encircling the disk $|s| \le |z|^{1/\alpha}$ counterclockwise.

For $0 < \alpha < 2$, the two parameter M-L function has the following asymptotics [10, 11]

$$E_{\alpha,\beta}(z) = \frac{1}{\alpha} z^{(1-\beta)/\alpha} e^{z^{1/\alpha}} - \sum_{k=1}^n \frac{z^{-k}}{\Gamma(\beta - \alpha k)} + O\left(|z|^{-1-n}\right), \quad |z| \to \infty, \tag{1.149}$$

for $|\arg z| \le \min\{\pi, \pi\alpha\}$, and

$$E_{\alpha,\beta}(z) = -\sum_{k=1}^n \frac{z^{-k}}{\Gamma(\beta - \alpha k)} + O\left(|z|^{-1-n}\right), \quad |z| \to \infty, \, \theta \le |\arg z| \le \pi, \tag{1.150}$$

for $\pi\alpha < |\arg z| < \pi$.

Therefore, we can use the following asymptotics of the two parameter M-L function

$$E_{\alpha,\beta}(z) \sim \frac{1}{\alpha} z^{(1-\beta)/\alpha} e^{z^{1/\alpha}}, \quad z \gg 1, \tag{1.151}$$

for $0 < \alpha < 2$. Accounting the series expansion of the M-L function [10, 11, 12],

$$E_{\alpha,\beta}(-z) = -\sum_{n=0}^\infty \frac{(-z)^{-n-1}}{\Gamma(\beta - \alpha(1+n))}$$

$$= -\sum_{n=1}^\infty \frac{(-z)^{-n}}{\Gamma(\beta - \alpha n)}, \quad z > 1, \tag{1.152}$$

for $0 < \alpha < 2$, one finds the asymptotic behavior of the two parameter M-L function. For large z ($z \gg 1$) one finds

$$E_{\alpha,\beta}(-z) \sim \frac{z^{-1}}{\Gamma(\beta - \alpha)} - \frac{z^{-2}}{\Gamma(\beta - 2\alpha)}, \quad z \gg 1. \tag{1.153}$$

Another important function in the theory of fractional differential equations is the associated two parameter M-L function, which is defined as follows

$$\mathcal{E}_{\alpha,\beta}(t; \lambda) = t^{\beta-1} E_{\alpha,\beta}(-\lambda t^\alpha), \tag{1.154}$$

with $(\alpha, \beta > 0; \lambda \in \mathbb{C})$. Its Laplace transform reads

$$\mathcal{L}\left[\mathcal{E}_{\alpha,\beta}(t; \mp\lambda)\right] = \frac{s^{\alpha-\beta}}{s^\alpha \mp \lambda}, \qquad (1.155)$$

where $\Re(s) > |\lambda|^{1/\alpha}$.

Thus, the asymptotic behavior of the associated two parameter M-L function reads

$$\mathcal{E}_{\alpha,\beta}(t; \lambda) = t^{\beta-1} E_{\alpha,\beta}(-\lambda t^\alpha)$$
$$\sim \frac{t^{\beta-\alpha-1}}{\lambda\Gamma(\beta-\alpha)} - \frac{t^{\beta-2\alpha-1}}{\lambda^2\Gamma(\beta-2\alpha)}, \qquad t \gg 1, \qquad (1.156)$$

for large t. For $\beta = \alpha$, the first term $(n = 0)$ in the series (1.152) vanishes, then one uses the next term with $n = 1$, which is

$$\mathcal{E}_{\alpha,\alpha}(t; \lambda) = t^{\alpha-1} E_{\alpha,\alpha}(-\lambda t^\alpha)$$
$$\sim -\frac{t^{\alpha-2\alpha-1}}{\lambda^2\Gamma(\alpha-2\alpha)} = -\frac{t^{-\alpha-1}}{\lambda^2\Gamma(-\alpha)}, \qquad t \gg 1. \qquad (1.157)$$

For $z \ll 1$, the two parameter M-L function behaves as

$$E_{\alpha,\beta}(-\lambda t^\alpha) \sim \frac{1}{\Gamma(\beta)} - \frac{\lambda t^\alpha}{\Gamma(\alpha+\beta)}$$
$$\simeq \frac{1}{\Gamma(\beta)} \exp\left(-\frac{\Gamma(\beta)}{\Gamma(\alpha+\beta)}\lambda t^\alpha\right), \qquad t \ll 1. \qquad (1.158)$$

For $0 < \alpha < 1$ it is stretched exponential, while for $1 < \alpha < 2$ it is compressed exponential.

Asymptotic properties of the M-L functions can be also defined by means of the Tauberian theorem, as in the example.

Example 1.12. Find the asymptotic behavior of the associated two parameter M-L function for $t \gg 1$, by using the Tauberian theorem.

From the Laplace transform formula (1.155) for the associated two parameter M-L function $\mathcal{L}\left[\mathcal{E}_{\alpha,\beta}(t; \lambda)\right] = \frac{s^{\alpha-\beta}}{s^\alpha+\lambda}$, and by its series expansion for $s \to 0$ we have

$$\mathcal{L}\left[\mathcal{E}_{\alpha,\beta}(t; \lambda)\right] = \frac{1}{\lambda}\frac{s^{\alpha-\beta}}{1+\frac{s^\alpha}{\lambda}} \underset{s\to0}{\sim} \frac{1}{\lambda}s^{\alpha-\beta}\left(1-\frac{s^\alpha}{\lambda}\right) \sim \frac{s^{\alpha-\beta}}{\lambda} - \frac{s^{2\alpha-\beta}}{\lambda^2}.$$

Next, if the values of parameters α and β satisfy the conditions of the application of the Laplace transform formula $\mathcal{L}\left[t^\sigma\right] = \frac{\Gamma(\sigma+1)}{s^{\sigma+1}}$, $\sigma > -1$, then we directly apply it to obtain the asymptotic behavior for $t \to \infty$, which reads

$$\mathcal{E}_{\alpha,\beta}(t;\lambda) \underset{t\to\infty}{\sim} \frac{t^{\beta-\alpha-1}}{\lambda\Gamma(\beta-\alpha)} - \frac{t^{\beta-2\alpha-1}}{\lambda^2\Gamma(\beta-2\alpha)}.$$

If the values of the parameters do not satisfy the condition of existence of the Laplace transform, we divide the both sides of the equation by s^n, $n = 1, 2, 3, \ldots$ in order to satisfy the condition. Then applying the inverse Laplace transform and n times differentiating with respect to t, $\frac{d^n}{dt^n}$, one observes the same result. $\qquad\square$

The definition (1.142) is also convenient to define an incomplete gamma function, as shown in the example.

Example 1.13. Show that

$$\gamma(a,t) = \Gamma(a)e^{-t}t^a E_{1,a+1}(t), \qquad (1.159)$$

where

$$\gamma(a,t) = \int_0^t \tau^{a-1}e^{-\tau}\,d\tau \qquad (1.160)$$

is the lower incomplete gamma function, which is related to the upper incomplete gamma function as

$$\Gamma(a,z) = \int_z^\infty \tau^{a-1}e^{-\tau}\,d\tau = \Gamma(a) - \gamma(a,z). \qquad (1.161)$$

We consider two ways to proof Eq. (1.159). In the first way, we use the Laplace transform, applied to the both sides of Eq. (1.159). For the lhs, we have

$$\mathcal{L}\left[\gamma(a,t)\right] = \mathcal{L}\left[\int_0^t \tau^{a-1}e^{-\tau}\,d\tau\right] = s^{-1}\mathcal{L}\left[\tau^{a-1}e^{-\tau}\right] = \frac{\Gamma(a)}{s(s+1)^a},$$
$$(1.162)$$

where the properties (1.27) and (1.20) of the Laplace transform are used. For the rhs, we have

$$\mathcal{L}\left[\Gamma(a)e^{-t}t^a E_{1,a+1}(t)\right] = \Gamma(a)\mathcal{L}\left[e^{-t}t^a E_{1,a+1}(t)\right]$$
$$= \Gamma(a)\frac{(s+1)^{1-(a+1)}}{(s+1)-1} = \frac{\Gamma(a)}{s(s+1)^a}, \qquad (1.163)$$

which is the same result of Eq. (1.162).

In the second way, we obtain Eq. (1.159) from the definition (1.160). Performing the variable change $1 - \tau/t = z$, we have

$$\gamma(a,t) = \int_0^t \tau^{a-1} e^{-\tau} \, d\tau = t^a e^{-t} \int_0^1 (1-z)^{a-1} e^{tz} \, dz. \qquad (1.164)$$

Expanding the exponential $e^{zt} = \sum_{n=0}^\infty (zt)^n [\Gamma(n+1)]^{-1}$, and taking the integration, which is the beta function (1.96), we obtain

$$\gamma(a,t) = t^a e^{-t} \Gamma(a) \sum_{n=0}^\infty \frac{t^n}{\Gamma(n+a+1)} = \Gamma(a) e^{-t} t^a E_{1,a+1}(t).$$

$$\square$$

The two parameter M-L function (1.142) is also implemented in the Wolfram Language as `MittagLefflerE[`α, β, z`]`.

1.3.3.3 *Three parameter Mittag-Leffler function*

The three parameter M-L function (also known as Prabhakar function) is defined by [18]

$$E_{\alpha,\beta}^\gamma(z) = \sum_{k=0}^\infty \frac{(\gamma)_k}{\Gamma(\alpha k + \beta)} \frac{z^k}{k!}, \qquad (1.165)$$

where $\beta, \gamma, z \in \mathbb{C}$, $\Re(\alpha) > 0$, $(\gamma)_k$ is the Pochhammer symbol (it is implemented in Wolfram Language as `Pochhammer[`γ, k`]`)

$$(\gamma)_0 = 1, \quad (\gamma)_k = \frac{\Gamma(\gamma+k)}{\Gamma(\gamma)}. \qquad (1.166)$$

This function is also entire function of order $\rho = 1/\Re(\alpha)$ and type 1. In its turn, it is a generalization of the two parameter M-L function, indeed

$$E_{\alpha,\beta}^1(z) = \sum_{k=0}^\infty \frac{\cancel{\Gamma(1+k)}}{\Gamma(1)} \frac{1}{\Gamma(\alpha k + \beta)} \frac{z^k}{\cancel{k!}} = \sum_{k=0}^\infty \frac{z^k}{\Gamma(\alpha k + \beta)} = E_{\alpha,\beta}(z). \qquad (1.167)$$

The three parameter M-L function can be presented by the following Mellin-Barnes integral

$$E_{\alpha,\beta}^\gamma(z) = \frac{1}{\Gamma(\gamma)} \frac{1}{2\pi\imath} \int_S \frac{\Gamma(s)\Gamma(\gamma-s)}{\Gamma(\beta-\alpha s)} (-z)^{-s} \, ds, \qquad (1.168)$$

where $|\arg z| < \pi$, and the contour of integration begins at $c - \imath\infty$ and ends at $c + \imath\infty$, $0 < \gamma < \Re(\gamma)$, and separates all poles of the integrand at

$s = -k$, $k = 0, 1, 2, \ldots$ to the left and all poles at $s = n + \gamma$ $n = 0, 1, \ldots$ to the right.

For $0 < \alpha < 2$ one obtains the following asymptotic behavior [10, 11]

$$E^{\gamma}_{\alpha,\beta}(z) = \frac{1}{\alpha^{\gamma}\Gamma(\gamma)} z^{(\gamma-\beta)/\alpha} e^{z^{1/\alpha}}, \quad z \to \infty. \tag{1.169}$$

Moreover, for $0 < \alpha < 2$, by means of the expression, see for example [10, 11, 12],

$$E^{\gamma}_{\alpha,\beta}(-z) = \frac{z^{-\gamma}}{\Gamma(\gamma)} \sum_{n=0}^{\infty} \frac{\Gamma(\gamma+n)}{\Gamma(\beta - \alpha(\gamma+n))} \frac{(-z)^{-n}}{n!}, \quad z > 1, \tag{1.170}$$

one finds the asymptotic behavior of the three parameter M-L function. For large z ($z \gg 1$) one finds

$$E^{\gamma}_{\alpha,\beta}(-z) \sim \frac{z^{-\gamma}}{\Gamma(\beta - \alpha\gamma)} - \gamma\frac{z^{-(\gamma+1)}}{\Gamma(\beta - \alpha(\gamma+1))}, \quad z \gg 1. \tag{1.171}$$

In the opposite case of $z \ll 1$, the three parameter M-L function behaves as follows

$$E^{\gamma}_{\alpha,\beta}(\pm\lambda t^{\alpha}) \sim \frac{1}{\Gamma(\beta)} \pm \gamma\frac{\lambda t^{\alpha}}{\Gamma(\alpha+\beta)}$$

$$\sim \frac{1}{\Gamma(\gamma)} \exp\left(\pm\gamma\frac{\Gamma(\beta)}{\Gamma(\alpha+\beta)}\lambda t^{\alpha}\right), \quad t \ll 1. \tag{1.172}$$

For $0 < \alpha < 1$ it is a stretched exponential, while for $1 < \alpha < 2$ it is a compressed exponential.

The associated three parameter M-L function has various applications in the theory of anomalous diffusion and non-exponential relaxation, and it is defined as follows

$$\mathcal{E}^{\gamma}_{\alpha,\beta}(t; \lambda) = t^{\beta-1} E^{\gamma}_{\alpha,\beta}(-\lambda t^{\alpha}), \tag{1.173}$$

with ($\min\{\alpha, \beta, \gamma\} > 0$; $\lambda \in \mathbb{R}$). Its Laplace transform yields

$$\mathcal{L}\left[\mathcal{E}^{\gamma}_{\alpha,\beta}(t; \mp\lambda)\right] = \frac{s^{\alpha\gamma-\beta}}{(s^{\alpha} \mp \lambda)^{\gamma}}, \tag{1.174}$$

where $|\lambda/s^{\alpha}| < 1$.

From the asymptotic behavior of the three parameter M-L function one finds the asymptotic behavior of the associated three parameter M-L function, which reads

$$\mathcal{E}^{\gamma}_{\alpha,\beta}(t; \lambda) = t^{\beta-1} E^{\gamma}_{\alpha,\beta}(-\lambda t^{\alpha}) \sim \frac{1}{\lambda^{\gamma}} \frac{t^{\beta-\alpha\gamma-1}}{\Gamma(\beta - \alpha\gamma)}, \quad t \gg 1, \tag{1.175}$$

for large t. We note that in the case $\beta = \alpha\gamma$, the first term ($n = 0$) vanishes in the series (1.170), then the main contribution is due to the next term with $n = 1$. Thus, one finds

$$\mathcal{E}_{\alpha,\beta}^{\gamma}(t; \lambda) \sim -\frac{\gamma}{\lambda^{\gamma+1}} \frac{t^{\beta-\alpha(\gamma+1)-1}}{\Gamma(\beta - \alpha(\gamma+1))} = -\frac{\gamma}{\lambda^{\gamma+1}} \frac{t^{-\alpha-1}}{\Gamma(-\alpha)}, \quad z \gg 1, \quad (1.176)$$

Then from the definition (1.176) one finds the asymptotic behavior of the associated three parameter M-L function for $t \ll 1$, which reads

$$\mathcal{E}_{\alpha,\beta}^{\gamma}(t; \lambda) \sim t^{\beta-1} \left[\frac{1}{\Gamma(\beta)} - \gamma \frac{\lambda t^{\alpha}}{\Gamma(\alpha+\beta)} \right]$$

$$\sim \frac{t^{\beta-1}}{\Gamma(\gamma)} \exp\left(-\gamma \frac{\Gamma(\beta)}{\Gamma(\alpha+\beta)} \lambda t^{\alpha} \right). \quad (1.177)$$

Example 1.14. [19] Show that the three parameter M-L function is associated with the nth derivative of the two parameter M-L function, by the relation

$$\left(\frac{\mathrm{d}}{\mathrm{d}z} \right)^n E_{\alpha,\beta}(z) = n! \, E_{\alpha,\beta+\alpha n}^{n+1}(z), \quad n \in \mathbb{N}. \quad (1.178)$$

Accounting that $\frac{\mathrm{d}^n z^k}{\mathrm{d}z^n} = \frac{k!}{(k-n)!} z^{k-n}$ for $n \leq k$, $n, k \in \mathbb{N}$, we have

$$\left(\frac{\mathrm{d}}{\mathrm{d}z} \right)^n E_{\alpha,\beta}(z) = \left(\frac{\mathrm{d}}{\mathrm{d}z} \right)^n \sum_{k=0}^{\infty} \frac{z^k}{\Gamma(\alpha k + \beta)}$$

$$= \sum_{k=n}^{\infty} \frac{k!}{(k-n)!} \frac{z^{k-n}}{\Gamma(\alpha k + \beta)} = n! \sum_{k=n}^{\infty} \binom{k}{n} \frac{z^{k-n}}{\Gamma(\alpha k + \beta)}$$

$$= n! \sum_{r=0}^{\infty} \binom{n+r}{n} \frac{z^r}{\Gamma(\alpha(r+n) + \beta)}$$

$$= n! \sum_{r=0}^{\infty} \frac{(n+1)_r}{\Gamma(\alpha r + \beta + \alpha n)} \frac{z^r}{r!} = n! \, E_{\alpha,\beta+\alpha n}^{n+1}(z),$$

where we use

$$\frac{(n)_k}{k!} = \binom{n+k-1}{k}. \quad (1.179)$$

$\square$

Example 1.15. Show that

$$r^{1-a}\frac{\gamma(a-1,rt)}{\Gamma(a-1)} = t^{a-1}E_{1,a}^{a-1}(-rt) \qquad (1.180)$$

for $a > 0$, where $\gamma(a,t)$ is the lower incomplete gamma function (1.160).

By the Laplace transform, we have

$$\frac{r^{1-a}}{\Gamma(a-1)}\mathcal{L}\left[\int_0^{rt}\tau^{a-2}e^{-\tau}\,d\tau\right] = \frac{1}{\Gamma(a-1)}\mathcal{L}\left[\int_0^t\tau^{a-2}e^{-r\tau}\,d\tau\right]$$

$$= \frac{r^{2-2a}}{\Gamma(a-1)}s^{-1}\mathcal{L}\left[\tau^{a-2}e^{-r\tau}\right] = s^{-1}(s+r)^{1-a},$$

where we first applied the property (1.27), and then the shift rule (1.20). From the Laplace transform (1.174) of the associated three parameter M-L function we have

$$\mathcal{L}^{-1}\left[\frac{(s+r)^{1-a}}{s}\right] = \mathcal{L}^{-1}\left[\frac{s^{-1}}{(s+1)^{a-1}}\right] = t^{a-1}E_{1,a}^{a-1}(-rt),$$

which establishes Eq. (1.180).

One can also find the following relation, which can be proven by the Laplace transform,

$$t^{a-1}E_{1,a}^{a-1}(-rt) = e^{-rt}\frac{t^{a-1}}{\Gamma(a)} + r^{1-a}\frac{\gamma(a,rt)}{\Gamma(a)}, \qquad 0 < a < 1. \qquad (1.181)$$

For $a = 3/2$, this expression relates to the error function $\mathrm{erf}(z)$ as follows

$$r^{-1/2}\frac{\gamma(1/2,rt)}{\Gamma(1/2)} = \frac{\mathrm{erf}(\sqrt{rt})}{\sqrt{r}} = t^{1/2}E_{1,3/2}^{1/2}(-rt), \qquad (1.182)$$

where the relation $\gamma(1/2,z) = \sqrt{\pi}\,\mathrm{erf}(z)$ is used. $\qquad\qquad\square$

1.4   Fractional calculus

Fractional calculus was inspired in 1695 by L'Hopital's query in a letter to Leibniz about a fractional exponent ν in derivative $d^\nu y/dx^\nu$ of a function $y(x)$. However, the modern theory of fractional calculus was started by Riemann. The textbook by Oldham and Spanier [20] presents a detailed, comprehensive chronology of the mathematical results. A historical survey with interesting mathematical examples can also be found in a monograph by Miller and Ross [21].

To have a first insight into the problem of fractional calculus, we start from the n-th integer derivative, defined by means of a contour integral [21]. Let us consider the Cauchy integral

$$\frac{d^n}{dz^n} f(z) \equiv D^n f(z) = \frac{n!}{2\pi i} \int_C \frac{f(x)}{(x-z)^{n+1}} dx, \qquad (1.183)$$

where C is a suitable contour around z and $f(z)$ is a test function. We assume that the test function satisfies all necessary conditions for the integral (1.183), such as being single-valued and analytical, and is suitable for the further description of fractional integro-differentiation. The residue theorem ensures the existence of the n-th derivative of $f(z)$ at point z. Replacing n by an arbitrary number, say μ, the point z is a branch point, and we take a branch cut along the real axis as we did in the case of the gamma function. We replace the factorial $n!$ by $\mu! = \Gamma(\mu+1)$, and the contour integral has a branch point at $z = x$.

We define the integral $_aD_t^\mu f(t)$, where the subscripts define the limit of integration and μ is the order of integral by

$$_aD_t^\mu f(t) \equiv D_{a+}^\mu f(t) = \frac{\Gamma(\mu+1)}{2\pi i} \int_a^t (\zeta - t)^{-\mu-1} f(\zeta) \, d\zeta$$

$$= \frac{\Gamma(\mu+1)}{2\pi i} \int_C (\zeta - t)^{-\mu-1} f(\zeta) \, d\zeta, \qquad (1.184)$$

where the last expression is an analytical continuation in the complex plain $\zeta \to z = t + it'$. C is a contour around t that starts and ends at $\zeta = a$. Cutting the real axis from $t - 0$, we obtain

$$(\zeta - t)^{-\mu-1} = e^{-(\mu+1)[\ln|\zeta-t|+i\phi]}, \qquad (1.185)$$

where $\phi = \arg(\zeta - t)$. For the upper edge of the cut, $\phi = \pi$, while $\phi = -\pi$ for the lower edge of the cut. The contour integral (1.184) consists of three parts,

$$\int_C (\zeta - t)^{-\mu-1} f(\zeta) \, d\zeta = \int_{C(\phi=\pi)} + \int_{C(\phi=-\pi)} + \int_{C_\epsilon}$$

$$= e^{i(\mu+1)\pi} \int_a^{t-\epsilon} (t-\tau)^{-\mu-1} f(\tau) \, d\tau$$

$$+ e^{-i(\mu+1)\pi} \int_{t-\epsilon}^a (t-\tau)^{-\mu-1} f(\tau) \, d\tau$$

$$+ \int_{C_\epsilon} (\zeta - t)^{-\mu-1} f(\zeta) \, d\zeta. \qquad (1.186)$$

The last integral vanishes in the limit $\epsilon \to 0$, and we obtain

$$D_{a+}^{\mu} f(t) = \frac{\Gamma(\mu+1)}{2\pi \imath} \int_C (\zeta - t)^{-\mu-1} f(\zeta) d\zeta$$

$$= \frac{\Gamma(\mu+1) \sin[(\mu+1)\pi]}{\pi \imath} \int_a^t (t - \tau)^{-\mu-1} f(\tau)\, d\tau. \qquad (1.187)$$

Using Euler's reflection formula (1.107) in the form $\Gamma(1 + \mu)\Gamma(-\mu) = \pi/\sin[(\mu+1)\pi]$ and performing the reflection $\mu \to -\mu$, we have

$$_a D_t^{-\mu} f(t) \equiv \left(I_{a+}^{\mu} f \right)(t) = \frac{1}{\Gamma(\mu)} \int_a^t (t - \tau)^{\mu-1} f(\tau)\, d\tau. \qquad (1.188)$$

After this brief introduction according to Refs. [5, 13, 21], we follow Ref. [22] and present formal definitions and properties of the fractional integration and differentiation.

1.4.1 *Fractional integrals*

The Riemann-Liouville (R-L) fractional integral of order $\mu > 0$ with the lower limit a is defined by [22]

$$\left(I_{a+}^{\mu} f \right)(t) = \frac{1}{\Gamma(\mu)} \int_a^t \frac{f(\tau)}{(t - \tau)^{1-\mu}}\, d\tau, \quad t > a, \quad \Re(\mu) > 0. \qquad (1.189)$$

To complete the definition, for $\mu = 0$ it is used

$$\left(I_{a+}^{0} f \right)(t) = f(t). \qquad (1.190)$$

The basic properties of the R-L fractional integral are

$$I_{0+}^{\gamma} I_{0+}^{\delta} = I_{0+}^{\gamma+\delta} = I_{0+}^{\delta} I_{0+}^{\gamma}, \quad \text{(semi-group property)} \qquad (1.191)$$

$$I_{0+}^{\gamma} t^s = \frac{\Gamma(s+1)}{\Gamma(s+1+\gamma)} t^{s+\gamma}, \quad \gamma \geq 0, \quad s > -1, \quad t > 0. \qquad (1.192)$$

The Laplace transform of the R-L fractional integral reads

$$\mathcal{L}\left[I_{0+}^{\mu} f(t) \right] = s^{-\mu} \mathcal{L}[f(t)]. \qquad (1.193)$$

1.4.2 *Fractional derivatives*

The R-L fractional derivative of order $\mu > 0$ with the lower limit a is defined by [22]

$$\left(_{RL}D_{a+}^{\mu} f \right)(t) = \left(\frac{d}{dt} \right)^n \left(I_{a+}^{n-\mu} f \right)(t)$$

$$= \frac{1}{\Gamma(n-\mu)} \frac{d^n}{dt^n} \int_a^t (t - \tau)^{n-\mu-1} f(\tau)\, d\tau, \qquad (1.194)$$

where $\Re(\mu) > 0$, $n = [\Re(\mu)] + 1$ is the integer part of the real number $\Re(\mu)$. By definition it follows

$$\left({}_{\mathrm{RL}}D_{a+}^0 f \right)(t) = f(t).$$

For $0 < \mu < 1$, i.e., $n = 1$, the R-L derivative becomes

$$\left({}_{\mathrm{RL}}D_{a+}^\mu f \right)(t) = \frac{d}{dt}\left(I_{a+}^{1-\mu} f \right)(t) = \frac{1}{\Gamma(1-\mu)}\frac{d}{dt}\int_a^t (t-\tau)^{-\mu} f(\tau)\, d\tau.$$

$$(1.195)$$

The Caputo fractional derivative of order $\mu > 0$ with the lower limit a is defined by [22]

$$_C D_{a+}^\mu f(t) = \left(I_{a+}^{n-\mu}\left(\frac{d}{dt}\right)^n f \right)(t)$$

$$= \frac{1}{\Gamma(n-\mu)}\int_a^t (t-\tau)^{n-\mu-1}\frac{d^n}{d\tau^n} f(\tau)\, d\tau, \qquad (1.196)$$

where the order of fractional integral and ordinary derivative is exchanged. For $0 < \mu < 1$ $(n = 1)$, it becomes

$$_C D_{a+}^\mu f(t) = \left(I_{a+}^{1-\mu}\frac{d}{dt} f \right)(t) = \frac{1}{\Gamma(1-\mu)}\int_a^t (t-\tau)^{-\mu}\frac{d}{d\tau} f(\tau)\, d\tau. \quad (1.197)$$

The Laplace transform of the R-L and Caputo fractional derivatives reads

$$\mathcal{L}\left[{}_{\mathrm{RL}}D_{0+}^\mu f(t) \right] = s^\mu \mathcal{L}\left[f(t) \right] - \sum_{k=0}^{n-1}\left[\lim_{t\to 0+}\frac{d^k}{dt^k}\left(I_{0+}^{n-\mu} f \right)(t) \right] s^{n-k-1}, \quad (1.198)$$

$$\mathcal{L}\left[{}_C D_{0+}^\mu f(t) \right] = s^\mu \mathcal{L}\left[f(t) \right] - \sum_{k=0}^{n-1}\left[\lim_{t\to 0+}\frac{d^k}{dt^k} f(t) \right] s^{\mu-k-1}, \qquad (1.199)$$

respectively, which for the case $0 < \mu < 1$ reduces to

$$\mathcal{L}\left[{}_{\mathrm{RL}}D_{0+}^\mu f(t) \right] = s^\mu \mathcal{L}\left[f(t) \right] - \left(I_{0+}^{1-\mu} f \right)(t = 0+), \qquad (1.200)$$

$$\mathcal{L}\left[{}_C D_{0+}^\mu f(t) \right] = s^\mu \mathcal{L}\left[f(t) \right] - s^{\mu-1} f(t = 0+). \qquad (1.201)$$

When $a = -\infty$, the fractional derivative is the Weyl derivative,

$$_W D_{-\infty}^\mu f(x) = \frac{1}{\Gamma(-\mu)}\int_{-\infty}^x \frac{f(x')\, dx'}{(x-x')^{1+\mu}}. \qquad (1.202)$$

If we impose the physically reasonable condition $f(-\infty) = 0$ together with its n derivatives, where $n - 1 < \nu < n$, then

$$_W D_{-\infty}^\mu f(x) = {}_{\mathrm{RL}}D_{-\infty}^\mu f(x) = {}_C D_{-\infty}^\mu f(x). \qquad (1.203)$$

One also has $_{\mathrm{W}}D_{-\infty}^{\mu}e^{\lambda x} = \lambda^{\mu}e^{\lambda x}$. This property is convenient for the Fourier transform $\mathcal{F}[f(x)] = \tilde{f}(k)$, which yields

$$\mathcal{F}\left[_{\mathrm{W}}D_{-\infty}^{\mu}f(x)\right] = (-ik)^{\mu}\tilde{f}(k). \tag{1.204}$$

The asymmetric Riesz-Feller derivative $_{\mathrm{RF}}D_{\theta}^{\alpha}$ of order α and skewness θ, is defined by its Fourier transform, see for example [3, 23, 24]

$$\mathcal{F}\left[_{\mathrm{RF}}D_{\theta}^{\alpha}f(x)\right](k) = -\psi_{\alpha}^{\theta}(k)\mathcal{F}[f(x)](k), \tag{1.205}$$

where

$$\psi_{\alpha}^{\theta}(k) = |k|^{\alpha}e^{\imath(\mathrm{sign}k)\theta\pi/2}, \quad 0 < \alpha \leq 2, \quad |\theta| \leq \min\{\alpha, 2 - \alpha\}. \tag{1.206}$$

Its integral representation is given by [3, 23, 24]

$$_{\mathrm{RF}}D_{\theta}^{\alpha}f(x) = \frac{\Gamma(1+\alpha)}{\pi}\left[\sin\frac{(\alpha+\theta)\pi}{2}\int_{0}^{\infty}\frac{f(x+\xi)-f(x)}{\xi^{1+\alpha}}d\xi\right.$$
$$\left. + \sin\frac{(\alpha-\theta)\pi}{2}\int_{0}^{\infty}\frac{f(x-\xi)-f(x)}{\xi^{1+\alpha}}d\xi\right]. \tag{1.207}$$

The symmetric Riesz-Feller derivative is obtained for $\theta = 0$. Thus, it is defined by the Fourier transform

$$\mathcal{F}\left[_{\mathrm{RF}}D_{0}^{\alpha}f(x)\right] = -\psi_{\alpha}^{0}(k)\mathcal{F}[f(x)] = -|k|^{\alpha}\mathcal{F}[f(x)]. \tag{1.208}$$

Its integral representation then becomes

$$_{\mathrm{RF}}D_{0}^{\alpha}f(x) = \frac{\Gamma(1+\alpha)}{\pi}\sin\frac{\alpha\pi}{2}\int_{0}^{\infty}\frac{f(x+\xi)-2f(x)+f(x-\xi)}{\xi^{1+\alpha}}d\xi. \tag{1.209}$$

The Riesz fractional integral is defined as follows

$$\frac{\partial^{\alpha}}{\partial|x|^{\alpha}} = \frac{1}{2\Gamma(\alpha)\cos(\alpha\pi/2)}\int_{a}^{b}\frac{f(y)}{|x-y|^{1-\alpha}}dy. \tag{1.210}$$

For $\alpha = 1$ it relates to the Hilbert transform [25], i.e.,

$$_{\mathrm{RF}}D_{0}^{1}f(x) = -\frac{1}{\pi}\frac{d}{dx}\int_{-\infty}^{\infty}\frac{f(\xi)}{x-\xi}d\xi. \tag{1.211}$$

1.4.3 *Generalized operators*

The generalized integral operator can be defined as follows [26]

$$(\mathbf{I}_{\zeta,t}f)(t) = \int_{0}^{t}\zeta(t-t')f(t')dt', \tag{1.212}$$

where $\zeta(t)$ is a kernel function, and it represents a convolution integral. Therefore, its Laplace transform reads

$$\mathcal{L}\left[\mathbf{I}_{\zeta,t}f(t)\right] = \hat{\zeta}(s)\hat{f}(s), \tag{1.213}$$

where we apply the Laplace transform of convolution (1.22).

From the definition (1.212), we conclude that for $\zeta(t) = \frac{t^{\mu-1}}{\Gamma(\mu)}$, $\mu > 0$ it turns to the R-L fractional integral (1.189). Moreover, for $\zeta(t) = e^{-rt}\frac{t^{\mu-1}}{\Gamma(\mu)}$ with $\mu > 0$ and $r > 0$, we have the tempered R-L fractional integral

$$\left(_{\mathrm{T}}I_{0+}^{\mu}f\right)(t) = \frac{1}{\Gamma(\mu)}\int_0^t e^{-r(t-t')}\frac{f(t')}{(t-t')^{1-\mu}}\,dt'. \tag{1.214}$$

A general form of "derivative" operators can be introduced as well [12, 26]. In particular, the generalized "derivative" operator in the R-L form is defined by

$$\left(_{\mathrm{RL}}\mathbf{G}_{\eta,t}f\right)(t) = \frac{d^n}{dt^n}\int_0^t \eta(t-t')f(t')\,dt', \tag{1.215}$$

where $\eta(t)$ is a memory kernel. The generalized "derivative" operator in the Caputo form is defined by

$$\left(_{\mathrm{C}}\mathbf{G}_{\gamma,t}f\right)(t) = \int_0^t \gamma(t-t')\frac{d^n}{dt'^n}f(t')\,dt', \tag{1.216}$$

where $\gamma(t)$ is also a memory kernel.

The Laplace transform of the generalized "derivative" operator in the R-L form reads

$$\mathcal{L}\left[\left(_{\mathrm{RL}}\mathbf{G}_{\eta,t}f\right)(t)\right](s) = s^n\hat{\eta}(s)\mathcal{L}\left[f(t)\right](s)$$
$$- \sum_{k=0}^{n-1}\left[\lim_{t\to 0+}\frac{d^k}{dt^k}\int_0^t \eta(t-t')f(t')\,dt'\right]s^{n-k-1}, \tag{1.217}$$

while for the generalized "derivative" operator in the Caputo form one finds

$$\mathcal{L}\left[\left(_{\mathrm{C}}\mathbf{G}_{\gamma,t}f\right)(t)\right](s) = s^n\hat{\gamma}(s)\mathcal{L}\left[f(t)\right](s) - \sum_{k=0}^{n-1}\left[\lim_{t\to 0+}\frac{d^k}{dt^k}f(t)\right]\hat{\gamma}(s)s^{n-k-1}. \tag{1.218}$$

Note that many known fractional derivatives are special cases of these generalized operators. These derivatives will be used throughout the book:

(1) R-L fractional derivative: $\eta(t) = \frac{t^{n-\mu-1}}{\Gamma(n-\mu)}$, $n = [\Re(\mu)] + 1$

$$\frac{1}{\Gamma(n-\mu)}\frac{d^n}{dt^n}\int_0^t (t-t')^{n-\mu-1}f(t')\,dt' \equiv \left(_{\mathrm{RL}}D_{0+}^{\mu}f\right)(t); \tag{1.219}$$

(2) Caputo fractional derivative: $\gamma(t) = \frac{t^{n-\mu-1}}{\Gamma(n-\mu)}$, $n = [\Re(\mu)] + 1$

$$\frac{1}{\Gamma(n-\mu)} \int_0^t (t-t')^{n-\mu-1} \frac{d^n}{dt'^n} f(t') \, dt' \equiv \left({}_{\mathrm{C}}D_{0+}^\mu f \right)(t); \qquad (1.220)$$

(3) Tempered R-L fractional derivative: $\eta(t) = e^{-rt} \frac{t^{n-\mu-1}}{\Gamma(n-\mu)}$, $n = [\Re(\mu)] + 1$, $r > 0$ is the tempering parameter

$$\frac{1}{\Gamma(n-\mu)} \frac{d^n}{dt^n} \int_0^t e^{-r(t-t')}(t-t')^{n-\mu-1} f(t') \, dt' \equiv \left({}_{\mathrm{TRL}}D_{0+}^\mu f \right)(t); \qquad (1.221)$$

(4) Tempered Caputo fractional derivative: $\gamma(t) = e^{-rt} \frac{t^{n-\mu-1}}{\Gamma(n-\mu)}$, $n = [\Re(\mu)] + 1$, $r > 0$ is the tempering parameter

$$\frac{1}{\Gamma(n-\mu)} \int_0^t e^{-r(t-t')}(t-t')^{n-\mu-1} \frac{d^n}{dt'^n} f(t') \, dt' \equiv \left({}_{\mathrm{TC}}D_{0+}^\mu f \right)(t); \qquad (1.222)$$

(5) Distributed order derivative in the R-L form:
$\eta(t) = \int_{n-1}^n p(\mu) \frac{t^{n-\mu-1}}{\Gamma(n-\mu)} \, d\mu$, $n = [\Re(\mu)] + 1$

$$\frac{1}{\Gamma(n-\mu)} \frac{d^n}{dt^n} \int_0^t \left[\int_{n-1}^n p(\mu)(t-t')^{n-\mu-1} f(t') \, d\mu \right] dt' \equiv \left({}_{\mathrm{DRL}}D_{0+}^\mu f \right)(t); \qquad (1.223)$$

(6) Distributed order derivative in the Caputo form:
$\gamma(t) = \int_{n-1}^n p(\mu) \frac{t^{n-\mu-1}}{\Gamma(n-\mu)} \, d\mu$, $n = [\Re(\mu)] + 1$

$$\frac{1}{\Gamma(n-\mu)} \int_0^t \left[\int_{n-1}^n p(\mu)(t-t')^{n-\mu-1} \frac{d^n}{dt'^n} f(t') \, d\mu \right] dt' \equiv \left({}_{\mathrm{DC}}D_{0+}^\mu f \right)(t); \qquad (1.224)$$

(7) Prabhakar derivative (in the R-L form):
$\eta(t) = t^{n-\mu-1} E_{\rho,n-\mu}^{-\gamma}(-\omega t^\rho)$, and $n = \lceil \mu \rceil$ is the ceiling function, which gives the smallest integer greater than or equal to μ

$$\frac{d^n}{dt^n} \int_0^t (t-t')^{n-\mu-1} E_{\rho,n-\mu}^{-\gamma}(-\omega[t-t']^\rho) f(t') \, dt' \equiv \left({}_{\mathrm{RL}}\mathcal{D}_{\rho,-\omega,0+}^{\gamma,\mu} f \right)(t); \qquad (1.225)$$

(8) Regularized Prabhakar derivative (in the Caputo form):
$\eta(t) = t^{n-\mu-1} E_{\rho,n-\mu}^{-\gamma}(-\omega t^\rho)$, and $n = \lceil \mu \rceil$

$$\int_0^t (t-t')^{n-\mu-1} E_{\rho,n-\mu}^{-\gamma}(-\omega[t-t']^\rho) \frac{d^n}{dt'^n} f(t') \, dt' \equiv \left({}_{\mathrm{C}}\mathcal{D}_{\rho,-\omega,0+}^{\gamma,\mu} f \right)(t). \qquad (1.226)$$

In this book, we shall mainly use the case with $n = 1$. Then, the Laplace transform of some of the aforementioned operators read:

$$\mathcal{L}\left[_{\mathrm{TRL}}D_{0+}^{\mu}f(t)\right] = (s+r)^{\mu-1}s\hat{f}(s) - \left(_{\mathrm{T}}I_{0+}^{1-\mu}f\right)(0+), \qquad (1.227)$$

$$\mathcal{L}\left[_{\mathrm{TC}}D_{0+}^{\mu}f(t)\right] = (s+r)^{\mu-1}\left[s\hat{f}(s) - f(0+)\right], \qquad (1.228)$$

$$\mathcal{L}\left[_{\mathrm{C}}\mathcal{D}_{\rho,-\nu,0+}^{\gamma,\mu}f(t)\right] = s^{-\rho\gamma+\mu-1}(s^{\rho}+\nu)^{\gamma}\left[s\hat{f}(s) - f(0+)\right]. \qquad (1.229)$$

References

[1] J. Bertrand, P. Bertrand, and J. P. Ovarlez. "Chapter 12 - Mellin Transform". In: *Transforms and Applications Handbook*. Ed. by A. D. Poularikas. CRC Press, Boca Raton: Cambridge University Press, 2010, pp. 12–1–12–36 (cit. on pp. 11, 12).

[2] Y. Brychkov, O. Marichev, and N. Savischenko. *Handbook of Mellin Transforms*. CRC, Boca Raton, 2018 (cit. on p. 12).

[3] W. Feller. *An introduction to probability theory and its applications, vol 2*. John Wiley & Sons, 2008 (cit. on pp. 13, 15, 38).

[4] B. D. Hughes. *Random Walks and Random Environments: Volume 1: Random Walks*. Oxford: Clarendon Press, 1995 (cit. on p. 15).

[5] A. Iomin, V. Mèndez, and W. Horsthemke. *Fractional Dynamics in Comblike Structures*. Singapore: World Scientific, 2018 (cit. on pp. 15, 36).

[6] G. B. Arfken and H. J. Weber. *Mathematical methods for physicists, 6th edn*. Elsevier, 2005 (cit. on p. 16).

[7] F. Olver. *Asymptotics and Special Functions*. CRC Press, 1997 (cit. on p. 16).

[8] H. Bateman and A. Erdélyi. *Higher Transcendental Functions [Volumes I-III]*. New York: McGraw-Hill, 1953–1955 (cit. on pp. 19, 27).

[9] G. M. Mittag-Leffler. "Sur la nouvelle fonction $E_\alpha(x)$". In: *Comptes rendus de l'Académie des Sciences Paris* 137.2 (1903), pp. 554–558 (cit. on p. 22).

[10] R. Gorenflo et al. *Mittag-Leffler functions, related topics and applications*. Vol. 2. Springer, 2014 (cit. on pp. 24, 28, 32).

[11] R. Garra and R. Garrappa. "The Prabhakar or three parameter Mittag–Leffler function: Theory and application". In: *Communications in Nonlinear Science and Numerical Simulation* 56 (2018), pp. 314–329 (cit. on pp. 24, 28, 32).

[12] T. Sandev and Z. Tomovski. *Fractional Equations and Models*. Springer Nature, 2019 (cit. on pp. 24, 28, 32, 39).

[13] I. Podlubny. *Fractional differential equations: an introduction to fractional derivatives, fractional differential equations, to methods of their solution and some of their applications*. Elsevier, 1998 (cit. on pp. 25, 36).

[14] A. Wiman. "Über den Fundamentalsatz in der Teorie der Funktionen $E_a(x)$". In: *Acta Mathematica* 29.1 (1905), pp. 191–201 (cit. on p. 26).

[15] R. P. Agarwal. "A propos d'une note de M. Pierre Humbert". In: *Comptes Rendus de l'Académie des Sciences Paris* 236.21 (1953), pp. 2031–2032 (cit. on p. 26).

[16] P. Humbert. "Quelques résultats relatifs à la fonction de Mittag-Leffler". In: *Comptes Rendus Hebdomadaires des Séances de l'Académie des Science* 236.15 (1953), pp. 1467–1468 (cit. on p. 26).

[17] P. Humbert and R. P. Agarwal. "Sur la fonction de Mittag-Leffler et quelques-unes de ses généralisations". In: *Bulletin of Mathematical Sciences* 77.2 (1953), pp. 180–185 (cit. on p. 26).

[18] T. R. Prabhakar. "A singular integral equation with a generalized Mittag-Leffler function in the kernel". In: *Yokohama Mathematical Journal* 19 (1971), pp. 7–15 (cit. on p. 31).

[19] A. A. Kilbas, H. M. Srivastava, and J. J. Trujillo. *Theory and applications of fractional differential equations.* Elsevier, 2006 (cit. on p. 33).

[20] K. B. Oldham and J. Spanier. *The Fractional Calculus: Theory and Applications of Differentiation and Integration to Arbitrary Order.* New York: Academic Press, 1974 (cit. on p. 34).

[21] K. S. Miller and B. Ross. *An Introduction to the Fractional Calculus and Fractional Differential Equations.* New York: John Wiley & Sons, 1993 (cit. on pp. 34–36).

[22] A. P. Prudnikov, Yu. A. Brychkov, and O. I. Marichev. *Integrals and series Volume 3: More special functions.* Taylor and Francis, Oxford, UK, 2003 (cit. on pp. 36, 37).

[23] F. Mainardi, G. Pagnini, and R. Gorenflo. "Mellin transform and subodination laws in fractional diffusion processes". In: *Fractional Calculus and Applied Analysis* 6.4 (2003), pp. 441–459 (cit. on p. 38).

[24] F. Mainardi, G. Pagnini, and R. K. Saxena. "Fox H functions in fractional diffusion". In: *Journal of Computational and Applied Mathematics* 178.1-2 (2005), pp. 321–331 (cit. on p. 38).

[25] W. Feller. *On a generalization of Marcel Riesz' potentials and the semi-groups generated by them.* Meddelanden Lunds Universitets Matematiska Seminarium, Lund, 1952 (cit. on p. 38).

[26] A. N. Kochubei. "General fractional calculus, evolution equations, and renewal processes". In: *Integral Equations and Operator Theory* 71.4 (2011), pp. 583–600 (cit. on pp. 38, 39).

Chapter 2

Fox *H*-function and related functions

The Fox *H*-function has been introduced on the bases of the Mellin transform and the Meijer's G-functions in the seminal paper by Charles Fox [1]. Apparently, the first mention of the Fox *H*-functions in physical literature was in Refs. [2, 3] in connection with applications in fractional diffusion [3, 4] and wave-diffusion [5] processes. In some extend, the Meijer's G-functions and the Fox *H*-functions are further developments and generalization of the generalized hypergeometric function. Thereof it is instructive to start this discussion from recalling some features of the generalized hypergeometric functions.

2.1 Introduction: *F* and *G* higher transcendental functions

The generalized hypergeometric functions are presented in the form of the generalized hypergeometric series [6],

$$
{}_pF_{q-1}(x) = {}_pF_{q-1}\left[\begin{matrix} a_1, \ldots, a_p \\ b_1, \ldots, b_q \end{matrix} \middle| x \right].
\tag{2.1}
$$

As is well known[1] the series ${}_pF_{q-1}(x)$ converges for all values of x when $p \leq q$ and diverges for $p > q$. It also converges for $|x| < 1$ if $p = q+1$. The series (2.1) can be obtained from a generalized hypergeometric equation as follows.

Example 2.1 (Generalized hypergeometric function).
Following Refs. [6], we consider a generalized hypergeometric

---

[1]An extended discussion with an extended list of references can be also find in Ref. [7].

equation, which formally reads

$$\left[x\frac{d}{dx}\prod_{j=1}^{q}\left(x\frac{d}{dx}+b_j-1\right)-x\prod_{j=1}^{p}\left(x\frac{d}{dx}+a_j\right)\right]u(x)=0. \quad (2.2)$$

Considering a formal power series solution $u(x)=\sum_{n=0}^{\infty}u_n x^n$ and taking into account that x^n is an eigenfunction of the dilatation operator, $x\frac{d}{dx}x^n=nx^n$, one obtains that Eq. (2.2) reduces to the recurrence relation,

$$n\prod_{j=1}^{q}(n-1+b_j)u_n=\prod_{j=1}^{p}(n-1+a_j)u_{n-1}, \quad (2.3)$$

which yields

$$u_n=\frac{\prod_{j=1}^{p}(a_j)_n}{n!\,\prod_{j=1}^{q}(b_j)_n}u_0\,, \quad (2.4)$$

where $(a)_n$ is the Pochhammer symbol (1.166). Taking $u_0=1$, one obtains the solution of the hypergeometric Eq. (2.2) in the form of the power series

$$u(x)=\sum_{n=0}^{\infty}\frac{\prod_{j=1}^{p}(a_j)_n}{n!\,\prod_{j=1}^{q}(b_j)_n}x^n \quad (2.5)$$

with the definition of the generalized hypergeometric functions

$$u(x)={}_pF_{q-1}\left[\begin{matrix}a_1,\ldots,a_p\\b_1,\ldots,b_q\end{matrix}\,\middle|\,x\right]. \quad (2.6)$$

□

2.1.1   *Meijer G-functions*

Let us consider a hypergeometric equation in slightly modified form for a function $G(x)$. Namely, now Eq. (2.2) can be rewritten as follows [8]

$$\left[\prod_{j=1}^{q}\left(x\frac{d}{dx}-b_j\right)-x\prod_{j=1}^{p}\left(x\frac{d}{dx}+1-a_j\right)\right]G(x)=0. \quad (2.7)$$

We consider this simplified Meijer's equation (see Ref. [6], Eq. 5.4(1)) to understand the structure of the Meijer's G-function. Then looking for the solution in the integral form

$$G_{p,q}(x)=\frac{1}{2\pi\imath}\int_{\Omega}g(s)x^s\,ds \quad (2.8)$$

with a suitable contour Ω of the integration over the complex s plane and substituting Eq. (2.8) in Eq. (2.7), we obtain

$$0 = \frac{1}{2\pi\imath} \int_\Omega g(s) \left[\prod_{j=1}^q (s - b_j)x^s + \prod_{j=1}^p (s + 1 - a_j)x^{s+1} \right] ds$$

$$= \frac{1}{2\pi\imath} \int_\Omega \left[g(s) \prod_{j=1}^q (s - b_j) + g(s-1) \prod_{j=1}^p (s - a_j) \right] x^s \, ds, \qquad (2.9)$$

where integrands with the same power x^s are collected. It is supposed that the integration contours for $g(s)$ and $g(s\pm1)$ are topologically equivalent *i.e.* can be deformed to each other without crossing the integrand singularities. One obtains thereof that

$$\frac{g(s)}{g(s-1)} = \frac{\prod_{j=1}^p (s - a_j)}{\prod_{j=1}^q (s - b_j)}. \qquad (2.10)$$

To understand the structure of $g(s)$, we consider several examples with different values of p and q.

Example 2.2 ($G_{p,q}(x)$ for $p = q = 1$). For $p = q = 1$, Eq. (2.10) reads

$$\frac{g(s)}{g(s-1)} = \frac{s - a_1}{s - b_1} = \frac{\Gamma(s - a_1 + 1)}{\Gamma(s - a_1)} \cdot \frac{\Gamma(s - b_1)}{\Gamma(s - b_1 + 1)}. \qquad (2.11)$$

Therefore $g(s) = \frac{\Gamma(s-a_1+1)}{\Gamma(s-b_1+1)}$ and integration in Eq. (2.8) is determined by poles $s = a_1 - n$ due to the singularities of the gamma function $\Gamma(s - a_1 + 1)$ with $n = 1, 2, \ldots$ and $|x| > 1$. Then $G_{1,1}(x)$ reads

$$G_{1,1}(x) = \frac{1}{2\pi i} \int_\Omega \frac{\Gamma(s - a_1 + 1)}{\Gamma(s - b_1 + 1)} x^s \, ds$$

$$= x^{a_1 - 1} \sum_{n=0}^\infty \lim_{\xi \to -n} (\xi + n)\Gamma(\xi) \cdot \frac{x^\xi}{\Gamma(\xi + a_1 - b_1)}$$

$$= x^{a_1 - 1} \sum_{n=0}^\infty \lim_{\xi \to -n} \frac{(\xi + n)(\xi + n - 1)\ldots\xi}{(\xi + n - 1)\ldots\xi}\Gamma(\xi) \cdot \frac{x^\xi}{\Gamma(\xi + a_1 - b_1)}$$

$$= x^{a_1 - 1} \sum_{n=0}^\infty \lim_{\xi \to -n} \frac{\Gamma(\xi + n + 1)}{(\xi + n - 1)\ldots\xi} \cdot \frac{x^{\xi-1}}{\Gamma(\xi + a_1 - b_1)}$$

$$= x^{a_1 - 1} \sum_{n=0}^\infty \frac{(-1)^n}{n!} \cdot \frac{x^{-n}}{\Gamma(a_1 - b_1 - n)}. \qquad (2.12)$$

Now let us treat $\Gamma(a_1 - b_1 - n) \equiv \Gamma(z)$ by means of the Euler's reflection formula (1.107). Using it twice, we have

$$\frac{1}{\Gamma(a_1 - b_1 - n)} \equiv \frac{1}{\Gamma(z)} = \Gamma(1 - z)\frac{\sin(\pi z)}{\pi}$$

$$= \Gamma(n + 1 - a_1 + b_1)\frac{(-1)^n \sin \pi(a_1 - b_1)}{\pi}$$

$$= (-1)^n \frac{\Gamma(n + 1 - a_1 + b_1)}{\Gamma(a_1 - b_1)\Gamma(1 - a_1 + b_1)}$$

$$= (-1)^n \frac{(1 - a_1 + b_1)_n}{\Gamma(a_1 - b_1)}.$$

Substituting this result into Eq. (2.12), we obtain $G_{1,1}(x)$ as follows

$$G_{1,1}(x) = \frac{x^{a_1 - 1}}{\Gamma(a_1 - b_1)} \sum_{n=0}^{\infty} \frac{(1 - a_1 + b_1)_n}{n!} \left(\frac{1}{x}\right)^n$$

$$= \frac{x^{a_1 - 1}}{\Gamma(a_1 - b_1)} \left(1 - \frac{1}{x}\right)^{a - 1 - b_1 - 1}, \qquad (2.13)$$

where $|x| > 1$. Note that for $|x| < 1$, the integral is zero, $G_{1,1}(x) = 0$, see discussion in Ref. [8]. □

However, our aim in this example is understanding the structure of $g(s)$ for arbitrary p and q. Therefore, we proceed with $p = 1$ and $q = N$. Then, Eq. (2.10) reads

$$\frac{g(s)}{g(s - 1)} = \frac{s - a_1}{(s - b_1)(s - b_2)\ldots(s - b_N)}$$

$$= \frac{\Gamma(s - a_1 + 1)}{\Gamma(s - a_1)} \cdot \frac{\Gamma(s - b_1)}{\Gamma(s - b_1 + 1)} \cdot \frac{\Gamma(s - b_2)}{\Gamma(s - b_2 + 1)} \cdots \frac{\Gamma(s - b_N)}{\Gamma(s - b_N + 1)},$$

which yields $g(s) = \frac{\Gamma(s - a_1 + 1)}{\prod_{j=1}^{N} \Gamma(s - b_j + 1)}$. Analogously, for $p = N$ and $q = 1$, we have $g(s) = \frac{\prod_{j=1}^{N} \Gamma(s - a_j + 1)}{\Gamma(s - b_1 + 1)}$. Eventually the integrand $g(s)$ in Eq. (2.10) reads

$$g(s) = \frac{\prod_{j=1}^{p} \Gamma(s - a_j + 1)}{\prod_{j=1}^{q} \Gamma(s - b_j + 1)}. \qquad (2.14)$$

Defining $G_{p,q}(x)$ in terms of the Meijer's G-function $G_{p,q}^{m,n}(x)$, where $0 < m \leq q$ and $0 < n \leq p$. When $n = p$ are indexes, which determine the number of gamma functions in numerator, we have

$$G_{p,q}(x) = G_{p,q}^{0,p}(x) = \frac{1}{2\pi\imath} \int_{\Omega} \frac{\prod_{j=1}^{p} \Gamma(s - a_j + 1)}{\prod_{j=1}^{q} \Gamma(s - b_j + 1)} x^s \, ds$$

$$= G_{p,q}^{0,p} \left[x \, \middle| \begin{array}{c} a_1, ..., a_p \\ b_1, ..., b_q \end{array} \right], \tag{2.15}$$

where Ω corresponds to the Mellin-Barnes integral [6]. Eventually, further generalization of the Meijer's G-function is determined by the integrand $g(s)$ as follows

$$g(s) = \frac{\prod_{j=1}^{m} \Gamma(b_j - s) \prod_{j=1}^{n} \Gamma(1 - a_j + s)}{\prod_{j=m+1}^{q} \Gamma(1 - b_j + s) \prod_{j=n+1}^{p} \Gamma(a_j - s)}. \tag{2.16}$$

The next step in the direction to the Fox H-function is a generalization of the arguments of the gamma functions by involving new parameters A_j and B_j [1, 9]. This general approach for the Fox H-function, including its properties is presented in the next section.

The Meijer G-function is also implemented in Wolfram Language by `MeijerG[{{a₁,...,aₙ},{aₙ₊₁,...,aₚ}},{{b₁,...,bₘ},{bₘ₊₁,...,b_q}},z]`.

2.2 Fox H-function

The Fox H-function (or H-function) is defined by means of the following Mellin-Barnes integral [1, 9, 10]

$$H_{p,q}^{m,n}(z) = H_{p,q}^{m,n} \left[z \, \middle| \begin{array}{c} (a_1, A_1), ..., (a_p, A_p) \\ (b_1, B_1), ..., (b_q, B_q) \end{array} \right]$$

$$= H_{p,q}^{m,n} \left[z \, \middle| \begin{array}{c} (a_p, A_p) \\ (b_q, B_q) \end{array} \right] = \frac{1}{2\pi\imath} \int_{\Omega} \theta(s) z^{-s} \, ds, \tag{2.17}$$

where

$$\theta(s) = \frac{\prod_{j=1}^{m} \Gamma(b_j + B_j s) \prod_{j=1}^{n} \Gamma(1 - a_j - A_j s)}{\prod_{j=m+1}^{q} \Gamma(1 - b_j - B_j s) \prod_{j=n+1}^{p} \Gamma(a_j + A_j s)}, \tag{2.18}$$

$0 \le n \le p$, $1 \le m \le q$, $a_i, b_j \in C$, $A_i, B_j \in \mathrm{R}^+$, $i = 1, ..., p$, $j = 1, ..., q$. Contour integration Ω starts at $c - \imath\infty$ and finishes at $c + \imath\infty$ separating the poles of the function $\Gamma(b_j + B_j s)$, $j = 1, ..., m$ with those of the function $\Gamma(1 - a_i - A_i s)$, $i = 1, ..., n$. It plays an important role in the theory of fractional differential equations enabling a closed form representation of the solutions of fractional diffusion-wave equations. It is a very general function, which reduces to many special cases in the form of well known special functions.

The Fox H-function (2.17) is implemented in the Wolfram Language as

FoxH$[\{\{\{a_1, A_1\}, \ldots, \{a_n, A_n\}\}, \{\{a_{n+1}, A_{n+1}\}, \ldots, \{a_p, A_p\}\}\},$
$\{\{\{b_1, B_1\}, \ldots, \{b_m, B_m\}\}, \{\{b_{m+1}, B_{m+1}\}, \ldots, \{b_q, B_q\}\}\}, z]$.

2.2.1 *Properties of the Fox H-function*

By using different properties of the gamma function, due to the relations (2.17) and (2.18), one can find many useful properties of the Fox H-function, and we present them as examples.

Proposition 2.1 (Symmetry property). *The Fox H-function is symmetric in the pairs $(a_1, A_1), \ldots, (a_n, A_n)$, as well as in the pairs $(a_{n+1}, A_{n+1}), \ldots, (a_p, A_p)$. The Fox H-function is symmetric also in the pairs $(b_1, B_1), \ldots, (b_m, B_m)$, as well as in the pairs $(b_{m+1}, B_{m+1}), \ldots, (b_q, B_q)$.*

Example 2.3. Show that

$$H_{2,2}^{2,1}\left[z \left|\begin{array}{l}(a_1, A_1), (a_2, A_2) \\ (b_1, B_1), (b_2, B_2)\end{array}\right.\right] = H_{2,2}^{2,1}\left[z \left|\begin{array}{l}(a_1, A_1), (a_2, A_2) \\ (b_2, B_2), (b_1, B_1)\end{array}\right.\right]. \quad (2.19)$$

This is a direct consequence of the Proposition 2.1, where we have $m = 2$, $n = 1$, $p = q = 2$. Since $m = 2$, the Fox H-function is symmetric in the pairs $(b_1, B_1), (b_2, B_2)$, which means that they can change the order to $(b_2, B_2), (b_1, B_1)$.

The same can be observed if one uses the definition of the Fox H-function. Thus, we have

$$H_{2,2}^{2,1}\left[z \left|\begin{array}{l}(a_1, A_1), (a_2, A_2) \\ (b_1, B_1), (b_2, B_2)\end{array}\right.\right]$$

$$= \frac{1}{2\pi\imath} \int_\Omega \frac{\Gamma(b_1 + B_1 s)\Gamma(b_2 + B_2 s)\Gamma(1 - a_1 - A_1 s)}{\Gamma(a_2 + A_2 s)} z^{-s}\, ds$$

$$= \frac{1}{2\pi\imath} \int_\Omega \frac{\Gamma(b_2 + B_2 s)\Gamma(b_1 + B_1 s)\Gamma(1 - a_1 - A_1 s)}{\Gamma(a_2 + A_2 s)} z^{-s}\, ds$$

$$= H_{2,2}^{2,1}\left[z \left|\begin{array}{l}(a_1, A_1), (a_2, A_2) \\ (b_2, B_2), (b_1, B_1)\end{array}\right.\right]. \quad (2.20)$$

$\square$

Example 2.4. Show that

$$
H_{3,3}^{3,2} \left[z \left| \begin{array}{l} (a_1, A_1), (a_2, A_2), (a_3, A_3) \\ (b_1, B_1), (b_2, B_2), (b_3, B_3) \end{array} \right. \right]
$$

$$
= H_{3,3}^{3,2} \left[z \left| \begin{array}{l} (a_2, A_2), (a_1, A_1), (a_3, A_3) \\ (b_3, B_3), (b_2, B_2), (b_1, B_1) \end{array} \right. \right]. \tag{2.21}
$$

This also directly follows from Proposition 2.1 where we use $m = 3$, $n = 2$, $p = q = 3$. Since $m = 3$, the Fox H-function is symmetric in the pairs (b_1, B_1), (b_2, B_2), (b_3, B_3), which means that they can change the order, for example, to (b_3, B_3), (b_2, B_2), (b_1, B_1). Moreover, since $n = 2$, the Fox H-function is symmetric in the pairs (a_1, A_1), (a_2, A_2), which means that they can change the order to (a_2, A_2), (a_1, A_1).

Indeed, from the definition of the Fox H-function, one finds

$$
H_{3,3}^{3,2} \left[z \left| \begin{array}{l} (a_1, A_1), (a_2, A_2), (a_3, A_3) \\ (b_1, B_1), (b_2, B_2), (b_3, B_3) \end{array} \right. \right]
$$

$$
= \frac{1}{2\pi\imath} \int_\Omega \frac{\prod_{j=1}^3 \Gamma(b_j + B_j s) \prod_{j=1}^2 \Gamma(1 - a_j - A_j s)}{\Gamma(a_3 + A_3 s)} z^{-s}\, ds
$$

$$
= \frac{1}{2\pi\imath} \int_\Omega \frac{\prod_{j=3}^1 \Gamma(b_j + B_j s) \prod_{j=2}^1 \Gamma(1 - a_j - A_j s)}{\Gamma(a_3 + A_3 s)} z^{-s}\, ds
$$

$$
= H_{3,3}^{3,2} \left[z \left| \begin{array}{l} (a_2, A_2), (a_1, A_1), (a_3, A_3) \\ (b_3, B_3), (b_2, B_2), (b_1, B_1) \end{array} \right. \right]. \tag{2.22}
$$

□

Following Ref. [10], we present another properties of the Fox-H-function.

Proposition 2.2 (Reduction formulas). *The following reduction formulas*

$$
H_{p,q}^{m,n}\left[z \left|\begin{array}{l} (a_1, A_1), ..., (a_p, A_p) \\ (b_1, B_1), ..., (b_{q-1}, B_{q-1}), (a_1, A_1) \end{array}\right.\right]
$$

$$
= H_{p-1,q-1}^{m,n-1}\left[z \left|\begin{array}{l} (a_2, A_2), ..., (a_p, A_p) \\ (b_1, B_1), ..., (b_{q-1}, B_{q-1}) \end{array}\right.\right], \tag{2.23}
$$

and

$$
H_{p,q}^{m,n}\left[z \left|\begin{array}{l} (a_1, A_1), ..., (a_{p-1}, A_{p-1}), (b_1, B_1) \\ (b_1, B_1), ..., (b_q, B_q) \end{array}\right.\right]
$$

$$
= H_{p-1,q-1}^{m-1,n}\left[z \left|\begin{array}{l} (a_1, A_1), ..., (a_{p-1}, A_{p-1}) \\ (b_2, B_2), ..., (b_q, B_q) \end{array}\right.\right], \tag{2.24}
$$

hold true, for $n \geq 1$, $q > m$.

Proof. *From the definition of the Fox H-function (2.17), we have*

$$
H_{p,q}^{m,n}\left[z \left|\begin{array}{l} (a_1, A_1), ..., (a_p, A_p) \\ (b_1, B_1), ..., (b_{q-1}, B_{q-1}), (a_1, A_1) \end{array}\right.\right]
$$

$$
= \frac{1}{2\pi\imath} \int_\Omega \frac{\prod_{j=1}^m \Gamma(b_j + B_j s)\,\Gamma(1-a_1-A_1 s)\prod_{j=2}^n \Gamma(1 - a_j - A_j s)}{\prod_{j=m+1}^{q-1} \Gamma(1-b_j-B_j s)\,\Gamma(1-a_1-A_1 s)\prod_{j=n+1}^{p} \Gamma(a_j+A_j s)} z^{-s}\, ds
$$

$$
= \frac{1}{2\pi\imath} \int_\Omega \frac{\prod_{j'=1}^m \Gamma(b_{j'} + B_{j'} s)\prod_{j'=1}^{n-1} \Gamma(1 - a_{j'} - A_{j'} s)}{\prod_{j'=m+1}^{q-1} \Gamma(1 - b_{j'} - B_{j'} s)\prod_{j'=n}^{p-1} \Gamma(a_{j'} + A_{j'} s)} z^{-s}\, ds
$$

$$
= H_{p-1,q-1}^{m,n-1}\left[z \left|\begin{array}{l} (a_2, A_2), ..., (a_p, A_p) \\ (b_1, B_1), ..., (b_{q-1}, B_{q-1}) \end{array}\right.\right].
$$

In a same way, we show

$$
H_{p,q}^{m,n}\left[z \left|\begin{array}{l} (a_1, A_1), ..., (a_p, A_p) \\ (b_1, B_1), ..., (b_{q-1}, B_{q-1}), (a_1, A_1) \end{array}\right.\right]
$$

$$
= \frac{1}{2\pi\imath} \int_\Omega \frac{\Gamma(b_1 + B_1 s)\prod_{j=2}^m \Gamma(b_j + B_j s)\prod_{j=1}^n \Gamma(1 - a_j - A_j s)}{\prod_{j=m+1}^{q} \Gamma(1 - b_j - B_j s)\prod_{j=n+1}^{p-1} \Gamma(a_j + A_j s)\,\Gamma(b_1 + B_1 s)} z^{-s}\, ds
$$

$$
= \frac{1}{2\pi\imath} \int_\Omega \frac{\prod_{j'=1}^{m-1} \Gamma(b_{j'} + B_{j'} s)\prod_{j'=1}^n \Gamma(1 - a_{j'} - A_{j'} s)}{\prod_{j'=m+1}^{q-1} \Gamma(1 - b_{j'} - B_{j'} s)\prod_{j'=n+1}^{p-1} \Gamma(a_{j'} + A_{j'} s)} z^{-s}\, ds
$$

$$
= H_{p-1,q-1}^{m-1,n}\left[z \left|\begin{array}{l} (a_1, A_1), ..., (a_{p-1}, A_{p-1}) \\ (b_2, B_2), ..., (b_q, B_q) \end{array}\right.\right].
$$

$\square$

Proposition 2.3. *For $\delta > 0$, it holds*

$$H_{p,q}^{m,n}\left[z^{\delta}\left|\begin{array}{c}(a_p, A_p)\\(b_q, B_q)\end{array}\right.\right] = \frac{1}{\delta}H_{p,q}^{m,n}\left[z\left|\begin{array}{c}(a_p, A_p/\delta)\\(b_q, B_q/\delta)\end{array}\right.\right]. \tag{2.25}$$

Proof. *From the definition (2.17), one finds*

$$H_{p,q}^{m,n}\left[z^{\delta}\left|\begin{array}{c}(a_p, A_p)\\(b_q, B_q)\end{array}\right.\right] = \frac{1}{2\pi\iota}\int_{\Omega}\theta(s)\left(z^{\delta}\right)^{-s}ds$$

$$\underset{s\delta=s'}{=}\frac{1}{\delta}\frac{1}{2\pi\iota}\int_{\Omega}\theta\left(\frac{s'}{\delta}\right)z^{-s'}\frac{ds'}{\delta} = \frac{1}{\delta}H_{p,q}^{m,n}\left[z\left|\begin{array}{c}(a_p, A_p/\delta)\\(b_q, B_q/\delta)\end{array}\right.\right]. \tag{2.26}$$

$\square$

Proposition 2.4. *The Fox H-function has the property*

$$z^{\sigma}H_{p,q}^{m,n}\left[z\left|\begin{array}{c}(a_p, A_p)\\(b_q, B_q)\end{array}\right.\right] = H_{p,q}^{m,n}\left[z\left|\begin{array}{c}(a_p + \sigma A_p, A_p)\\(b_q + \sigma B_q, B_q)\end{array}\right.\right]. \tag{2.27}$$

Proof. *From the definition (2.17), we have*

$$z^{\sigma}H_{p,q}^{m,n}\left[z\left|\begin{array}{c}(a_p, A_p)\\(b_q, B_q)\end{array}\right.\right] = \frac{1}{2\pi\iota}$$

$$\times\int_{\Omega}\frac{\prod_{j=1}^{m}\Gamma(b_j + B_j s)\prod_{j=1}^{n}\Gamma(1 - a_j - A_j s)}{\prod_{j=m+1}^{q}\Gamma(1 - b_j - B_j s)\prod_{j=n+1}^{p}\Gamma(a_j + A_j s)}z^{-s+\sigma}\,ds$$

$$\underset{s'=s-\sigma}{=}\frac{1}{2\pi\iota}$$

$$\times\int_{\Omega}\frac{\prod_{j=1}^{m}\Gamma(b_j + B_j(s'+\sigma))\prod_{j=1}^{n}\Gamma(1 - a_j - A_j(s'+\sigma))}{\prod_{j=m+1}^{q}\Gamma(1 - b_j - B_j(s'+\sigma))\prod_{j=n+1}^{p}\Gamma(a_j + A_j(s'+\sigma))}z^{-s'}\,ds'$$

$$=\frac{1}{2\pi\iota}$$

$$\times\int_{\Omega}\frac{\prod_{j=1}^{m}\Gamma([b_j + \sigma B_j] + B_j s')\prod_{j=1}^{n}\Gamma(1 - [a_j + \sigma A_j] - A_j s')}{\prod_{j=m+1}^{q}\Gamma(1 - [b_j+\sigma B_j] - B_j s')\prod_{j=n+1}^{p}\Gamma([a_j+\sigma A_j]+A_j s')}z^{-s'}\,ds'$$

$$= H_{p,q}^{m,n}\left[z\left|\begin{array}{c}(a_p + \sigma A_p, A_p)\\(b_q + \sigma B_q, B_q)\end{array}\right.\right]. \tag{2.28}$$

$\square$

In the similar way one can also show the following properties:

$$H_{p,q}^{m,n}\left[z\left|\begin{array}{c}(a_p, A_p)\\(b_q, B_q)\end{array}\right.\right] = H_{q,p}^{n,m}\left[\frac{1}{z}\left|\begin{array}{c}(1 - b_q, B_q)\\(1 - a_p, A_p)\end{array}\right.\right]. \tag{2.29}$$

$$H_{p+1,q+1}^{m,n+1}\left[z\left|\begin{array}{l}(0,\alpha),(a_p,A_p)\\(b_q,B_q),(r,\alpha)\end{array}\right.\right]=(-1)^r H_{p+1,q+1}^{m+1,n}\left[z\left|\begin{array}{l}(a_p,A_p),(0,\alpha)\\(r,\alpha),(b_q,B_q)\end{array}\right.\right],$$

$$(2.30)$$

The k-th derivative ($k \in \mathrm{N}$) of H-function is given by

$$\frac{d^k}{dz^k}\left\{z^\alpha H_{p,q}^{m,n}\left[(az)^\beta\left|\begin{array}{l}(a_p,A_p)\\(b_q,B_q)\end{array}\right.\right]\right\}$$

$$=z^{\alpha-k}H_{p+1,q+1}^{m,n+1}\left[(az)^\beta\left|\begin{array}{l}(-\alpha,\beta),(a_p,A_p)\\(b_q,B_q),(k-\alpha,\beta)\end{array}\right.\right],\qquad(2.31)$$

where $\beta > 0$. All these properties and relations are useful for treating fractional diffusion equations.

2.2.2 *Integral transforms of the Fox H-function*

The integral transforms constitute an important technique in solving fractional diffusion equations and treating the Fox H functions considered in ensuing chapters. We start the consideration from the Mellin transform, which is immediately obtained from the definition (2.17). Then the Mellin transform of the Fox H-function is

$$\int_0^\infty x^{\xi-1}H_{p,q}^{m,n}\left[ax\left|\begin{array}{l}(a_1,A_1),...,(a_p,A_p)\\(b_1,B_1),...,(b_q,B_q)\end{array}\right.\right]dx=a^{-\xi}\,\theta(\xi),\qquad(2.32)$$

where

$$\theta(\xi)=\frac{\prod_{j=1}^m\Gamma(b_j+B_j\xi)\prod_{j=1}^n\Gamma(1-a_j-A_j\xi)}{\prod_{j=m+1}^q\Gamma(1-b_j-B_j\xi)\prod_{j=n+1}^p\Gamma(a_j+A_j\xi)}$$

is defined in Eq. (2.18). Note in passing, it will be used to obtain the fractional moments of the fundamental solutions of fractional diffusion equations.

Supposing that the arguments of the gamma functions in $\theta(\xi)$ corresponds to the validity of the Laplace transform of the H-function (see Sec. 2.2.8 of Ref. [10]), the latter reads

$$\mathcal{L}\left\{t^{\rho-1}H_{p,q}^{m,n}\left[zt^\sigma\left|\begin{array}{l}(a_p,A_p)\\(b_q,B_q)\end{array}\right.\right]\right\}=s^{-\rho}H_{p+1,q}^{m,n+1}\left[zs^{-\sigma}\left|\begin{array}{l}(1-\rho,\sigma),(a_p,A_p)\\(b_q,B_q)\end{array}\right.\right],$$

$$(2.33)$$

where the validity conditions follow from the integration

$$\int_0^\infty t^{\rho+\sigma\xi-1}e^{-st}dt.$$

In the analogous way, the cosine Mellin transform of the Fox H-function yields

$$\int_0^\infty k^{\rho-1}\cos(kx)H_{p,q}^{m,n}\left[ak^\delta\left|\begin{array}{c}(a_p,A_p)\\(b_q,B_q)\end{array}\right.\right]dk$$

$$=\frac{\pi}{x^\rho}H_{q+1,p+2}^{n+1,m}\left[\frac{x^\delta}{a}\left|\begin{array}{c}(1-b_q,B_q),(\frac{1+\rho}{2},\frac{\delta}{2})\\(\rho,\delta),(1-a_p,A_p),(\frac{1+\rho}{2},\frac{\delta}{2})\end{array}\right.\right],$$
(2.34)

where the validity conditions are: $\Re\left(\rho+\delta\min_{1\leq j\leq m}\left(\frac{b_j}{B_j}\right)\right)>1$, $x^\delta>0$, $\Re\left(\rho+\delta\max_{1\leq j\leq n}\left(\frac{a_j-1}{A_j}\right)\right)<\frac{3}{2}$, $|\arg(a)|<\pi\theta/2$, $\theta>0$, $\theta=\sum_{j=1}^n A_j-\sum_{j=n+1}^p A_j+\sum_{j=1}^m B_j-\sum_{j=m+1}^q B_j$.

2.2.3 *Series and asymptotic expansions*

Series expansion of the Fox H-function (2.17) is given by [10, 11, 12]

$$H_{p,q}^{m,n}\left[z\left|\begin{array}{c}(a_1,A_1),...,(a_p,A_p)\\(b_1,B_1),...,(b_q,B_q)\end{array}\right.\right]$$

$$=\sum_{h=1}^m\sum_{k=0}^\infty\frac{\prod_{j=1,j\neq h}^m\Gamma\left(b_j-B_j\frac{b_h+k}{B_h}\right)\prod_{j=1}^n\Gamma\left(1-a_j+A_j\frac{b_h+k}{B_h}\right)}{\prod_{j=m+1}^q\Gamma\left(1-b_j+B_j\frac{b_h+k}{B_h}\right)\prod_{j=n+1}^p\Gamma\left(a_j-A_j\frac{b_h+k}{B_h}\right)}$$

$$\times\frac{(-1)^k z^{(b_h+k)/B_h}}{k!B_h},$$
(2.35)

when the poles $\prod_{j=1}^m\Gamma(b_j-B_j s)$ are simple, i.e., $B_h(b_j+l)\neq B_j(b_h+k)$ for $j\neq h$, $h=1,\ldots,m$, $l,k=0,1,2,\ldots$. This expansion exists for all $z\neq 0$ if $m^*>0$ and for $0<|z|<1/C$ if $m^*=0$, where

$$m^*=\sum_{j=1}^q B_j-\sum_{j=1}^p A_j,$$
(2.36)

and

$$C=\prod_{k=1}^p(A_k)^{A_k}\prod_{k=1}^q(B_k)^{-B_k}.$$
(2.37)

When the poles $\prod_{j=1}^n\Gamma(1-a_j+A_j s)$ are simple, i.e., $A_h(1-a_j+l)\neq A_j(1-a_h+k)$ for $j\neq h$, $h=1,\ldots,m$, $l,k=0,1,2,\ldots$, the Fox H-function has the following series expansion [10, 11]

$$
H_{p,q}^{m,n} \left[z \left| \begin{array}{c} (a_1, A_1), ..., (a_p, A_p) \\ (b_1, B_1), ..., (b_q, B_q) \end{array} \right. \right]
$$

$$
= \sum_{h=1}^{m} \sum_{k=0}^{\infty} \frac{\prod_{j=1, j \neq h}^{n} \Gamma \left(1 - a_j - A_j \frac{1 - a_h + k}{A_h} \right) \prod_{j=1}^{m} \Gamma \left(b_j + B_j \frac{1 - a_h + k}{A_h} \right)}{\prod_{j=n+1}^{p} \Gamma \left(a_j + A_j \frac{1 - a_h + k}{A_h} \right) \prod_{j=m+1}^{q} \Gamma \left(1 - b_j - B_j \frac{1 - a_h + k}{A_h} \right)}
$$

$$
\times \frac{(-1)^k \left(\frac{1}{z} \right)^{(1 - a_h + k)/A_h}}{k! A_h}. \tag{2.38}
$$

This expansion exists for all $z \neq 0$ if $m^* < 0$ and for $|z| > 1/C$ if $m^* = 0$, where m^* is given by (2.36) and C by (2.37).

The asymptotic expansion of the Fox H-function $H_{p,q}^{m,0}(z)$ for large z is [5]

$$
H_{p,q}^{m,0}(z) \sim B z^{(1-\alpha)/m^*} \exp \left(-m^* C^{1/m^*} z^{1/m^*} \right), \tag{2.39}
$$

where

$$
\alpha = \sum_{k=1}^{p} a_k - \sum_{k=1}^{q} b_k + \frac{1}{2}(q - p + 1), \tag{2.40}
$$

$$
B = (2\pi)^{\frac{q-p-1}{2}} C^{(1-\alpha)/m^*} (m^*)^{-1/2} \prod_{k=1}^{p} (A_k)^{-a_k + 1/2} \prod_{k=1}^{m} (B_k)^{b_k - 1/2}, \tag{2.41}
$$

while m^* is given by (2.36) and C by (2.37). This formula, as we will see in the next chapters, is very important in the analysis of the solution of fractional diffusion and Fokker-Planck equations (we also use the abbreviation FPE).

2.2.4 Relation to other functions

To gain more insights into the theory of the H-function, we also come to grips with some details of its relation to other functions. To this end we shall resort to the inverse Mellin transform by means of the Mellin-Barnes integral (2.17). Evaluating the integration of the gamma functions in $\theta(\xi)$ as a sum of residues at the poles $\xi = 0, -1, -2, \ldots$, we show how the Fox H-function relates to other functions. We start from the simplest examples of the exponential and the M-L functions [10].

Example 2.5 (H-function vs exponential and M-L functions).
Let us first consider $\theta(-\xi) = \Gamma(\xi)$, then we have

$$H_{0,1}^{1,0}\left[z\,\middle|\,(0,1)\right] = \frac{1}{2\pi i}\int_{c-i\infty}^{c+i\infty}\Gamma(\xi)z^{-\xi}\,d\xi$$

$$= \sum_{k=0}^{\infty}\lim_{\xi\to-k}(\xi+k)\Gamma(\xi)z^{-\xi}$$

$$= \sum_{k=0}^{\infty}\lim_{\xi\to-k}\frac{(\xi+k)(\xi+k-1)\cdots\xi}{(\xi+k-1)\cdots\xi}\Gamma(\xi)z^{-\xi}$$

$$= \sum_{k=0}^{\infty}\lim_{\xi\to-k}\frac{\Gamma(\xi+k+1)}{(\xi+k-1)\cdots\xi}z^{-\xi}$$

$$= \sum_{k=0}^{\infty}\frac{(-1)^k}{k!}z^k = e^{-z}. \tag{2.42}$$

Again, evaluating the Mellin-Barnes integral as a sum of residues of the two gamma functions for $\theta(\xi) = \frac{\Gamma(\xi)\Gamma(1-\xi)}{\Gamma(\beta-\alpha\xi)}$, we have the two parameter M-L function,

$$H_{2,1}^{1,1}\left[z\,\middle|\,\begin{matrix}(0,1)\\(0,1),(1-\beta,\alpha)\end{matrix}\right] = \frac{1}{2\pi i}\int_{c-i\infty}^{c+i\infty}\frac{\Gamma(\xi)\Gamma(1-\xi)}{\Gamma(\beta-\alpha\xi)}z^{-\xi}\,d\xi$$

$$= \sum_{k=0}^{\infty}\lim_{\xi\to-k}\frac{(\xi+k)\pi}{\sin(\pi\xi)\Gamma(\beta-\alpha\xi)}z^{-\xi}$$

$$= \sum_{k=0}^{\infty}\frac{(-z)^k}{\Gamma(k\alpha+\beta)} = E_{\alpha,\beta}(-z). \tag{2.43}$$

Here, in the second line we have used the limit $\lim_{\xi\to-k}[(\xi+k)\pi]/\sin(\xi\pi) = \lim_{\xi\to-k}(-1)^k[(\xi+k)\pi]/\sin[(\xi+k)\pi]$ and the property (1.107). Correspondingly, the one parameter M-L function is

$$E_\alpha(-z) = H_{1,2}^{1,1}\left[z\,\middle|\,\begin{matrix}(0,1)\\(0,1),(0,\alpha)\end{matrix}\right], \tag{2.44}$$

The three parameter M-L function is a special case of the Fox H-function, as well

$$E_{\alpha,\beta}^{\delta}(-z) = \frac{1}{\Gamma(\delta)}H_{1,2}^{1,1}\left[z\,\middle|\,\begin{matrix}(1-\delta,1)\\(0,1),(1-\beta,\alpha)\end{matrix}\right]. \tag{2.45}$$

$\square$

Thus, by using relations (2.43) and (2.34), the cosine transform of the two parameter M-L function is given in terms of the Fox H-function,

$$\int_0^\infty \cos(kx) E_{\alpha,\beta}\left(-ak^2\right) dk = \int_0^\infty \cos(kx) H_{2,1}^{1,1}\left[ak^2 \left| \begin{array}{c} (0,1) \\ (0,1),(1-\beta,\alpha) \end{array}\right.\right] dk$$

$$= \frac{\pi}{x} H_{3,3}^{2,1}\left[\frac{x^2}{a} \left| \begin{array}{c} (1,1),(\beta,\alpha),(1,1) \\ (1,2),(1,1),(1,1) \end{array}\right.\right]$$

$$= \frac{\pi}{x} H_{1,1}^{1,0}\left[\frac{x^2}{a} \left| \begin{array}{c} (\beta,\alpha) \\ (1,2) \end{array}\right.\right], \qquad (2.46)$$

where we apply the reduction formula (2.23) in Proposition 2.2. In the same way, from Eq. (2.34), we find the cosine transform of the three parameter M-L function

$$\int_0^\infty \cos(kx) E_{\alpha,\beta}^\delta\left(-ak^\lambda\right) dk$$

$$= \frac{1}{\Gamma(\delta)} \int_0^\infty \cos(kx) H_{1,2}^{1,1}\left[ak^\lambda \left| \begin{array}{c} (1-\delta,1) \\ (0,1),(1-\beta,\alpha) \end{array}\right.\right] dk$$

$$= \frac{1}{\Gamma(\delta)} \frac{\pi}{x} H_{2,3}^{2,1}\left[\frac{x^\lambda}{a} \left| \begin{array}{c} (1,1),(\beta,\alpha),(1,\lambda/2) \\ (1,\lambda),(\delta,1),(1,\lambda/2) \end{array}\right.\right], \qquad (2.47)$$

which for $\lambda = 2$, reduces to

$$\int_0^\infty \cos(kx) E_{\alpha,\beta}^\delta\left(-ak^2\right) dk$$

$$= \frac{1}{\Gamma(\delta)} \frac{\pi}{x} H_{2,3}^{2,1}\left[\frac{x^2}{a} \left| \begin{array}{c} (1,1),(\beta,\alpha),(1,1) \\ (1,2),(\delta,1),(1,1) \end{array}\right.\right]$$

$$= \frac{1}{\Gamma(\delta)} \frac{\pi}{x} H_{1,2}^{2,0}\left[\frac{x^2}{a} \left| \begin{array}{c} (\beta,\alpha),(1,1) \\ (1,2),(\delta,1) \end{array}\right.\right], \qquad (2.48)$$

due to the reduction property (2.2).

Note also that for a special case of parameters of the Fox H-function, one obtains

$$H_{0,1}^{1,0}\left[z \left| \begin{array}{c} \\ (b,B) \end{array}\right.\right] = B^{-1} z^{b/B} \exp\left(-z^{1/B}\right), \qquad (2.49)$$

which for $B = 1$ turns to

$$H_{0,1}^{1,0}\left[z \left| \begin{array}{c} \\ (b,1) \end{array}\right.\right] = z^b \exp\left(-z\right), \qquad (2.50)$$

as well as

$$\frac{1}{(1 \pm z)^a} = \frac{1}{\Gamma(a)} H_{1,1}^{1,1}\left[\pm z \left| \begin{array}{c} (1-a,1) \\ (0,1) \end{array}\right.\right]. \qquad (2.51)$$

The modified Bessel function (of the third kind) $K_\nu(z)$ is a special case of the Fox H-function, as well. Using the same technique of Example 2.5, we establish this in the example.

Example 2.6 (*H*-function vs Bessel function).
Consider the Mellin-Barnes integral, one arrives at the following chain of transformations

$$H_{0,2}^{2,0}\left[\frac{z^2}{4}\;\middle|\;(\tfrac{a+\nu}{2},1),(\tfrac{a-\nu}{2},1)\right]$$

$$=\frac{1}{2\pi\imath}\int_{c-\imath\infty}^{c+\imath\infty}\Gamma\left(\tfrac{a+\nu}{2}+\xi\right)\Gamma\left(\tfrac{a-\nu}{2}+\xi\right)\left(\tfrac{z}{2}\right)^{-2\xi}\mathrm{d}\xi$$

$$=\left(\frac{z}{2}\right)^a\left[\sum_{k=0}^{\infty}\lim_{\xi+\frac{\nu}{2}\to-k}\left(\xi+\tfrac{\nu}{2}+k\right)\Gamma\left(\tfrac{\nu}{2}+\xi\right)\Gamma\left(\tfrac{-\nu}{2}+\xi\right)\left(\tfrac{z}{2}\right)^{-2\xi}\right.$$

$$\left.+\sum_{k=0}^{\infty}\lim_{\xi-\frac{\nu}{2}\to-k}\left(\xi-\tfrac{\nu}{2}+k\right)\Gamma\left(\tfrac{\nu}{2}+\xi\right)\Gamma\left(\tfrac{-\nu}{2}+\xi\right)\left(\tfrac{z}{2}\right)^{-2\xi}\right]$$

$$=\left(\frac{z}{2}\right)^a\left[\sum_{k=0}^{\infty}\lim_{\xi+\frac{\nu}{2}\to-k}\frac{\Gamma\left(\xi+k+\tfrac{\nu}{2}+1\right)}{(\xi+k-1+\nu/2)\cdots(\xi+\nu/2)}\Gamma\left(\tfrac{-\nu}{2}+\xi\right)\left(\tfrac{z}{2}\right)^{-2\xi}\right.$$

$$\left.+\sum_{k=0}^{\infty}\lim_{\xi i-\frac{\nu}{2}\to-k}\frac{\Gamma\left(\xi+k-\tfrac{\nu}{2}+1\right)}{(\xi+k-1-\nu/2)\cdots(\xi-\nu/2)}\Gamma\left(\tfrac{\nu}{2}+\xi\right)\left(\tfrac{z}{2}\right)^{-2\xi}\right]$$

$$\times\left(\frac{z}{2}\right)^a\frac{\pi}{\sin(\nu\pi)}\left[\left(\frac{z}{2}\right)^{-\nu}\sum_{k=0}^{\infty}\frac{\left(\frac{z}{2}\right)^{2k}}{k!\Gamma(k-\nu+1)}-\left(\frac{z}{2}\right)^{\nu}\sum_{k=0}^{\infty}\frac{\left(\frac{z}{2}\right)^{2k}}{k!\Gamma(k+\nu+1)}\right]$$

$$=\left(\frac{z}{2}\right)^a\frac{\pi}{\sin(\nu\pi)}[I_{-\nu}(z)-I_\nu(z)]=2\left(\frac{z}{2}\right)^a K_\nu(z),\qquad(2.52)$$

where $I_\nu(z)$ and $K_\nu(z)$ are the modified Bessel functions [13]. The series representation of the result for $z\to0$, $\nu\notin Z$ is

$$K_\nu(z)\simeq\frac{\Gamma(\nu)}{2}\left(\frac{z}{2}\right)^{-\nu}\left[1+\frac{z^2}{4(1-\nu)}+\cdots\right]$$

$$+\frac{\Gamma(-\nu)}{2}\left(\frac{z}{2}\right)^{\nu}\left[1+\frac{z^2}{4(\nu+1)}+\cdots\right].\qquad(2.53)$$

$\square$

The Bessel and modified Bessel functions are implemented in Wolfram Language as `BesselJ[n, z]`, `BesselY[n, z]`, `BesselI[n, z]` and `BesselK[n, z]`.

2.2.4.1 *Wright function*

Another type of transcendental/hypergeometric functions is the Wright function, which intimately relates to the fractional calculus and the Fox H-function and is a particular case of the latter.[2]

The Fox-Wright function is defined by [10, 14, 15]

$$_p\Psi_q(z) = {}_p\Psi_q \left[\begin{matrix} (a_1, A_1), ..., (a_p, A_p); \\ (b_1, B_1), ..., (b_q, B_q); \end{matrix} z \right] = \sum_{k=0}^{\infty} \frac{\prod_{j=1}^{p} \Gamma(a_j + A_j k)}{\prod_{j=1}^{q} \Gamma(b_j + B_j k)} \cdot \frac{z^k}{k!},$$

(2.54)

where $a_j, A_j \in C$, $\Re[A_j] > 0$, for $j = 1, ..., p$ i $b_j, B_j \in C$, $\Re[B_j] > 0$, for $j = 1, ..., q$, $1 + \Re\left(\sum_{j=1}^{q} B_j - \sum_{j=1}^{p} A_j\right) \geq 0$. For a special case of the Wright function ($p = 0$, $q = 1$, $b_1 = \beta$, $B_1 = \alpha$), the following notation is used [10]

$$\varphi(\alpha, \beta; z) = \sum_{n=0}^{\infty} \frac{1}{\Gamma(\alpha n + \beta)} \frac{z^n}{n!} = H_{0,2}^{1,0} \left[-z \left| \begin{matrix} - \\ (0,1), (1 - \beta, \alpha) \end{matrix} \right. \right], \quad (2.55)$$

where $\Re(\alpha) > -1$, $\beta \in C$.

It is easily seen from the definition that

$$E_{\alpha,\beta}^{\gamma}(z) = \frac{1}{\Gamma(\gamma)} {}_1\Psi_1 \left[\begin{matrix} (\gamma, 1); \\ z \\ (\beta, \alpha); \end{matrix} \right]. \quad (2.56)$$

Let us return to Eq. (2.55), which is a generalization of the Bessel functions [16]. It is also known as the Wright function of the first kind for real $\alpha > -1$ [14, 15]. For $\alpha = 1$ and $\beta = \nu + 1$, it reduces to the Bessel functions (see Example 2.6)

$$(\varphi(1, 1 \pm \nu; z^2/4) = \sum_{n=0}^{\infty} \frac{1}{\Gamma(n \pm \nu + 1)} \frac{(z/2)^{2n}}{n!} = (z/2)^{\mp\nu} I_{\pm\nu}(z) \quad (2.57a)$$

$$\varphi(1, 1 + \nu; z^2/4) = \sum_{n=0}^{\infty} \frac{(-1)^n}{\Gamma(n + \nu + 1)} \frac{(z/2)^{2n}}{n!} = (z/2)^{-\nu} J_{\nu}(z). \quad (2.57b)$$

The Laplace transform immediately yields the two parameter M-L function that reads

$$\mathcal{L}[\varphi(\alpha, \beta; z)] = s^{-1} E_{\alpha,\beta}(s^{-1}). \quad (2.58)$$

Taking into account the integral representation of the reciprocal gamma function (1.117), the Hankel integral representation of the Wright function (2.55) reads

$$\varphi(\alpha, \beta; z) = \frac{1}{2\pi\imath} \int_{\{Ha\}} \sum_{n=0}^{\infty} \frac{z^n}{n!} \tau^{-(\alpha n + \beta)} e^{\tau} d\tau = \frac{1}{2\pi\imath} \int_{\{Ha\}} \tau^{-\beta} e^{\tau + z\tau^{-\alpha}} d\tau.$$

(2.59)

---

[2]An extended review of the Wright function can be found in Ref. [14].

2.2.5 *Stable distributions*

In this section, we briefly touch some mathematical aspects of the Lévy (or stable) distributions. An extended discussion can be found in Refs. [12, 17, 18, 19, 20, 21]. Lévy distribution $L(x; \alpha, \beta)$ is defined by its Fourier transform

$$L(x; \alpha, \beta) = \frac{1}{2\pi} \int_{-\infty}^{\infty} \tilde{L}(k; \alpha, \beta) e^{ikx} \, dk, \tag{2.60a}$$

$$\tilde{L}(k; \alpha, \beta) = \exp\{-|k|^{\alpha}[1 - i\beta\,\mathrm{sign}(k)\tan(\pi\alpha/2)]\}, \tag{2.60b}$$

where $0 < \alpha \le 2$ and $\beta \in [-1, 1]$. When the skewness parameter $\beta = 0$ the distribution is symmetrical. We used here two sub-classes, which are mostly popular in applications [21]. The first one is a symmetrical stable distribution formed by the stable characteristic function

$$\tilde{L}(k; \alpha, 0) \equiv \tilde{L}(k; \alpha) = e^{-|k|^{\alpha}}. \tag{2.61}$$

In particular,

$$L(x; 2) = \frac{1}{\sqrt{4\pi}} e^{-\frac{x^2}{4}} \tag{2.62}$$

is the Gaussian (normal) distribution, when all moments are finite, while

$$L(x; 1) = \frac{1}{\pi\,(1 + x^2)} \tag{2.63}$$

is the Cauchy distribution, when all moments diverge.

Another example is the Lévy-Smirnov distribution

$$\tilde{L}(x; 1/2) = \frac{1}{\sqrt{4\pi x^3}} \exp\left(-\frac{1}{4x}\right), \quad x > 0. \tag{2.64}$$

As shown in Eq. (2.50), the characteristic function (2.61) corresponds to the Lévy distribution, which is determined by the Fox H-function

$$e^{-|k|^{\alpha}} = H^{1,0}_{0,1}\left[|k|^{\alpha}\,\middle|\, \begin{matrix} - \\ (0, 1) \end{matrix}\right]. \tag{2.65}$$

Then the cosine Mellin transformation (2.34) yields

$$L(x; \alpha) = \frac{1}{2\pi} \int_{-\infty}^{\infty} e^{-|k|^{\alpha}} e^{ikx} \, dx = \frac{1}{\pi} \int_{0}^{\infty} \cos(kx) \, e^{-k^{\alpha}} \, dx$$

$$= \frac{1}{\pi} \int_{0}^{\infty} \cos(kx) \, H^{1,0}_{0,1}\left[|k|^{\alpha}\,\middle|\, \begin{matrix} - \\ (0, 1) \end{matrix}\right] dx$$

$$= \frac{1}{|x|} H^{1,1}_{2,2}\left[|x|^{\alpha}\,\middle|\, \begin{matrix} (1.1), (1, \alpha/2) \\ (1, \alpha), (1, \alpha/2) \end{matrix}\right] = \frac{1}{\pi\alpha} \sum_{l=0}^{\infty} \frac{(-1)^l \Gamma(\frac{2l+1}{\alpha})}{(2l)!} x^{2l},$$
$$\tag{2.66}$$

where the expansion of the Fox H-function results from the integration [17].

Another approach is through the auxiliary functions of the Wright type, related to the Mainardi function [14, 22, 23], which is defined as follows

$$M_\alpha(y) = \sum_{n=0}^{\infty} \frac{1}{\Gamma(-\alpha n + 1 - \alpha)} \frac{(-y)^n}{n!}. \tag{2.67}$$

Correspondingly, its relation to the Fox H-function is [24]

$$M_\alpha(y) = H_{1,1}^{1,0} \left[y \left| \begin{matrix} (1-\alpha, \alpha) \\ (0,1) \end{matrix} \right. \right]. \tag{2.68}$$

Then the one-sided Lévy stable probability density $L_\alpha(y)$ in Eq. (2.66) can be expressed through the $M_\alpha(y)$ as follows [25]

$$L_\alpha(t) = \frac{\alpha}{t^{\alpha+1}} M_\alpha \left(\frac{1}{t^\alpha} \right). \tag{2.69}$$

In this case, the Fourier transform is replaced by the Laplace transform [26]

$$L_\alpha(t) = \mathcal{L}^{-1} \left[e^{-s^\alpha} \right]. \tag{2.70}$$

The Lévy distribution (2.60a) is implemented in Wolfram Language by StableDistribution$[\alpha, \beta]$, while its characteristic function (2.60b) by CharacteristicFunction$[$StableDistribution$[\alpha, \beta], \mathtt{k}]$.

References

[1]   C. Fox. "The G and H functions as symmetrical Fourier kernels". In: *Transactions of the American Mathematical Society* 98.3 (1961), pp. 395–429 (cit. on pp. 45, 49).

[2]   W. R. Schneider. "Stable distributions: Fox function representation and generalization". In: *Stochastic processes in classical and quantum systems.* Berlin: Springer, 1986, pp. 497–511 (cit. on p. 45).

[3]   W. Wyss. "The fractional diffusion equation". In: *Journal of Mathematical Physics* 27.11 (1986), pp. 2782–2785 (cit. on p. 45).

[4]   W. R. Schneider. "Fractional diffusion". In: *Dynamics and Stochastic Processes Theory and Applications.* Berlin: Springer, 1990, pp. 276–286 (cit. on p. 45).

[5]   W. R. Schneider and W. Wyss. "Fractional diffusion and wave equations". In: *Journal of Mathematical Physics* 30.1 (1989), pp. 134–144 (cit. on pp. 45, 56).

[6]   H. Bateman and A. Erdélyi. *Higher Transcendental Functions [Volumes I-III].* New York: McGraw-Hill, 1953–1955 (cit. on pp. 45, 46, 49).

[7]   R. A. Askey and A. B. O. Daalhuis. "Generalized hypergeometric functions and Meijer G-Function". In: *NIST Handbook of Mathematical Functions.* Ed. by F. W. J. Olver, D. M. Lozier, and R. F. Boisvert. Cambridge University Press, 2010. Chap. 16 (cit. on p. 45).

[8]   R. Beals and J. Szmigielski. "Meijer G–Functions: A Gentle Introduction". In: *Notices of the American Mathematical Society* 60.7 (2013), pp. 866–872 (cit. on pp. 46, 48).

[9]   B. L. J. Braaksma. "Asymptotic expansions and analytic continuations for a class of Barnes-integrals". In: *Compositio Mathematica* 15 (1962–1964), pp. 239–341 (cit. on p. 49).

[10] A. M. Mathai, R. K. Saxena, and H. J. Haubold. *The H-function: theory and applications.* Springer Science & Business Media, 2009 (cit. on pp. 49, 52, 54–56, 60).

[11] A. M. Mathai and R. K. Saxena. *The H-function with applications in statistics and other disciplines.* New Delhi: John Wiley & Sons, 1978 (cit. on pp. 55, 56).

[12] R. Metzler and J. Klafter. "The random walk's guide to anomalous diffusion: a fractional dynamics approach". In: *Physics Reports* 339.1 (2000), pp. 1–77 (cit. on pp. 55, 61).

[13] M. Abramovitz and I. A. Stegun. *Handbook of Mathematical Functions with Formulas, Graphs, and Mathematical Tables.* New York: Dover Publications, 1972 (cit. on p. 59).

[14] Yu. Luchko. "The Wright function and its applications". In: *Volume 1 Basic Theory.* Ed. by Anatoly Kochubei and Yuri Luchko. Berlin, Boston: De Gruyter, 2019, pp. 241–268 (cit. on pp. 60, 62).

[15] R. Garra and F. Mainardi. "Some applications of Wright functions in fractional differential equations". In: *Reports on Mathematical Physics* 87.2 (2021), pp. 265–273. ISSN: 0034-4877 (cit. on p. 60).

[16] E. M. Wright. "The asymptotic expansion of the generalized Bessel function." English. In: *Proceedings of London Mathematical Society* 2 (1934), pp. 257–270. ISSN: 0024-6115. DOI: 10.1112/plms/s2-38.1.257 (cit. on p. 60).

[17] E. W. Montroll and J. T. Bendler. "On Lévy (or stable) distributions and the Williams-Watts model of dielectric relaxation." In: *Journal of Statistical Physics* 34 (1984), pp. 129–162 (cit. on p. 61).

[18] B. J. West et al. "Fractional diffusion and Lévy stable processes". In: *Physical Review E* 55 (1997), pp. 99–106 (cit. on p. 61).

[19] J. P. Bouchaud and A. Georges. "Anomalous Diffusion in Disordered Media: Statistical Mechanisms, Models and Physical Applications". In: *Physics Reports* 195.4-5 (1990), pp. 127–293 (cit. on p. 61).

[20] S. Jespersen, R. Metzler, and H. C. Fogedby. "Lévy flights in external force fields: Langevin and fractional Fokker-Planck equations and their solutions". In: *Physical Review E* 59 (3 1999), pp. 2736–2745 (cit. on p. 61).

[21] V. Uchaikin and R. Sibatov. *Fractional Kinetics in Solids*. Singapore: World Scientific, 2013 (cit. on p. 61).

[22] F. Mainardi. *Fractional Calculus and Waves in Linear Viscoelasticity*. London: Imperial College Press, 2010 (cit. on p. 62).

[23] T. Sandev and Z. Tomovski. *Fractional Equations and Models*. Springer Nature, 2019 (cit. on p. 62).

[24] F. Mainardi, G. Pagnini, and R. K. Saxena. "Fox H functions in fractional diffusion". In: *Journal of Computational and Applied Mathematics* 178.1-2 (2005), pp. 321–331 (cit. on p. 62).

[25] R. Gorenflo and F. Mainardi. "Parametric Subordination in Fractional Diffusion Processes". In: *Fractional Dynamics: Recent Advances*. World Scientific, 2012, pp. 229–263 (cit. on p. 62).

[26] E. Barkai. "Fractional Fokker-Planck equation, solution, and application". In: *Physical Review E* 63.4 (2001), p. 046118 (cit. on p. 62).

Chapter 3

Elements of random walk theory

The term "random walk" was invented by Karl Pearson in his letter to the English journal Nature, which appeared on 27 July 1905. An extended theoretical studies of random walk phenomena have been started from three fundamental works by: (1) Louis Bachelier in his thesis devoted to the theory of financial speculations in 1900; (2) Albert Einstein's work (1905); and (3) independent result of Marian Smoluchowski (1906) on the theory and explanation of Brownian motion.

In this chapter we present some basic material on a random walk theory. We follow a concise introduction to the topics in Refs. [1, 2]. The literature on this subject is vast and the reader has an abundance of books to chose from, we mention but few [3, 4, 5, 6] used here.

3.1 Stochastic variables and characteristics

A random walk as a stochastic process is described by random variables, say $X(t)$, indexed by the time variable $t \geq 0$, which can be a set of integer ($t \in \mathbb{Z}_+$) or real ($t \in \mathbb{R}_+$)) numbers. Then random variables $X(t)$ describe a stochastic process, which can be either discrete-time stochastic process or continuous-time stochastic one. Being a random value, $X(t)$ corresponds to a family of realizations, and for different times $t = t_1, t_2, \ldots, t_n$, the realization is $X(t_1) = x_1$, $X(t_2) = x_2, \ldots, X(t_n) = x_n$. Each realization appears with the weight, which is described by the probability density function (PDF) $P(x, t)$, which contains the basic information of realization of $X(t)$ for each t, such that

$$P(x_j, t_j) \, dx_j = \Pr(x_j < X(t_j) < x_j + dx_j) \tag{3.1}$$

is the probability that $X(t_j) \in (x_j, x_j + dx_j)$ at $t = t_j$, where $j = 1, 2, \ldots, n$ and each x_j varies in its own region, or a field of realizations: $x_j \in \Omega_j$.

One also considers the joint PDF $P(x_1, t_1; x_2, t_2)$, which defines the probability that for $t = t_1$, $x_1 < X(t_1) < x_1 + dx_1$ and for $t = t_2$, $x_2 < X(t_2) < x_2 + dx_2$. Continuing this way for all t_j we arrive at the joint PDF $P(x_1, t_1; x_2, t_2; \ldots; x_n, t_n)$, such that all possible realizations of the process, defined in the n dimensional space $\Omega_1 \times \Omega_2 \times \cdots \times \Omega_n$, has the probability one,

$$\int_{\Omega_1} \cdots \int_{\Omega_n} P(x_n, t_n; \ldots; x_2, t_2; x_1, t_1)\, dx_1 \cdots dx_n = 1 \qquad (3.2)$$

with the property

$$\int_{\Omega_j} P(x_n, t_n; \ldots; x_j, t_j; \ldots; x_2, t_2; x_1, t_1)\, dx_j$$

$$= P(x_n, t_n; \ldots; x_{j-1}, t_{j-1}; x_{j+1}, t_{j+1}; \ldots; x_2, t_2; x_1, t_1). \qquad (3.3)$$

Another important function is the conditional PDF, which provides $X(t_i) = x_i$ for $j+1 \le i \le n$ given that $X(t_i) = x_i$ for $1 \le i \le j$ and which is defined through the Bayes theorem for the probabilities

$$\Pr(x_n, t_n; \ldots; x_{j+1}, t_{j+1} \mid x_j, t_j; \ldots; x_1, t_1) = \frac{P(x_n t_n; \ldots; x_1, t_1)}{P(x_j, t_j; \ldots; x_1, t_1)}, \qquad (3.4)$$

where we use that the PDFs are invariant with respect to the index permutations and $1 < j < n$. This expression allows us to connect the joint PDF for the interval (t_1, t_j) to that one for the interval (t_1, t_n). An important relation also follows from Eqs. (3.3) and (3.4) that

$$P(x_2, t_2) = \int_{\Omega_1} P(x_2, t_2 \mid x_1, t_1) P(x_1, t_1)\, dx_1, \qquad (3.5)$$

where the conditional PDF plays the role of the *transition PDF* or the Green's function for the evolution of the PDF.

3.1.1   *Averaged characteristics*

The random variables $X(t)$ and their PDF $P(x, t)$ supply the information about the random process and can translate the information from the PDF to regular, average quantities or moments. The moment of order n of the random variable $X(t)$ is defined as

$$\langle X^n(t) \rangle = \int_{\Omega} x^n P(x, t)\, dx. \qquad (3.6)$$

The first moment $\langle X(t) \rangle$ is also called the mean, or expectation value. The second moment is $\langle X^2 \rangle$, and the combination of the first and the second moments defines the variance, or dispersion,

$$\sigma^2 = \langle (X - \langle X \rangle)^2 \rangle = \langle X^2 \rangle - \langle X \rangle^2. \tag{3.7}$$

We also call it the mean squared displacement (MSD).

Introducing the Fourier transform of the PDF, which is the expectation value of the exponential, the n-th moment can be obtained from a generating function, which also known as the *characteristic function* of the variable $X(t)$:

$$\langle e^{ikX(t)} \rangle = \int_\Omega e^{ikx} P(x,t)\, dx = \tilde{P}(k,t). \tag{3.8}$$

Differentiating it n times with respect to k at $k = 0$,

$$\left. \frac{d^n}{d\,k^n} \tilde{P}(k,t) \right|_{k=0}, \tag{3.9}$$

one obtains Eq. (3.6).

Another important average value is the correlation function of two random variables $X(t_1)$, and $X(t_2)$, which is defined as

$$\mathrm{Cor}\,(X(t_1), X(t_2)) = \langle [X(t_1) - \langle X(t_1) \rangle][X(t_2) - \langle X(t_2) \rangle] \rangle. \tag{3.10}$$

If the correlation function vanishes $\mathrm{Cor}\,(X(t_1), X(t_2)) = 0$, the variables are *uncorrelated*. If the two variables are *independent* then $P(x_1, t_1; x_2, t_2) = P(x_1, t_1) \cdot P(x_1, t_2)$, which also implies that the two variables are uncorrelated:

$$\mathrm{Cor}\,(X(t_1), X(t_2)) = \langle X(t_1) - \langle X(t_1) \rangle \rangle \cdot \langle X(t_2) - \langle X(t_2) \rangle \rangle = 0. \tag{3.11}$$

Note that uncorrelated does not imply independent. A random process with independent values at every instant of time,

$$P(x_n t_n; \ldots; x_1, t_1) = \prod_{i=1}^{n} P(x_i, t_i), \tag{3.12}$$

is known as a purely or *completely random process*.

3.2 Markov processes

The dependence relation between the random variables $X(t_1), \ldots, X(t_n)$ characterizes the random process. Processes with a short memory, where the dependence relation remains relatively simple, namely, when two variables are independent, are the so-called Markov processes. That is, the process is Markovian if the conditional (transition) PDF possesses the property

$$P(x_n, t_n \mid x_{n-1}, t_{n-1}; \ldots; x_1, t_1) = P(x_n, t_n \mid x_{n-1}, t_{n-1}). \qquad (3.13)$$

Therefore, a Markov process is completely determined by the "initial" PDF $P(x_1, t_1)$ and the transition PDF $P(x_j, t_j \mid x_{j-1}, t_{j-1})$. Repeated use of Eqs. (3.13) and (3.4) leads to

$$P(x_n, t_n; x_{n-1}, t_{n-1}; \ldots; x_2, t_2; x_1, t_1)$$
$$= P(x_n, t_n \mid x_{n-1}, t_{n-1}) \times P(x_{n-1}, t_{n-1} \mid x_{n-2}, t_{n-2})$$
$$\cdots \times P(x_2, t_2 \mid x_1, t_1) \times P(x_1, t_1). \qquad (3.14)$$

3.2.1   *Smoluchowski–Chapman–Kolmogorov equation*

Let us consider the three-point joint PDF in Eq. (3.14), which reads

$$P(x_3, t_3; x_2, t_2; x_1, t_1) = P(x_3, t_3 \mid x_2, t_2)P(x_2, t_2 \mid x_1, t_1)P(x_1, t_1), \qquad (3.15)$$

where $t_1 < t_2 < t_3$. Integrating Eq. (3.15) with respect to x_2 and taking into account property (3.3), we find

$$\int_{\Omega_2} P(x_3, t_3; x_2, t_2; x_1, t_1)\, dx_2$$
$$= \int_{\Omega_2} P(x_3, t_3 \mid x_2, t_2)P(x_2, t_2 \mid x_1, t_1)P(x_1, t_1)\, dx_2 \qquad (3.16)$$

and

$$P(x_3, t_3; x_1, t_1) = \int_{\Omega_2} P(x_3, t_3 \mid x_2, t_2)P(x_2, t_2 \mid x_1, t_1)P(x_1, t_1)\, dx_2. \qquad (3.17)$$

Since

$$P(x_3, t_3; x_1, t_1) = P(x_3, t_3 \mid x_1, t_1)P(x_1, t_1), \qquad (3.18)$$

we conclude that the transition PDFs of a Markov process obey the equation

$$P(x_3, t_3 \mid x_1, t_1) = \int_{\Omega_2} P(x_3, t_3 \mid x_2, t_2)P(x_2, t_2 \mid x_1, t_1)\, dx_2, \qquad (3.19)$$

which is commonly known as the Smoluchowski, or the Chapman–Kolmogorov equation.[1] This equation states that for a Markov process

---

[1]In a paper by Montroll and West, this equation is called the chain equation of Bachelier, Smoluchowski, Chapman, and Kolmogorov [7]: "Equation [...] was first introduced by Bachelier (1900) in his thesis (under the direction of Poincaré) on market speculation. It was later discussed, independently by Smoluchowski (1906), Chapman (1916), and Kolmogorov (1931)."

the transition from the state $X(t_1) = x_1$ to the state $X(t_3) = x_3$ can be decomposed into two stages. First, a transition from the state x_1 at time t_1 to an arbitrary state x_2 at an arbitrary intermediate time t_2. Then, collect the transitions form all possible intermediate states to the state x_3 at time t_3.

Equation (3.19) is a consistency equation for the conditional or transition PDFs of a Markov process and its solution gives a complete description of any Markov process. The Chapman–Kolmogorov equation is a nonlinear functional equation, and a general solution is not known. For some classes of Markov processes, either an explicit solution can be obtained, or Eq. (3.19) can be cast into the form of a partial differential equation for the transition PDF, which is easier to deal with.

3.2.2 *Markov chains*

The Smoluchowski–Chapman–Kolmogorov equation can be written also for a process $X(t)$ with discrete state space and discrete time t. Thereof, Eqs. (3.5) and (3.19) for the PDF $P(x_n, t) \equiv P(n, t)$ and the transition PDF $P(x_n|x_m, t) \equiv P(n|m, t)$ are

$$P(n, t) = \sum_m P(n, t|m, t-1)P(m, t-1), \qquad (3.20a)$$

$$P(n, t|n_0, t_0) = \sum_m P(n, t|m, t-1)P(m, t-1|n_0, t_0), \qquad (3.20b)$$

respectively. If the state space is finite or countable, the process is called a Markov chain, and Eqs. (3.20a) and (3.20b) are known as the Markov equations. The transition probability $P(n, t|m, t-1)$ contains the necessary information about the transition mechanism from the state m to n, or position x_m to x_n. One can thus define the *transition* or Markov matrix $\mathbf{Q}(t-1)$ whose (m, n)-th element is

$$Q_{n,m}(t) = P(n, t|m, t-1). \qquad (3.21)$$

For a time-homogeneous process, the matrix is time independent, and the PDF that $X(t)$ has a value x_n at time t is

$$P(n, t) = \sum_m Q_{n,m} P(m, t-1). \qquad (3.22)$$

A more detailed discussion of Markov chains and their applications can be found, e.g., in the book by Karlin and Taylor [8]. Here we consider only some examples of random walks on the one dimensional lattice.[2]

Example 3.1 (One dimensional random walk). As the example, we formulate the one dimensional realization of this problem in the framework of the Markov chain. Therefore, we consider Eq. (3.22) for a random walk in one dimension, where a particle moves to the right with probability p and to the left with probability $q = 1 - p$. Then the elements of the transition matrix are

$$Q_{m,n} = p\delta_{n,m+1} + q\delta_{n,m-1}, \tag{3.23}$$

and Eq. (3.20b) reads

$$P(n,t) = pP(n-1,t-1) + qP(n+1,t-1), \tag{3.24}$$

where $t = 1, 2, \ldots$ is the number of steps taken by the random walker. The walker starts at the site l, i.e., the initial condition is $P(n,0) = \delta_{n,l}$.

We apply the method of characteristic functions, which is the discrete Fourier transform on the interval $[-\pi, \pi]$,

$$\tilde{P}(\theta,t) = \sum_{n=-\infty}^{\infty} e^{in\theta} P(t,n), \quad P(n,t) = \frac{1}{2\pi} \int_{-\pi}^{\pi} e^{-in\theta} \tilde{P}(\theta,t)\, d\theta. \tag{3.25}$$

Equations (3.23) and (3.25) imply that the characteristic function is given by

$$\tilde{P}(\theta,t) = (pe^{i\theta} + qe^{-i\theta})^t \tilde{P}(\theta,0), \tag{3.26}$$

where $\tilde{P}(\theta,0) = e^{il\theta}$. Consequently, the solution for the PDF reads

$$P(n,t) = \frac{1}{2\pi} \int_{-\pi}^{\pi} \sum_{j=1}^{t} \binom{t}{j} p^j q^{t-j} e^{i\theta(2j-t-n+l)}\, d\theta = \binom{t}{\frac{t+n-l}{2}} p^{\frac{t+n-l}{2}} q^{\frac{t-n+l}{2}},$$
$$\tag{3.27}$$

where $\binom{n}{m} = \frac{n!}{m!(n-m)!}$ is a binomial coefficient.   □

---

[2]Now we are at the position to consider Karl Pearson's problem posed in Nature as the following question: "Can any of your readers refer me to a work wherein I should find a solution of the following problem, or failing the knowledge of any existing solution provide me with an original one? I should be extremely grateful for aid in the matter. A man starts from a point O and walks l yards in a straight line; he then turns through any angle whatever and walks another l yards in a second straight line. He repeats this process n times. I require the probability that after n stretches he is at a distance between r and $r + \delta r$ from his starting point O."

Example 3.2 (Asymptotic limit). In the asymptotic limit $t \to \infty$, the binomial distribution (3.27) is approximately a Gaussian distribution, which can be immediately obtained from Stirling's formula for the gamma function $\Gamma(z)$,

$$\Gamma(z) \approx \sqrt{\frac{2\pi}{z}} e^{-z} z^z . \tag{3.28}$$

Therefore for $p = q = 1/2$, the solution (3.27) reduces to the Gaussian distribution

$$P(n, t) \approx \frac{1}{\sqrt{\pi t}} e^{\frac{(n-l)^2}{2t}} . \tag{3.29}$$

In the case of general p and q, this asymptotic behavior is known as the de Moivre–Laplace theorem [6], which is a special case of the Central Limit Theorem. $\qquad\square$

3.2.3 *Steady state, spectral decomposition and KS entropy*

Let us return to the Markov chain (3.22). Then the Markov dynamics for the probability $P(x_m, t) \equiv P(m, t)$ to be at position x_m at time t is according to the equation

$$P(n, t+1) = \sum_m Q_{n,m} P(m, t) \tag{3.30}$$

with the Markov *stochastic* matrix, which is determined by the condition [9]

$$\sum_n Q_{n,m} = 1 . \tag{3.31}$$

Let us assume that the normalized steady state $P_{st}(x_n)$ of equation (3.30) exists and reads as follows[3]

$$P_{st}(x_n) = \sum_m Q_{n,m} P_{st}(x_m), \tag{3.32}$$

or equivalently it reads as an eigenvalue equation $\mathbf{Q}P_{st} = \lambda P_{st}$ with the eigenvalue $\lambda = 1$. The Perron–Frobenius theorem[4] ensures that $\lambda = 1$ is the

---

[3]We borrow this discussion and notations from the very detailed paper [10], adjusting this issue for the very concise presentation. An extended discussion of this issue on the nonequilibrium steady states also can be found in Refs. [11] and [12].

[4]We are citing the corresponding text from Ref. [9]: "We can now state and prove the first part of the Perron-Frobenius theorem for irreducible matrices. Theorem 1. If the matrix $A \in R^{n \times n}$ is not negative and irreducible, then (a) The matrix A has a positive eigenvalue, r, equal to the spectral radius of A; (b) There is a positive (right) eigenvector associated with the eigenvalue r; (c) The eigenvalue r has algebraic multiplicity 1."

highest eigenvalue of the positive Markov matrix with the right eigenvector $|r(x_n)\rangle = P_{st}(x_n)$ being the steady state, while the left eigenvector is unit $\langle l(x_n)| = 1$. Here we use the Dirac notation of "bra" $\langle\cdot|$ and "ket" $|\cdot\rangle$ vectors, which are in use in quantum mechanics.

By means of this approach, one also defines the spectral decomposition of the stochastic Markov matrix. Following Ref. [10] we have

$$\mathbf{Q} = |r\rangle\langle l| + \sum_m \lambda_m |\lambda_m^r\rangle\langle\lambda_m^l| \equiv |r\rangle\langle l| + \sum_m e^{-\xi_m}|\xi_m^r\rangle\langle\xi_m^l|, \qquad (3.33)$$

where the other eigenvalues are presented in the exponential form $\lambda_m = e^{-\xi_m} < 1$ and all the left and right eigenvectors satisfy the closure relation

$$\mathbf{1} = |r\rangle\langle l| + \sum_m |\xi_m^r\rangle\langle\xi_m^l|. \qquad (3.34)$$

We can also redefine the finite time propagator in terms of the bra and ket vectors with the matrix elements $Q_{n,m}(\tau) = P(n,\tau|m,\tau-1) = \langle x_n|\mathbf{Q}|x_m\rangle$, where it is supposed that $x(\tau) = x_n$ and $x(\tau-1) = x_m$. Then Eq. (3.22) reads

$$Q(\tau) \equiv Q(\tau|\tau-1) = P(x(\tau),\tau|x(\tau-1),\tau-1) = \langle x(\tau)|\mathbf{Q}|x(\tau-1)\rangle. \qquad (3.35)$$

Then the Markov dynamics is according to the propagator

$$\langle x(t)|\mathbf{Q}^t|x(0)\rangle = \prod_{\tau=1}^{t} Q(\tau) = \langle x(t)|r\rangle\langle l|x(0)\rangle$$
$$+ \sum_m e^{-\xi_m t}\langle x(t)|\xi_m^r\rangle\langle\xi_m^l|x(0)\rangle. \qquad (3.36)$$

In the limit $t \to \infty$ the Markov dynamics reduces to the stationary equation (3.32) for the steady state.

Let us consider a random trajectory $x(\tau)$, with $0 \leq \tau \leq t$ according to the Markov matrix (or propagator) (3.36) and where the initial position is associated with the steady state $\langle x(0)|r\rangle = P_{st}(x(0))$. Therefore, the PDF at time t is

$$P(x(0 \leq \tau \leq t)) \left[\prod_{\tau=1}^{t}\langle x(\tau)|\mathbf{Q}|x(\tau-1)\rangle\right]\langle x(0)|r\rangle. \qquad (3.37)$$

From a generic definition of entropy per unit time $-\frac{1}{t}\langle\ln[P(x(0 \leq \tau \leq t))]\rangle$ let us compute the Kolmogorov-Sinai (KS) entropy per unite time h_{KS}, defined in the framework of the steady state, and which determines chaotic properties of the random work [11]. From Eq. (3.37), we have

$$h_{KS} = \lim_{t \to \infty} -\frac{1}{t} \langle \ln [P(x(0 \le \tau \le t))] \rangle$$

$$= \lim_{t \to \infty} \left\{ -\frac{1}{t} \sum_{\tau=1}^{t} \langle \ln [Q(\tau)] \rangle - \frac{1}{t} \ln [P_{st}(x(0))] \right\}$$

$$= \sum_m P_{st}(x_m) \left[-\sum_n Q_{n,m} \ln (Q_{n,m}) \right]. \tag{3.38}$$

Here, we took into account that in the limit $t \to \infty$, the stationary distribution is determined by Eq. (3.32) and the Markov matrix $Q(\tau)$ reduces to the operator $|r\rangle\langle l|$.

3.2.4 *Random Markov chain*

In the section, we suppose that the transition matrix of Eq. (3.22) is random. For example, Eq. (3.22) can be presented in the form

$$f(n+1) + f(n-1) = Ef(n) - V(n)f(n), \tag{3.39}$$

where $V(n)$ is a random delta-correlated process. A random walker now is a quantum particle, jumping on a lattice with a random potential $V(n)$, which is described by a stationary wave function f_n, such that $|f_n|^2$ is the PDF of finding a particle at the lattice site n. Then Eq. (3.22) takes the form of the discrete stationary Schrödinger equation, where E is the eigenspectrum, while the stationary wave functions satisfy the normalization condition $\sum_{n=-\infty}^{\infty} |f_n|^2 = 1$, that is the probability to find a quantum particle inside the lattice is equal to unity. The boundary condition $f_n = 0$ is taken at infinity for $n = \pm\infty$.

Note that Eq. (3.39) is the discrete variant of an Anderson model, within which one considers statistical properties of Anderson localization[5] [14] in a linear one dimensional lattice with random potential $V(n)$.

Let us present Eq. (3.39) in the matrix form, in such a way, that the transition matrix $\mathbf{Q}(n)$ is the 2×2 unimodular matrix, which determines the transition from a vector state $\vec{f}(n) = (f_n, f_{n-1})^T$ for any n to the state $\vec{f}(n+1) = (f_{n+1}, f_n)^T$:

$$\begin{pmatrix} f_{n+1} \\ f_n \end{pmatrix} = \begin{bmatrix} E - V(n) & -1 \\ 1 & 0 \end{bmatrix} \begin{pmatrix} f_n \\ f_{n-1} \end{pmatrix}. \tag{3.40}$$

---

[5]Philip W. Anderson (1923–2020) won a share of the Nobel Prize in Physics in 1977 for his discovery of electron localization, whereby disordered metals become insulators, and for his pioneering work on magnetism, see Ref. [13].

The iteration procedure yields

$$\vec{f}(n) = \prod_{j=1}^{n} \mathbf{Q}(j)\vec{f}(0). \tag{3.41}$$

Here $\vec{f}(0) = (f_1, f_0)^T$ is an initial state in Eq. (3.41). According to the Furstenberg-Kesten theorem[6] [16], the product of random matrices $\mathbf{Q}^N \equiv \prod_{j=1}^{N} \mathbf{Q}(j)$ for $N \to \infty$ exists with the probability 1 and limits to the value $\lambda = \lambda(E)$, which is called maximum Lyapunov characteristic exponent[7]:

$$\lambda = \lim_{N \to \infty} \frac{1}{N} \ln \left\| \prod_{j=1}^{N} \mathbf{Q}(j) \right\|. \tag{3.42}$$

Note, that the Lyapunov exponent is not a random value. This follows from the asymptotic behavior of the norm of product of matrices [16]. For example, for two matrices $\mathbf{A}$ and $\mathbf{B}$ we have $\|\mathbf{AB}\| \le \|\mathbf{A}\|\|\mathbf{B}\|$. Keeping in mind the exact result of Ref. [16], and for the sake of simplicity, we make a crude estimation, taking just the equality sign. The operator norm $\|\mathbf{Q}\|$ is determined by the modulus of the spectrum $e(E) = \text{spec}(\mathbf{Q})$. Therefore, for any vector $\vec{f}$, the operator norm of $\mathbf{Q}(j)$ reads $\|\mathbf{Q}\vec{f}\| \sim |e_j(E)|\|\vec{f}\|$ for any E. Then we have

$$\|\mathbf{Q}^N \vec{f}_E\| = \exp\left(\sum_{j=1}^{N} \ln |e_j(E)| \right) \|\vec{f}\| = \exp(N\lambda_N), \tag{3.43a}$$

$$\lambda_N = \frac{1}{N} \sum_{j}^{N} \ln |e_j(E)| = \langle \ln |e(E)| \rangle. \tag{3.43b}$$

For $N \gg 1$, $\lambda_N \to \lambda$. Then, this unimodular asymptotic matrix $\mathbf{Q}^N$ has two eigenvalues $e^{\pm \lambda N}$. As shown above, the exponential growth takes place for almost all values of E (with an accuracy of measure 0). Therefore, due to this exponential growth, there is no physical solutions for almost all energies E. However, there is a zero measure of E_n, which is the eigenvalues of $\mathbf{Q}(n)$ with the initial eigenfunctions $\vec{f}_{E_n}$, when the physical solution takes place and corresponds to the exponential decay $\vec{f}_E(j) = e^{-\lambda n}\vec{f}_E(0)$ according to the asymptotic matrix $\mathbf{Q}^N$ for $E = E_n$. This zero measure is a discrete

---

[6]See also detail of the discussion in Ref. [15].
[7]Usually it is denoted by λ_1. According to an Oseledets' multiplicative ergodic theorem [17] the asymptotic product of $D \times D$ random matrices has a set of Lyapunov exponents $\lambda_1 \ge \lambda_2 \ge \cdots \ge \lambda_D$, which are called characteristic Lyapunov exponents.

spectrum of the Schrödinger operator (3.39), when the eigenfunctions $\vec{f}_{E_n}$ are exponentially localized keeping the normalization condition. Therefore, the random walk of a quantum particle in the one dimensional lattice with a random potential is exponentially localized for large n with the localization length $1/\lambda$ [18].

3.3 The first passage

Another important characteristic of a random walk is the first-passage probability, which is the probability that a particle visits a point n for the first time at step t. A complete description can be found in Refs. [3, 19]. We however, discuss this example as an implementation of the generating function formalism, see Sec. 1.2.2 and we borrow this material from Ref. [2]. For our example of the discrete random walk in one dimension, the fist-passage probability density $F(n, t)$ is defined in complete analogy with the PDF $P(n, t)$. These two quantities are related each other as follows

$$P(n, t) = \delta_{t,0} \delta_{n,0} + \sum_{t'=1}^{t} F(n, t') P(0, t - t').$$ (3.44)

The rhs of the expression has the following interpretation: the first term determines the initial condition of a particle which starts at the origin $n = 0$ at step 0. After t steps it finishes at the point n. However, it can visit the point n at some earlier step t' with the probability $F(n, t')$ and for the other $t - t'$ steps it makes a loop (or loops) by leaving and returning to n with the transition probability $P(n - n, t - t') = P(0, t - t')$.

To find $F(n, t)$, we use a *generating function*, or z-transform, which can be introduced by analogy with the characteristic function, see Sec. 1.2.2. We multiply Eq. (3.44) by z^t and sum over t,

$$\sum_t P(n, t) z^t = \delta_{n,0} + \sum_t z^t \sum_{t'=1}^{t} F(n, t') P(0, t - t').$$ (3.45)

Using the definitions of the z-transforms,

$$P(n, z) = \sum_t P(n, t) z^t, \quad F(n, z) = \sum_t F(n, t) z^t,$$ (3.46)

and using the convolution rule, also known as the Cauchy rule,

$$\sum_{k=0}^{\infty} \sum_{l=0}^{k} a_l b_{k-l} = \left(\sum_{k=0}^{\infty} a_k \right) \left(\sum_{k=0}^{\infty} b_k \right),$$ (3.47)

we arrive at the expression

$$F(n, z) = \frac{P(n, z) - \delta_{n,0}}{P(0, z)}. \qquad (3.48)$$

We have reduced the problem to finding the generating function $P(n, z)$. To that end, we return to the simplified Eq. (3.23) with $p = q = 1/2$ and the initial condition $P(n, 0) = \delta_{n,0}$. Performing both the Fourier and z-transforms, we obtain from Eq. (3.23)

$$\tilde{P}(\theta, z) = z \cos(\theta) \tilde{P}(\theta, z) + 1, \qquad (3.49)$$

with the solution in the form of the integral

$$P(n, z) = \frac{1}{2\pi} \int_{-\pi}^{\pi} \frac{e^{-\imath \theta n} \, d\theta}{1 - z \cos(\theta)}. \qquad (3.50)$$

The result of the integration [19] is discussed in the next example.

Example 3.3. We perform the variable change $w = e^{\mp \imath \theta}$ for $n \geqslant 0$. Due to symmetry, the result is the same for both positive and negative n, and we take $e^{-\imath \theta n} = w^{|n|}$. In the complex w-plane the integral (3.50) corresponds to the contour integral with a contour around a unit circle:

$$P(n, z) = \frac{1}{2\pi \imath z} \oint \frac{w^{|n|} dw}{(w - w_1)(w - w_2)}, \qquad (3.51)$$

where $w_{1,2} = \frac{1}{z}(1 \pm \sqrt{1 - z^2})$ are the roots of the quadratic equation. Since $|w_2| < 1$, it is the only pole of the integrand, and the result of the residue calculation is

$$P(n, z) = \frac{\left[\frac{1}{z}\left(1 - \sqrt{1 - z^2}\right)\right]^{|n|}}{\sqrt{1 - z^2}}. \qquad (3.52)$$

$\square$

The probability of the walker reaching n at some time during the random walk is

$$F(n) = \sum_{t=0}^{\infty} F(n, t) = F(n, z = 1). \qquad (3.53)$$

Consequently, the probability for the walker to return to the origin is given by

$$F(0) = \sum_{t=0}^{\infty} F(0, t) = F(0, z = 1). \qquad (3.54)$$

According to Eq. (3.48)

$$F(0, z) = 1 - \frac{1}{P(0, z)}. \tag{3.55}$$

Let us define $\chi(t) = 1$ if $X(t) = 0$ and $\chi(t) = 0$ otherwise. Then $N = \sum_t \chi(t)$ is the number of times that the walker is at the origin, $n = 0$. The average value of N is given by

$$\langle N \rangle = \left\langle \sum_t \chi(t) \right\rangle = \sum_t \langle \chi(t) \rangle$$

$$= \sum_t \sum_n \chi(t) P(n, t) = \sum_t P(0, t) = P(0, z = 1). \tag{3.56}$$

Equation (3.29) implies that $P(0, t) \approx \frac{1}{\sqrt{\pi t}}$ and consequently that $P(0, z = 1) = \sum_t P(0, t) = \infty$. In other words, the walker visits the origin on average infinitely often. Substituting this result into Eq. (3.55), we find that $F(0, z = 1) = 1$. The nearest-neighbour lattice random walk in one dimension is recurrent, since the probability to return to the origin is one. In general, any one dimensional random walk is recurrent.[8]

The question then arises as to the distribution of number of steps for this recurrent walk, or what is the *first return probability* to the origin $F(0, t)$. This question is answered in the next example.

Example 3.4. Using Eqs. (3.46) and (3.52), we find

$$\sum_{t=0}^{\infty} F(0, t) z^t = 1 - \sqrt{1 - z^2}. \tag{3.57}$$

We write the rhs as the binomial series,

$$1 - \sqrt{1 - z^2} = -\sum_{k=1}^{\infty} \frac{\Gamma\left(\frac{3}{2}\right)(-1)^k}{\Gamma(k+1)\Gamma\left(\frac{3}{2} - k\right)} z^{2k}. \tag{3.58}$$

Taking into account the relation for the gamma function,

$$\frac{-1}{\Gamma\left(\frac{3}{2} - k\right)} = \frac{2\cos(\pi k)\Gamma\left(\frac{1}{2} + k\right)}{\pi(2k - 1)}, \tag{3.59}$$

and using the Stirling's formula, see Example 3.2, we obtain

$$F(0, t) = \begin{cases} \frac{1}{2}\sqrt{\frac{e}{\pi}} k^{-\frac{3}{2}}, & t \text{ is even, } t = 2k \\ 0, & t \text{ is odd.} \end{cases} \tag{3.60}$$

---

[8]See the discussion in Ref. [3].

The last line is obvious, since the return to the origin in one dimension is possible only for an even number of steps. Of course, the first passage through the origin coincides with the first return to the origin for the nearest-neighbour lattice random walk in one dimension.   □

The result in Eq. (3.60) is a special case of the Sparre Andersen theorem [20]. The latter is discussed in Ref. [3], and we quote: "For any symmetric random walk (independent of the existence of moments) the first passage probability through the origin behaves [...] asymptotically" as

$$F(0,t) = \frac{1}{2\sqrt{\pi}} t^{-3/2}. \tag{3.61}$$

3.4 Continuous time random walk

For the generalization of the discrete random walk (considered up to now) to the continuous time random walk (CTRW), it is supposed that both the length of jumps and the waiting time between any successive jumps are continuous random values distributed according to a PDF $\Upsilon(x,t)$. The latter is considered as the jump PDF, from which the jump length PDF $w(x)$ and the waiting time PDF $\psi(t)$ can be deduced. In our simplified consideration, we suppose that the relation $\Upsilon(x,t) = w(x)\psi(t)$ is hold, see Ref. [21] for a general description of the CTRW. One of the main assumptions of the CTRW is that the waiting times and jump lengths are independent identically distributed values with the PDFs $\psi(t)$ and $w(x)$, respectively.[9] Therefore, realizations of a CTRW result from a sequence of these independent jumps with the PDF $w(x)$ and delays with the waiting time PDF $\psi(t)$.

3.4.1 *Renewal theory*

We begin with a discussion of $\psi(t)$. Continuing discussion of Markov chains, we are asking the question how many steps or arrival events N take place during time t with the probability $\mathcal{P}_N(t)$, where the elapsed time t is the sum of the waiting times of the steps, $t = t_1 + t_2 + \cdots + t_N$. Therefore, our first task is establishing a relation between $\mathcal{P}_N(t)$ and $\psi(t)$ in the framework of renewal theory [22]. Accounting that the probability that no events have

---

[9]An instructive discussion of random walk theory and CTRWs can be found, e.g., in Refs. [3, 21].

occurred in the time interval $(0, t)$ is given by

$$P_{N=0}(t) = 1 - \int_0^t \psi(t')dt' = \int_t^\infty \psi(t')dt' = \Psi(t), \qquad (3.62)$$

we define the function $\Psi(t)$ as the *survival probability*. Correspondingly, it relates to the waiting time PDF

$$\psi(t) = -\frac{d\Psi(t)}{dt}. \qquad (3.63)$$

For a single event, $P_{N=1}(t)$ accounts for the probability that the first event occurs at time t', and there are no subsequent events until time t, which results in the convolution,

$$P_1(t) = \int_0^t \psi(t')\Psi(t - t')\, dt' \equiv \psi * \Psi. \qquad (3.64)$$

Since the waiting times are independent, the distribution for two events is obtained by iteration,

$$P_2(t) = \psi * P_1(t). \qquad (3.65)$$

This eventually results in a chain of iterations

$$P_N(t) = \psi * P_{N-1} = \psi_N * \Psi, \qquad (3.66)$$

where ψ_N a chain for the PDFs of N steps during time t,

$$\psi_N(t) = \int_0^t \psi_{N-1}(t')\psi(t - t')\, dt' \equiv \psi * \psi_{N-1}(t) = \overbrace{\psi * \psi \cdots * \psi}^{N}. \qquad (3.67)$$

The Laplace transforms of $\psi_N(t)$ and $P_N(t)$ follow from the Laplace transform of integrals. From Eq. (3.62), we have

$$\hat{\Psi}(s) = \int_0^\infty e^{-st}\Psi(t)\, dt = \frac{1 - \hat{\psi}(s)}{s}, \qquad (3.68)$$

where

$$\hat{\psi}(s) = \int_0^\infty e^{-st}\psi(t)\, dt \equiv \mathcal{L}[\psi(t)], \qquad (3.69)$$

which also yields

$$\mathcal{L}\left[\overbrace{\psi * \psi \cdots * \psi}^{N}\right] = [\hat{\psi}(s)]^N. \qquad (3.70)$$

Eventually, we obtain

$$\hat{P}_N(s) = \mathcal{L}\left[P_N(t)\right] = [\hat{\psi}(s)]^N \frac{1 - \hat{\psi}(s)}{s}. \qquad (3.71)$$

3.4.2   *Random steps*

Now let us consider only the jumps, which are instantaneous events. In this case, x is the position of a walker after N jumps, $x = \sum_{n=1}^{N} x_n$. It is worth stressing that in this jumps-only process, the number of jumps $n \in [1, N]$ plays the role of an internal clock [3]. We introduce the PDF of being at the position x after N jumps as $w_N(x)$, where $w_1(x) = w(x)$. In complete analogy with the waiting times, it corresponds to the same Markov chain as in Eq. (3.67),

$$w_N(x) = \int_{-\infty}^{\infty} w_{N-1}(x')w(x-x')\,dx' \equiv w * w_{N-1} = \overbrace{w * w \cdots * w}^{N}. \quad (3.72)$$

This expression assumes that $x \equiv x_N$ and $x' \equiv x_{N-1}$ are positions of a walker after N and $N-1$ jumps respectively, where $x, x' \in \mathbb{R}$, and that all possible jump lengths $x - x'$ along the infinite axis are distributed according to $w(x - x')$.

According to Eq. (3.8) the characteristic function for the N steps reads

$$\tilde{w}_N(k) = \mathcal{F}[w_N(x)](k) = \int_{-\infty}^{\infty} w_N(x)e^{ikx}\,dx, \quad (3.73)$$

which is the average of the exponential, $\tilde{w}(k) = \langle e^{ikx} \rangle$, for $N = 1$. Applying the Fourier transformation to the chain (3.72) we obtain from the definition (3.73)

$$\tilde{w}_N(k) = \int_{-\infty}^{\infty} e^{ikx} \int_{-\infty}^{\infty} w_{N-1}(x')\frac{1}{2\pi} \int_{-\infty}^{\infty} \tilde{w}(k')e^{-ik'(x-x')}\,dk'dx'dx$$

$$\int_{-\infty}^{\infty} \tilde{w}_{N-1}(k')\tilde{w}(k')\delta(k'-k)\,dk' = \tilde{w}_{N-1}(k)\tilde{w}(k) = [\tilde{w}(k)]^N. \quad (3.74)$$

3.4.3   *Montroll-Weiss equation*

We now combine the two considerations of the previous section and consider a random walk in space and time. To this end we introduce the PDF $P(x,t)$ to find a particle or a walker at time t at position x. Taking into account all possible N steps with $N \in [0, \infty]$ that a walker made during time t to arrive at the position x, we obtain the PDF $P(x,t)$ as a subordination

$$P(x,t) = \sum_{N=0}^{\infty} \mathcal{P}_N(t)w_N(x), \quad (3.75)$$

which translates[10] the internal clock N of the random walk (3.72) to the physical time t. Performing the Fourier and Laplace transformations, we obtain Eq. (3.75) as follows,

$$\hat{\tilde{P}}(k,s) = \sum_{N=0}^{\infty} \hat{\mathcal{P}}_N(s)\tilde{w}_N(k). \tag{3.76}$$

We replace $\hat{\mathcal{P}}_N(s)$ and $\tilde{w}_N(k)$ by their expressions in Eqs. (3.71) and (3.74),

$$\hat{\tilde{P}}(k,s) = \frac{1-\hat{\psi}(s)}{s} \sum_{N=0}^{\infty} [\hat{\psi}(s)]^N [\tilde{w}(k)]^N$$

$$= \frac{1-\hat{\psi}(s)}{s} \cdot \frac{1}{1-\tilde{w}(k)\hat{\psi}(s)}. \tag{3.77}$$

This is the well known Montroll-Weiss equation [23], which is the central result of the CTRW theory. Here it is presented in a simplified decoupled form for a one dimensional random walk. It is also supposed that the initial condition for the PDF $P(x,t)$ is the Dirac delta function.

3.4.3.1 *Master equation*

Let us rewrite Eq. (3.77) as follows,

$$s\hat{\tilde{P}}(k,s)\left[1+\frac{\hat{\psi}(s)}{1-\hat{\psi}(s)}\right] - 1 = \frac{s\tilde{w}(k)\hat{\psi}(s)}{1-\hat{\psi}(s)}\hat{\tilde{P}}(k,s). \tag{3.78}$$

Here we used that

$$1/[1-\hat{\psi}(s)] = 1 + \hat{\psi}(s)/[1-\hat{\psi}(s)]. \tag{3.79}$$

Therefore the Master equation in Fourier-Laplace space is

$$s\hat{\tilde{P}}(k,s) - 1 = \hat{K}(s)[\tilde{w}(x) - 1]\hat{\tilde{P}}(k,s). \tag{3.80}$$

Here $\hat{K}(s)$ is the memory kernel

$$\hat{K}(s) = \frac{s\hat{\psi}(s)}{1-\hat{\psi}(s)}. \tag{3.81}$$

Performing the inverse Fourier and Laplace transformations, we obtain the Master equation for the PDF $P = P(x,t)$ for the one dimensional random walk

$$\frac{\partial P(x,t)}{\partial t} = \int_0^t K(t-t') \int_{-\infty}^{\infty} [w(x-x')P(x',t') - P(x,t')]\,dx'dt' . \tag{3.82}$$

Note that here $\Omega_1 \in (-\infty\,\infty)$. In general case, it is determined by boundary conditions, as well as the initial condition is $P(x,t=0) = P_0(x)$.

---

[10]A pedagogical explanation of subordination can be found in Ref. [3].

3.5   CTRW and fractional diffusion equations

In the section, we consider various forms of the diffusion (Fokker-Planck) equation, which describe various realizations of a random walk. As shown above, a CTRW is described by the master equation. Here we present a general form of a diffusion equation, borrowing the inferring from Ref. [21]. Thus, we start from the equation [24]

$$\mathcal{P}(x,t) = \int_{-\infty}^{\infty} \int_0^{\infty} \mathcal{P}(x',t')\Upsilon(x-x',t-t')\,dx'dt' + \delta(x)\delta(t), \qquad (3.83)$$

which present Eq. (3.75) in the iteration form and relates the PDF $\mathcal{P}(x,t)$ of arriving at position x at time t with the PDF $\mathcal{P}(x',t')$ of arriving at position x' at time t' (one should not confuse it with $\mathcal{P}_N(t)$). The second term in the rhs of Eq. (3.83) corresponds to the initial condition of the random walk. Then, the PDF $P(x,t)$ to be at x at time t is given by the convolution

$$P(x,t) = \int_0^t \mathcal{P}(x,t')\Psi(t-t')\,dt', \qquad (3.84)$$

which describes arrival at x at time $t' < t$ and not moving since. Then, in Fourier-Laplace space, we arrive at the Montroll-Weiss equation of the form

$$\hat{\tilde{P}}(k,s) = \frac{1-\hat{\psi}(s)}{s} \cdot \frac{\tilde{P}_0(k)}{1-\hat{\tilde{\Upsilon}}(k,s)}, \qquad (3.85)$$

where $\tilde{P}_0(k)$ is the Fourier image of the initial condition $P_0(x)$.

3.5.1   *Diffusion limit*

It should be admitted that different realizations of the CTRW described by Eq. (3.85), are determined by averaged characteristics of the waiting times and the length of jumps being finite or diverging values. Namely, it depends on the mean or characteristic waiting time $\mathcal{T}$ (do not confuse it with the integral transform operator)

$$\langle t \rangle = \mathcal{T} = \int_0^{\infty} \int_{-\infty}^{\infty} t\Upsilon(x,t)\,dtdx = \int_0^{\infty} t\psi(t)\,dt \qquad (3.86)$$

and the jump length variance

$$\langle x^2(t) \rangle = \Sigma^2 = \int_0^{\infty} \int_{-\infty}^{\infty} x^2\Upsilon(x,t)\,dtdx = \int_{-\infty}^{\infty} x^2 w(x)\,dx. \qquad (3.87)$$

Using this characteristics, let us account the asymptotic form of the Fourier-Laplace image of the jump PDF $\hat{\tilde{\Upsilon}}(k,s)$ in the diffusion limit, $s \to 0$ and $k \to 0$. In the lowest order of the expansion, we have

$$\hat{\tilde{\Upsilon}}(k,s) \sim 1 - \tau_\alpha s^\alpha - \sigma_\beta |k|^\beta + o(k^\beta s^\alpha), \quad 0 < \alpha \le 1, \quad 0 < \beta \le 2. \quad (3.88)$$

The values of $\alpha \in (0,1]$ and $\beta \in (0,2]$ depend on values of $\mathcal{T}$ and Σ^2.

3.5.1.1 *Brownian motion*

First we consider both $\mathcal{T}$ and Σ^2 being finite. Thus, we assume that the jump length PDF is a Gaussian, while the waiting time PDF is Poissonian,

$$w(x) = (4\pi\sigma^2)^{-1/2} \exp(-x^2/4\sigma^2),$$
$$\psi(t) = \tau^{-1} e^{-t/\tau}, \quad (3.89)$$

such that $\Sigma^2 = 2\sigma^2$ and $\mathcal{T} = \tau$. The Fourier and Laplace transforms yield

$$\tilde{w}(k) = \frac{1}{\sqrt{4\pi\sigma^2}} \int_{-\infty}^{\infty} e^{-\frac{x^2}{4\sigma^2}} e^{ikx}\, dx \approx e^{-\sigma^2 k^2} \approx 1 - \sigma^2 k^2, \quad (3.90a)$$

$$\hat{\psi}(s) = \tau^{-1} \int_0^\infty e^{-t/\tau} e^{-st}\, dt = \frac{1}{1+\tau s} \approx 1 - \tau s, \quad (3.90b)$$

$$\hat{\tilde{\Upsilon}}(k,s) \approx \tilde{w}(k)\hat{\psi}(s) \approx 1 - \tau s - \sigma^2 k^2. \quad (3.90c)$$

Substituting asymptotic expressions of Eq. (3.90) into Eq. (3.85), and defining a transport coefficient $\mathcal{D} = \sigma^2/\tau$, we obtain

$$s\hat{\tilde{P}}(k,s) - \tilde{P}_0(k) = -\mathcal{D}k^2 \hat{\tilde{P}}(k,s). \quad (3.91)$$

After the inverse Fourier and Laplace transformations, we obtain a standard Fokker-Planck equation (FPE)

$$\frac{\partial P(x,t)}{\partial t} = \mathcal{D}\frac{\partial^2 P(x,t)}{\partial x^2}. \quad (3.92)$$

The evolution of the initial PDF $P_0(x)$ is defined by the propagator, or Green's function $G(x,t \,|\, x',t') = G(x-x', t-t')$, according to the convolution $P(x,t) = \int G(x,t \,|\, x', t' = 0)P_0(x')\, dx'$. Thus, the FPE (3.92) is exactly the same for the Green's function with the initial condition $P_0(x) = \delta(x)$. Therefore, Eq. (3.92) describes Brownian motion with the Gaussian kernel

$$P(x,t) = \frac{1}{\sqrt{4\pi\mathcal{D}t}} \int_{-\infty}^{\infty} e^{-\frac{(x-x')^2}{4\mathcal{D}t}} P_0(x')dx', \quad (3.93)$$

and the initial condition $P(x, t=0) = P_0(x)$.

The MSD is

$$\langle x^2(t) \rangle = -\mathcal{L} \left[\frac{\partial^2}{\partial k^2} \frac{1}{s + \mathcal{D}k^2} \Big|_{k=0} \right] = 2\mathcal{D}t. \tag{3.94}$$

Here the transport coefficient D is the diffusion coefficient[11] of dimension $[\mathcal{D}] = [x]^2[t]^{-1} = m^2 s^{-1}$.

3.5.1.2   *Superdiffusion – Lévy flights*

Let the characteristic waiting time is finite, $\mathcal{T} = \tau$, while the jump length variance Σ diverges and is determined by a Lévy distribution for the jump length, i.e.

$$\tilde{w}(k) = \exp(-\sigma^\beta |k|^\beta) \approx 1 - \sigma^\beta |k|^\beta \tag{3.95}$$

then $\hat{\tilde{\Upsilon}} \approx 1 - \tau s - \sigma^\beta |k|^\beta$. Substituting this asymptotic expansion into Eq. (3.85), we obtain its propagator/Green's function as follows

$$\hat{\tilde{G}}(k, s) = \frac{1}{s + \mathcal{D}_\beta |k|^\beta}. \tag{3.96}$$

Here $\mathcal{D}_\beta = \sigma^\beta / \tau$. Then performing the Fourier–Laplace inversions, we obtain the fractional diffusion equation in the form of the fractional Fokker-Planck equation (FFPE) for the Green's function

$$\frac{\partial G(x, t)}{\partial t} = \mathcal{D}_\beta \, \partial^\beta_{|x|} G(x, t), \tag{3.97}$$

where

$$\partial^\beta_{|x|} G(x, t) \equiv \left[{}_W D^\beta_{-\infty} G(x, t) - {}_W D^\beta_{-\infty} G(-x, t) \right]$$

is the combination of the fractional Weyl derivatives with respect to x (see Eq. (1.210)). Note that the PDF is $P(x, t) = \int G(x - x', t) P_0(x') dx'$. Considering

$$\hat{G}(k, t) = \mathcal{L}^{-1} \left[\hat{\tilde{G}}(k, s) \right] = \exp[-\mathcal{D}_\beta t |k|^\beta] \tag{3.98}$$

as the characteristic function, we obtain divergence of the MSD, namely

$$\langle x^2(t) \rangle = -\frac{d^2}{dk^2} e^{-\mathcal{D}_\beta t |k|^\beta} \Big|_{k=0} = \infty, \quad 0 < \beta < 2. \tag{3.99}$$

Such anomalous diffusion is known as Lévy flights [21]. We do not discuss it here.

---

[11]Coefficient 2 in Eq. (3.94) should not wary us. It is a matter of definition of the diffusion coefficient. In some literature [6], the coefficient at the second derivative in the FPE is $\mathcal{D}/2$, then the MSD is $\mathcal{D}t$. In this case $\mathcal{D} = 2\sigma^2 \tau^{-1}$.

We show that Lévy flights as well as Brownian motion, does not break Markovian property of random walk. It means that the Green's function (3.98) corresponds to continuous Markov chain in the form of the Smoluchowski equation (3.19). Indeed, considering Markov processes being homogeneous in space and time, i.e.,

$$P(x, t \mid x', t') = G(x, t \mid x', t') = G(x - x', t - t'), \tag{3.100}$$

Eq. (3.19) can be rewritten in a form suitable for the Fourier transformation,

$$G(x - x_0, t - t_0) = \int_{-\infty}^{\infty} G(x - x', t - t')G(x' - x_0, t' - t_0) \, dx', \tag{3.101}$$

where we use $\Omega_2 \equiv (-\infty, \infty)$. Fourier transforming Eq. (3.101), we obtain

$$\tilde{G}(k, t - t_0) = \tilde{G}(k, t - t')\tilde{G}(k, t' - t_0), \tag{3.102}$$

which has the general solution for the conditional PDF, or Green's function

$$\tilde{G}(k, \tau) = \exp\left[-|k|^\beta \tau\right], \quad 0 < \beta \leq 2. \tag{3.103}$$

For $\beta = 1$, when we demand a simple symmetrical form of the solution, we obtain the Cauchy-Lorentz distribution,

$$P(x|x_0, \tau) \equiv G(x - x_0, \tau) = \frac{1}{\pi} \cdot \frac{\tau}{\tau^2 + (x - x_0)^2}, \quad \tau = \mathcal{D}_1(t - t_0), \tag{3.104}$$

which satisfies the correct initial condition as $\tau \to 0$, namely $P(x - x_0, 0) = \delta(x - x_0)$, and $\mathcal{D}_1$ is a kinetic coefficient.

3.5.1.3 Subdiffusion

The situation changes dramatically, when the characteristic waiting time diverges, $\mathcal{T} \to \infty$. In this case the random walk is not Markovian anymore. The normalized power law waiting time distribution is

$$\psi(t) = \alpha \tau^\alpha (\tau + t)^{-\alpha - 1}. \tag{3.105}$$

To obtain asymptotic – diffusion limit expression for the Laplace image of $\psi(t)$, we use two its properties: (i) from normalization condition it follows that $\hat{\psi}(s = 0) = \mathcal{L}[\psi(t)](s = 0) = 1$ and (ii) $\psi(t) \sim \tau_\alpha t^{-1-\alpha}$ for $t \to \infty$. Therefore, estimating the Laplace transform as integration by part and accounting the property (ii), we have

$$\hat{\psi}(s) = \alpha \tau_\alpha \int_0^\infty (\tau + t)t^{-1-\alpha}e^{-st} \, dt = 1 - s\tau_\alpha \int_0^\infty (\tau + t)^{-\alpha}e^{-st} \, dt$$

$$\approx 1 - \tau_\alpha \int_0^\infty t^{-\alpha}e^{-st} \, dt \approx 1 - \tau_\alpha s^\alpha. \tag{3.106}$$

The jump lengths are taken to be distributed by a Gaussian with the Fourier image (3.90a) and the finite variance Σ^2. Then we obtain[12] $\hat{\tilde{\Upsilon}}(k,s) \sim 1 - \tau^\alpha s^\alpha - \sigma^2 k^2$. Substituting it in Eq. (3.99), we have

$$\hat{\tilde{P}}(k,s) = \frac{P_0(k)/s}{1 + \mathcal{D}_\alpha k^2 s^{-\alpha}}, \qquad (3.107)$$

where $\mathcal{D}_\alpha = \sigma^2/\tau^\alpha$ is a transport coefficient. Performing the Fourier-Laplace inversion and employing the property (1.193) of the Laplace transform for the fractional R-L integral, we have

$$P(x,t) - P_0(x) = \mathcal{D}_\alpha \, I_{0+}^\alpha \frac{\partial^2 P(x,t)}{\partial x^2}. \qquad (3.108)$$

Application of the differential operator $\frac{\partial}{\partial t}$ yields the fractional Fokker-Plank equation (FFPE) as follows

$$\frac{\partial P(x,t)}{\partial t} = \mathcal{D}_\alpha \, {}_{\mathrm{RL}}D_{0+}^{1-\alpha} \frac{\partial^2 P(x,t)}{\partial x^2}. \qquad (3.109)$$

An alternative form of the FFPE with the Caputo fractional derivative can be obtain as well. To this end, multiplying Eq. (3.107) by s^α and accounting property of the Laplace transform of the Caputo fraction derivative one obtains

$$_{\mathrm{C}}D_{0+}^\alpha P(x,t) = \mathcal{D}_\alpha \frac{\partial^2 P(x,t)}{\partial x^2}. \qquad (3.110)$$

where $_{\mathrm{C}}D_{0+}^\alpha f(t)$ is the Caputo fractional derivative with respect to t (1.196). One also estimates the MSD, which reads

$$\langle x^2(t) \rangle = -\frac{d^2}{dk^2} \mathcal{L}^{-1} \left[\frac{s^{\alpha-1}}{s^\alpha + \mathcal{D}_\alpha k^2} \right] \Bigg|_{k=0}$$

$$= -\frac{d^2}{dk^2} \left[\sum_{j=1}^\infty \frac{(-k^2 \mathcal{D}_\alpha t^\alpha)^j}{\Gamma(1+j\alpha)} \right] \Bigg|_{k=0} = \frac{2\mathcal{D}_\alpha t^\alpha}{\Gamma(1+\alpha)}, \qquad (3.111)$$

and it is subdiffusion for $0 < \alpha < 1$.

3.5.2 A general case of FFPE

As another example, let us consider the case when both $\mathcal{T}$ and Σ^2 diverge, while the ratio $\mathcal{D}_{\alpha,\beta} = \sigma^\beta/\tau^\alpha \equiv \mathcal{D}$ is finite (do not mix it with the diffusion

---

[12]Note that this results from the crude inferring of $\hat{\psi}(s)$ in Eq. (3.106). Discussions of the accurate inferring of $\hat{\tilde{\Upsilon}}(k,s)$ based on the CTRW with spatiotemporal coupling can be found *e.g.*, in Ref. [24], see also the discussion and the literature list in Ref. [21].

coefficient of normal diffusion). Then, we consider the fractional in both space and time FPE [2], which reads

$$_CD_{0+}^\alpha G(x,t) = \mathcal{D} \, \partial_{|x|}^\beta G(x,t), \tag{3.112}$$

where Eq. (3.112) is a general form of the FFPE, with $0 < \alpha < 1$ and $0 < \beta < 2$ and a generalized diffusion coefficient $\mathcal{D}$. In Fourier-Laplace space, it reads

$$s^\alpha \hat{\tilde{G}}(k,s) = -\mathcal{D} |k|^\beta \hat{\tilde{G}}(k,s) + s^{-\alpha}. \tag{3.113}$$

This equation can be solved in terms of the Fox H-function. The Laplace inversion of Eq. (3.113) yields the one parameter M-L function (1.118), namely

$$\tilde{G}(k,t) = \mathcal{L}^{-1}\left[\frac{s^{\alpha-1}}{s^\alpha + \mathcal{D}|k|^\beta}\right] = E_\alpha\left(-\mathcal{D}t^\alpha |k|^\beta\right). \tag{3.114}$$

Let us establish this relation by writing the M-L function in the form of a Mellin transform $\mathcal{M}[f(z)](p) = f(p)$. Using the well known formula for the Mellin transform [25],

$$\mathcal{M}[f(z)](p) = \frac{1}{\Gamma(1-p)}\mathcal{M}\left[\mathcal{L}[f(z)](s)\right](1-p), \tag{3.115}$$

we represent the M-L function as the inverse Mellin transform via a chain of transformations (as an alternative approach, see also Example 2.5)

$$\begin{aligned} E_\alpha(-z^\alpha) &= \mathcal{M}^{-1}\left\{\frac{1}{\Gamma(1-p)}\mathcal{M}\left[\mathcal{L}[E_\alpha(-z^\alpha)](s)\right](1-p)\right\} \\ &= \mathcal{M}^{-1}\left\{\frac{1}{\Gamma(1-p)}\left[\int_0^\infty \frac{s^{\alpha-1}s^{p-1}\,ds}{s^\alpha+1}\right](1-p)\right\} \\ &= \frac{1}{2\pi i}\int_C \frac{\Gamma(1-p/\alpha)\Gamma(p/\alpha)}{\alpha\Gamma(1-p)}z^{-p}\,dp \\ &= \frac{1}{\alpha}H_{2,1}^{1,1}\left[z \,\middle|\, \begin{matrix}(0,1/\alpha)\\(0,1/\alpha),(1,1)\end{matrix}\right] = H_{2,1}^{1,1}\left[z^\alpha \,\middle|\, \begin{matrix}(0,1)\\(0,1),(1,\alpha)\end{matrix}\right]. \end{aligned} \tag{3.116}$$

Here $z^\alpha = \mathcal{D}t^\alpha |k|^\beta$ and the last two lines reflect the property of the Fox H-function. The Mellin transform integration is also considered as a form of Eq. (1.98) for the beta function. To perform the inverse Fourier transform, which reduces to the cosine Fourier transform, we employ the Mellin transform in combination with the cosine Fourier transform,

$$\mathcal{M}\left[\mathcal{F}_c[\phi(k)](x)\right](p) = \Gamma(p)\cos\left(\tfrac{p\pi}{2}\right)\mathcal{M}[\phi(k)](1-p). \tag{3.117}$$

We also use the following property of the Mellin transform [25],

$$\mathcal{M}\left[z^{\rho}\phi(ak^{\beta})\right](p) = \frac{1}{\beta}a^{-\frac{p+\rho}{\beta}}\mathcal{M}\left[\phi(k)\right]\left(\frac{p+\rho}{\beta}\right). \qquad (3.118)$$

Note that modulus of the argument is lifted due to the cosine. Taking these properties into account with $\rho = 0$ and $a = \mathcal{D}t^{\alpha}$, we obtain that the Mellin transform of the Fox H-function in Eq. (3.116) with the argument k instead of $z^{\alpha} = \mathcal{D}t^{\alpha}|k|^{\beta}$, is given by

$$\left[\frac{\Gamma(1-p/\beta)\Gamma(p/\beta)}{\Gamma(1-p\alpha/\beta)}\right](p \to 1-p) = \frac{\Gamma\left(\frac{1}{\beta}-\frac{p}{\beta}\right)\Gamma\left(1-\frac{1}{\beta}-\frac{p}{\beta}\right)}{\Gamma\left(1-\frac{\alpha}{\beta}+p\frac{\alpha}{\beta}\right)}. \qquad (3.119)$$

Therefore, the Fourier inversion of the Fox H-function in Eq. (3.116) reads,

$$G(x,t) = \mathcal{F}^{-1}[\tilde{G}(k,t)](x) = \frac{1}{\pi}\mathcal{F}_c[\tilde{G}(k,t)](x) = \frac{1}{\beta\pi}(\mathcal{D}t^{\alpha})^{-\frac{1}{\beta}}$$

$$\times \mathcal{M}^{-1}\left\{(t^{\alpha})^{\frac{p}{\beta}}\Gamma(p)\cos\left(\frac{p\pi}{2}\right)\frac{\Gamma\left(\frac{1}{\beta}-\frac{p}{\beta}\right)\Gamma\left(1-\frac{1}{\beta}-\frac{p}{\beta}\right)}{\Gamma\left(1-\frac{\alpha}{\beta}+p\frac{\alpha}{\beta}\right)}\right\}. \qquad (3.120)$$

Employing the expression $\pi^{-1}\cos(\pi\nu) = \left[\Gamma\left(\frac{1+\nu}{2}\right)\Gamma\left(\frac{1-\nu}{2}\right)\right]^{-1}$, we continue the chain of transformation of Eq. (3.120) as follows

$$G(x,t) = \frac{1}{\beta(\mathcal{D}t^{\alpha})^{\frac{1}{\beta}}}\mathcal{M}^{-1}\left\{(\mathcal{D}t^{\alpha})^{\frac{p}{\beta}}\frac{\Gamma(p)\Gamma\left(\frac{1}{\beta}-\frac{p}{\beta}\right)\Gamma\left(1-\frac{1}{\beta}-\frac{p}{\beta}\right)}{\Gamma\left(\frac{1}{2}+\frac{p}{2}\right)\Gamma\left(\frac{1}{2}-\frac{p}{2}\right)\Gamma\left(1-\frac{\alpha}{\beta}+p\frac{\alpha}{\beta}\right)}\right\}$$

$$= \frac{1}{\beta(\mathcal{D}t^{\alpha})^{\frac{1}{\beta}}}H^{2,1}_{3,3}\left[\frac{|x|}{(\mathcal{D}t^{\alpha})^{\frac{1}{\beta}}}\middle|\begin{array}{c}(1-\frac{1}{\beta},\frac{1}{\beta}),(1-\frac{\alpha}{\beta},\frac{\alpha}{\beta}),(\frac{1}{2},\frac{1}{2})\\(0,1),(1-\frac{1}{\beta},\frac{1}{\beta}),(\frac{1}{2},\frac{1}{2})\end{array}\right]$$

$$= \frac{1}{\beta|x|}H^{2,1}_{3,3}\left[\frac{|x|}{(\mathcal{D}t^{\alpha})^{\frac{1}{\beta}}}\middle|\begin{array}{c}(1,\frac{1}{\beta}),(1,\frac{\alpha}{\beta}),(1,\frac{1}{2})\\(1,1),(1,\frac{1}{\beta}),(1,\frac{1}{2})\end{array}\right]. \qquad (3.121)$$

For the special case $\beta = 2$ (time fractional diffusion equation), the solution becomes

$$G(x,t) = \frac{1}{2|x|}H^{2,1}_{3,3}\left[\frac{|x|}{\sqrt{\mathcal{D}t^{\alpha}}}\middle|\begin{array}{c}(1,1/2),(1,\alpha/2),(1,1/2)\\(1,1),(1,1/2),(1,1/2)\end{array}\right]$$

$$= \frac{1}{2|x|}H^{1,0}_{1,1}\left[\frac{|x|}{\sqrt{\mathcal{D}t^{\alpha}}}\middle|\begin{array}{c}(1,\alpha/2)\\(1,1)\end{array}\right], \qquad (3.122)$$

while for $\alpha = 1$ (space fractional diffusion equation), the solution reads

$$G(x,t) = \frac{1}{\beta|x|} H_{3,3}^{2,1} \left[\frac{|x|}{(Dt^\alpha)^{\frac{1}{\beta}}} \middle| \begin{array}{l} (1,1/\beta),(1,1/\beta),(1,1/2) \\ (1,1),(1,1/\beta),(1,1/2) \end{array} \right]$$

$$= \frac{1}{\beta|x|} H_{2,2}^{1,1} \left[\frac{|x|}{(Dt^\alpha)^{\frac{1}{\beta}}} \middle| \begin{array}{l} (1,1/\beta),(1,1/2) \\ (1,1),(1,1/2) \end{array} \right]. \tag{3.123}$$

3.6 Stochastic equations

A random process $\mathbf{g}(t)$, described by n functions $q_j(t)$, $j = 1,\ldots,n$ is controlled by a system of stochastic or Langevin equations

$$\dot{\mathbf{q}} = f(\mathbf{q},t) + g(\mathbf{q},t), \tag{3.124}$$

where $f(\mathbf{q},t)$ describes deterministic processes, while $g(\mathbf{q},t)$ is a random force. Our aim in the section is to show how Eq. (3.124) generates the probabilistic description of the random process in the framework of a PDF, which for the n dimensional process can be given as follows

$$P(\mathbf{q},t) = \left\langle \prod_{j=1}^{n} \delta(q_j(t) - q_j) \right\rangle, \tag{3.125}$$

where the average procedure is performed with respect to known distribution of the random force.

For the illustrative sake, we consider a simplified one dimensional Langevin equation[13]

$$\dot{X}(t) = -f[X(t)] + \xi(t). \tag{3.126}$$

Here $\xi(t)$ is an additive Gaussian white noise whose probability distribution functions is given by the functional form

$$[d\rho(\xi)] = \exp\left[-\frac{1}{4D}\int \xi^2(t)\,dt\right][d\xi(t)], \tag{3.127}$$

where $[d\xi(t)] = \prod_{\tau=0}^{t} d\xi(\tau)/\sqrt{4\pi D/d\tau}$. Since the noise ξ is Gaussian, it is completely described by its mean $\langle\xi(t)\rangle = 0$ and correlation function:

$$\langle\xi(t)\xi(t')\rangle = 2D\,\delta(t - t'), \tag{3.128}$$

---

[13]This formal form corresponds to a model with a well-defined velocity for the Brownian particle, which is due to Langevin [26, 27], $m\dot{V}(t) = -\gamma V(t) + \sigma\xi(t)$, where $X(t) = V(t)$ is the velocity of the Brownian particle, m is the mass of the particle, and $mf[X(t)] = \gamma$ is the damping or friction constant, while $\sigma/m = 1$ is the strength of the random force $\xi(t)$, which describes the impact of the fluid molecules on the particle.

see Appendix A. The spectral density $S(\omega) = D/\pi$ of the white noise correlation function is flat and given by

$$\langle \xi(t)\xi(t') \rangle = 2D\,\delta(t - t') = \int_{-\infty}^{\infty} S(\omega)e^{i\omega t}\,d\omega. \tag{3.129}$$

The integral form of the Langevin equation (3.126) is the solution of the equation in the functional form

$$X(t) = X_0 + \int_0^t f[X(t')]\,dt' + \int_0^t \xi(t')\,dt'. \tag{3.130}$$

3.6.1 Fokker-Planck equation

The one dimensional random process described by Eq. (3.126) generates the PDF

$$P(x,t) = \langle \delta(X(t) - x) \rangle, \tag{3.131}$$

which is the one dimensional realization of Eq. (3.125). To obtain the Fokker-Planck equation for the PDF, we differentiate Eq. (3.131) with respect to time. Accounting Eq. (3.126), we obtain

$$\dot{P} = \left\langle \left[-f[X(t)] + \xi(t)\right] \frac{\delta}{\delta X(t)} \delta\left(X(t) - x\right) \right\rangle. \tag{3.132}$$

Using the well known replacement of the functional derivative $\frac{\delta}{\delta X(t)}$ by the partial derivative $-\frac{\partial}{\partial x}$ due to the symmetry property of the delta function $\delta(X(t) - x)$, then taking this partial derivative out of the average integration, and taking into account that $f[X(t)]\delta(X(t) - x) = f(x)\delta(X(t) - x)$, one obtains

$$\dot{P} = \frac{\partial}{\partial x}\left[f(x)P(x,t) - \langle \xi(t)\delta\left(X(t) - x\right) \rangle \right]. \tag{3.133}$$

The next step of the treatment is averaging of two functionals according to Appendix A.2. Then we have a chain of transformations

$$\begin{aligned}
\langle \xi(t)\delta\left(X(t) - x\right) \rangle &= 2D\left\langle \frac{\delta}{\delta\xi(t)}\left[\delta\left(X(t) - x\right)\right] \right\rangle \\
&= 2D\frac{\partial}{\partial x}\left\langle \frac{\delta X(t)}{\delta\xi(t)}\delta\left(X(t') - x\right) \right\rangle \\
&= D\frac{\partial P(x,t)}{\partial x},
\end{aligned} \tag{3.134}$$

where the functional (3.130) is used in the last line as well as (see Appendix A) the following helpful identities

$$\int_0^t \delta(t' - t)\, dt' = \int_0^t \frac{\delta\xi(t)}{\delta\xi(t')} = \frac{\delta X[\xi]}{\delta\xi(t)} = \lim_{\Delta t \to 0} \int_{t-\Delta t/2}^t \frac{\delta\xi(t')}{\delta\xi(t)}\, dt' = \frac{1}{2}.$$

Not also that we omitted the zero functional derivative:

$$\frac{\delta}{\delta\xi(t)} \int_0^t f[x(t')]\, dt' = \lim_{\Delta t \to 0} \int_{t-\Delta t/2}^t \frac{\delta f[x(t')]}{\delta\xi(t)}\, dt' = 0.$$

We also take into account that x and t are independent variables that yields

$$\dot{P}(x,t) = \frac{\partial}{\partial t} P(x,t) + \frac{\partial}{\partial x} P(x,t)\dot{x} = \frac{\partial}{\partial t} P(x,t).$$

Collecting results of Eqs. (3.131)–(3.134), we obtain the Fokker Plank equation as follows

$$\frac{\partial P(x,t)}{\partial t} = \frac{\partial}{\partial x}\left[f(x)P(x,t)\right] + \mathcal{D}\frac{\partial^2 P(x,t)}{\partial x^2}. \tag{3.135}$$

3.6.1.1   *The Furutsu-Novikov theorem*

Let us return to the last step of the inferring of the FPE (3.135) based on expression

$$\langle \xi(t)f[X(t)]\rangle = 2\mathcal{D}\left\langle \int_0^t \frac{\delta\xi(\tau)}{\delta\xi(t)}\frac{\delta F[\xi(\tau)]}{\delta\xi(\tau)}\, d\tau\right\rangle$$

$$= \int_0^t \langle\xi(t)\xi(\tau)\rangle \left\langle \frac{\delta F[\xi(\tau)]}{\delta\xi(\tau)}\right\rangle d\tau. \tag{3.136}$$

The last expression is known as the Furutsu-Novikov theorem. The result has been derived independently by Furutsu [28, 29] and Novikov [30] in their works on electromagnetic waves and turbulence, respectively.

3.7   Langevin equation: Parametrization and FFPE

In the section, we consider the CTRW approach for the Langevin equations. As discussed above in Sec. 3.4, it is supposed that the length of jumps r and the waiting time between any successive jumps τ are continuous random values distributed according to PDFs $w(r)$ and $\psi(\tau)$, correspondingly. Both the random jump lengths and random waiting times can be parametrized by introducing internal, or *operational time u*, which can be *e.g.*, the arc length along the trajectory. For example, such a situation takes place in optical ray dynamics in Lévy glasses [31]. By means of this internal operational time u we parametrize the random walk jump length $r = r(u)$ and the delay time $\tau = \tau(u)$ [32]. Thus, the position of a particle $X(u)$ and physical times $t(u)$ are the functions of the operational time u, and these values are determined by the functionals as follows

$$X[r(u)] = \int_0^u r(v)\, dv, \quad t[\tau(u)] = \int_0^u \tau(v)\, dv, \qquad (3.137)$$

and the corresponding Langevin equations are

$$\frac{dX(u)}{du} = r(u) \qquad (3.138a)$$

$$\frac{dt(u)}{du} = \tau(u) \qquad (3.138b)$$

This parametrization of the one dimensional random process described by Eqs. (3.138) generates two PDFs: $f_1(x, u) = \langle \delta(X(u) - x) \rangle$ and $f_2(t, u) = \langle \delta(t(u) - t) \rangle$. The averaging procedure is due to the random processes described by functional distributions $w[r(u)]$ and $\psi[\tau(u)]$. However, to perform this averaging procedure, we need to know general forms of $w(r)$ and $\psi(\tau)$ as fractional stable distributions, whose properties are determined by the asymptotic, diffusion limit, discussed above. As discussed in Ref. [33], the symmetrical distribution $w(r)$ for the jump lengths $r \in \mathbb{R}$ and correspondingly $w[r(u)]$ can be presented by the (inverse) Fourier transform. For the time variable $\tau \in \mathbb{R}_+$, the distribution is expressed by the inverse Laplace transform, which however can be converted to the Fourier transform, as well. Eventually, the integral forms of the distributions are

$$w[r(u)] = \int \exp\left\{-\int_{-\infty}^{\infty} \left[\sigma_\beta |k(v)|^\beta - \imath k(v) r(v)\right] dv\right\} [dk(u)], \quad (3.139a)$$

$$\psi[\tau(u)] = \int \exp\left\{-\int \left[\tau_\alpha [\imath \omega(v)]^\alpha - \imath \omega(v) \tau(v)\right] dv\right\} [d\omega(u)]. \quad (3.139b)$$

Here both $[dk(u)]$ and $[d\omega(u)]$ have the form

$$[dz(u)] = \prod_{v=0}^{u} dz(v)/\sqrt{4\pi/adv}\,,$$

where $z(u) = [|k(u)|, \omega(u)]$ and $a = (\sigma_\beta, \imath^\alpha \tau_\alpha)$, while $0 < \alpha < 1$ and $0 < \beta \le 2$ are defined previously. The next step is the averaging procedure, which is presented in the next example.

Example 3.5. To perform averaging of the Dirac delta function $\delta(p(u) - q)$, we present it in the form of the Fourier integral:

$$\delta(p(u) - q) = \frac{1}{2\pi} \int \exp\{i\eta[p(u) - q]\} \, d\eta, \qquad (3.140)$$

where $p(u) = \int_0^u \lambda(v) \, dv$. Thus, we have

$$\langle \delta(p(u) - q) \rangle = \frac{1}{2\pi} \int e^{iq\eta} \langle [\dots] \rangle \, d\eta,$$

where the averaging term $\langle [\cdots] \rangle$ reads

$$\langle [\cdots] \rangle$$

$$= \int \int \exp\left\{ -\int [a(z(v))^\mu - iz(v)\lambda(v) - i\lambda(v)\eta] \, dv \right\} [dz(u)][d\lambda(u)]$$

$$= \int \exp\left[-\int a(z(v))^\mu \, dv \right] [\delta(z(u) - \eta) \, dz(u)] = e^{-a\eta^\mu u}. \qquad (3.141)$$

Note that $q = (x, t)$, while $p(u) = [X(u), t(u)]$. Eventually, we obtain

$$\langle \delta(p(u) - q) \rangle = \frac{1}{2\pi} \int e^{-a\eta^\mu u + iq\eta} \, d\eta. \qquad (3.142)$$

□

The result of Example 3.5 suggests the Fourier integral representation for both the PDF $f_1(r, u)$ and the PDF $f_2(t, u)$. However, as admitted above since t and u are non-negative values, $f_2(t, u)$ can be presented in the form of the inverse Laplace transform by performing the variable change $s = i\omega$. Note also that for $f_1(r, u)$, one should take modulus $|\eta|^\mu$ inside the exponential. Therefore, we have

$$f_1(x, u) = \frac{1}{2\pi} \int_{-\infty}^{\infty} e^{-\sigma_\beta |k|^\beta u + ikx} \, dk, \qquad (3.143\text{a})$$

$$f_2(t, u) = \frac{1}{2\pi} \int_{-\infty}^{\infty} e^{-\tau_\alpha (i\omega)^\alpha u + it\omega} \, d\omega \qquad (3.143\text{b})$$

$$= \frac{1}{2\pi i} \int_{-i\infty}^{i\infty} e^{-\tau_\alpha s^\alpha u + st} \, ds. \qquad (3.143\text{c})$$

Scaling arguments for $ku^{1/\beta} \sim x/u^{1/\beta}$ and $su^{1/\alpha} \sim t/u^{1/\alpha}$ provide the scaling forms [32]: $f_1(x, u) = u^{-1/\beta} g_1(xu^{-1/\beta})$ and $f_2 = u^{-1/\alpha} g_2(tu^{-1/\alpha})$. Note, that the case $\alpha = 1$ is also contained in Eqs. (3.143)(b,c). Indeed, it yields $\delta(t - \tau_1 u)$, which immediately leads to $P(x, t) = \int f_1(x, u) f_2(t, u) du$.

These scaling solutions correspond to the following equations

$$\frac{\partial f_1(x,u)}{\partial u} = -\frac{\sigma_\beta}{2\pi} \int_{-\infty}^{\infty} |k|^\beta e^{-\sigma_\beta |k|^\beta u + ikx}\, dk = \sigma_\beta \partial_{|x|}^\beta f_1(x,u), \quad (3.144a)$$

$$\frac{\partial f_2(t,u)}{\partial u} = -\frac{\tau_\alpha}{2\pi\imath} \int_{-\imath\infty}^{\imath\infty} s^\alpha e^{-\tau_\alpha s^\alpha u + st}\, ds = -\frac{\partial f_3(t,u)}{\partial t}, \quad (3.144b)$$

where $f_3(t,u)$ is a new PDF in the $t-u$ space:

$$f_3(t,u) = \frac{\tau_\alpha}{2\pi\imath} \int_{-\imath\infty}^{\imath\infty} s^{\alpha-1} e^{-\tau_\alpha s^\alpha u + st}\, ds. \quad (3.145)$$

In what follows analysis this function plays important role in the definition of the physical time FFPE.

First we consider two examples of the exact solutions for the PDFs in terms of the Fox H-functions.

Example 3.6. Let us obtain the exact expression for the PDF $f_1(x,u)$. To this end we present the exponential $e^{-\sigma_\beta |k|^\beta u}$ in Eq. (3.143a) in the form of the Fox H-function:

$$e^{-\sigma_\beta |k|^\beta u} \equiv e^z = \frac{1}{2\pi\imath} \int_{c-\imath\infty}^{c+\imath\infty} \Gamma(\eta) z^{-\eta}\, d\eta$$

$$= H_{0,1}^{1,0}\left[z \,\middle|\, (0,1) \right] = H_{0,1}^{1,0}\left[\sigma_\beta |k|^\beta u \,\middle|\, (0,1) \right]. \quad (3.146)$$

Then the Fourier transform of Eq. (3.146) reduces to the cosine Melline transform of Eq. (2.34), which yields

$$f_1(x,u) = \frac{2}{2\pi} \int_0^\infty \cos(k|x|) H_{0,1}^{1,0}\left[\sigma_\beta |k|^\beta u \,\middle|\, (0,1) \right] dk$$

$$= \frac{1}{|x|} H_{2,2}^{1,1}\left[\frac{|x|^\beta}{\sigma_\beta u} \,\middle|\, \begin{array}{l} (1,1),(1,\beta/2) \\ (1,\beta),(1,0),(1,\beta/2) \end{array} \right]. \quad (3.147)$$

$\square$

Example 3.7. In this example we obtain the exact expression for the PDF $f_2(t,u)$. To this end we first perform the Laplace inversion of the integral presentation of the Fox H function in Eq. (3.146). Accounting for Eq. (3.143c), and denoting

$$as^\alpha = u\tau_\alpha s^\alpha,$$

we have a chain of transformations

$$
f_2(t, u) = \frac{1}{2\pi\imath} \int_{-\imath\infty}^{\imath\infty} e^{-\tau_\alpha s^\alpha u + st} \, ds \equiv \mathcal{L}^{-1}\left[e^{-as^\alpha}\right]
$$

$$
= \int_{c-\imath\infty}^{c+\imath\infty} \Gamma(\eta)\mathcal{L}^{-1}\left[(as^\alpha)^{-\eta}\right] d\eta = \int_{c-\imath\infty}^{c+\imath\infty} \frac{\Gamma(\eta)}{\Gamma(\alpha\eta)} a^{-\eta} t^{\alpha\eta-1} \, d\eta
$$

$$
= t^{-1} H_{1,1}^{1,0}\left[\frac{a}{t^\alpha} \,\middle|\, \begin{matrix}(0,\alpha)\\(1,1)\end{matrix}\right] = t^{-1} H_{1,1}^{0,1}\left[\frac{t^\alpha}{\tau_\alpha u} \,\middle|\, \begin{matrix}(0,1)\\(1,\alpha)\end{matrix}\right]. \tag{3.148}
$$

$\square$

3.7.1 *The PDF in physical time*

Our next step is calculation of the PDF $P(x,t)$ in physical time t and space $x \in \mathbb{R}$. To this end we use PDFs $f_1(x,u)$ and $f_3(t,u)$ to define the PDF $P(x,t)$. Taking into account Eqs. (3.143a), and (3.145) we have

$$
P(x,t) = \int_0^\infty f_1(x,u) f_3(t,u) \, du
$$

$$
= \frac{\tau_\alpha}{2\pi} \int_{-\infty}^\infty e^{\imath kx} \, dk \frac{1}{2\pi\imath} \int_{-\imath\infty}^{\imath\infty} s^{\alpha-1} e^{st} \, ds \int_0^\infty e^{-\sigma_\beta |k|^\beta u} e^{-\tau_\alpha s^\alpha} \, du
$$

$$
= \mathcal{F}^{-1}\mathcal{L}^{-1}\left[\frac{s^{\alpha-1}}{s^\alpha + \mathcal{D}_{\alpha,\beta}|k|^\beta}\right]
$$

$$
= \mathcal{F}^{-1}\left[E_\alpha\left(-\mathcal{D}_{\alpha,\beta}|k|^\beta t^\alpha\right)\right], \tag{3.149}
$$

where $\mathcal{D}_{\alpha,\beta} = \sigma_\beta/\tau_\alpha$ is the generalized transport coefficient, discussed in Sec. 3.5.2 and the PDF (3.149) is presented in the form of the Fourier inversion of the M-L function $E_\alpha\left(-\mathcal{D}_{\alpha,\beta}|k|^\beta t^\alpha\right)$. Eventually, we arrived at the desired result for the PDF $P(x,t)$ in the form of the Fox H-function in physical time and space, which is the solution of the FFPE (3.112). It also should be stressed that $f_3(t,u)$ plays the role of the subordination function $h(t,u)$ in Eq. (3.149).

References

[1]   V. Méndez, D. Campos, and F. Bartumeus. *Stochastic Foundations in Movement Ecology: Anomalous Diffusion, Front Propagation and Random Searches*. Berlin: Springer, 2014 (cit. on p. 65).

[2]   A. Iomin, V. Mèndez, and W. Horsthemke. *Fractional Dynamics in Comblike Structures*. Singapore: World Scientific, 2018 (cit. on pp. 65, 75, 87).

[3]   J. Klafter and I. M. Sokolov. *First Steps in Random Walks: From Tools to Applications*. New York: Oxford University Press, 2011 (cit. on pp. 65, 75, 77, 78, 80, 81).

[4]   B. D. Hughes. *Random Walks and Random Environments: Volume 1: Random Walks*. Oxford: Clarendon Press, 1995 (cit. on p. 65).

[5]   N. G. Van Kampen. *Stochastic Processes in Physics and Chemistry*. 3rd. Amsterdam: North Holland, 2007 (cit. on p. 65).

[6]   S. M. Rytov, Yu. A. Kravtsov, and V. I. Tatarskii. *Principles of Statistical Radiophysics 1: Elements of Random Process Theory*. Berlin: Springer-Verlag, 1987 (cit. on pp. 65, 71, 84).

[7]   E. W. Montroll and B. J. West. "On an Enriched Collection of Stochastic Processes". In: *Fluctuation Phenomena*. Ed. by E. W. Montroll and J. L. Lebowitz. Amsterdam: Elsevier, 1979, pp. 61–175 (cit. on p. 68).

[8]   S. Karlin and H. M. Taylor. *A First Course in Stochastic Processes*. 2nd. Boston: Academic Press, 1975 (cit. on p. 70).

[9]   P. Lancaster and M. Tismenetsky. *The Theory of Matrices*. New York: Academic Press, 1985 (cit. on p. 71).

[10]   C. Monthus. "Revisiting the Ruelle thermodynamic formalism for Markov trajectories with application to the glassy phase of random trap models". In: *Journal of Statistical Mechanics: Theory and Experiment* 2021.6 (2021), p. 063301 (cit. on pp. 71, 72).

[11]   P. Gaspard. *Chaos, Scattering and Statistical Mechanics*. Cambridge: Cambridge UP, 1998 (cit. on pp. 71, 73).

[12]   V. Lecomte, C. Appert-Rolland, and F. Van Wijland. "Thermodynamic Formalism for Systems with Markov Dynamics". In: *Journal of Statistical Physics* 127 (2007), pp. 51–106 (cit. on p. 71).

[13]   P. Coleman. "Philip W. Anderson (1923–2020)". In: *Nature* 581 (2020), p. 29 (cit. on p. 73).

[14]   P. W. Anderson. "Absence of Diffusion in Certain Random Lattices". In: *Physical Review* 109 (5 1958), pp. 1492–1505 (cit. on p. 73).

[15]   A. Crisanti, G. Paladin, and A. Vulpiani. *Products of Random Matrices in Statistical Physics*. Berlin: Springer, 1993 (cit. on p. 74).

[16]   H. Furstenberg and H. Kesten. "Products of Random Matrices". In: *Annals of Mathematical Statistics* 31 (1960), pp. 457–469 (cit. on p. 74).

[17]   V. I. Oseledets. "A multiplicative ergodic theorem. Characteristic Ljapunov, exponents of dynamical systems". In: *Transactions of the Moscow Mathematical Society* 19 (1968). 07.11.21, pp. 179–210 (cit. on p. 74).

[18]   I. M. Lifshits, S. A. Gredeskul, and L. A. Pastur. *Introduction to the theory of disordered systems*. New York: Wiley-Interscience, 1988 (cit. on p. 75).

[19] S. Redner. *A guide to first-passage processes*. Cambridge University Press, 2001 (cit. on pp. 75, 76).

[20] E. S. Andersen. "On the Fluctuations of Sums of Random Variables". In: *Mathematica Scandinavica* 1.2 (1953), pp. 263–285 (cit. on p. 78).

[21] R. Metzler and J. Klafter. "The random walk's guide to anomalous diffusion: a fractional dynamics approach". In: *Physics Reports* 339.1 (2000), pp. 1–77 (cit. on pp. 78, 82, 84, 86).

[22] D. R. Cox. *Renewal Theory*. London: Methuen, 1967 (cit. on p. 78).

[23] E. W. Montroll and G. H. Weiss. "Random Walks on Lattices. II". In: *Journal of Mathematical Physics* 6.2 (1965), pp. 167–181 (cit. on p. 81).

[24] J. Klafter, A. Blumen, and M. F. Shlesinger. "Stochastic pathway to anomalous diffusion". In: *Physical Review A* 35 (7 1987), pp. 3081–3085 (cit. on pp. 82, 86).

[25] F. Oberhettinger and L. Badii. *Tables of Laplace transforms*. Springer Science & Business Media, 2012 (cit. on pp. 87, 88).

[26] P. Langevin. "Sur la théorie du mouvement brownien". In: *Comptes Rendus de l'Académie des Sciences Paris* 146 (1908), pp. 530–533 (cit. on p. 89).

[27] D. S. Lemons and A. Gythiel. "Paul Langevin's 1908 paper "On the Theory of Brownian Motion" ["Sur la théorie du mouvement brownien," C. R. Acad. Sci. (Paris) 146, 530–533 (1908)]". In: *American Journal of Physics* 65.11 (1997), pp. 1079–1081 (cit. on p. 89).

[28] K. Furutsu. "On the theory of radio wave propagation over inhomogeneous earth". In: *Journal of Research of the National Bureau of Standards, Section D: Radio Propagation* 67 (1963), pp. 39–62 (cit. on p. 91).

[29] K. Furutsu. "On the statistical theory of electromagnetic waves in a fluctuating medium (I)". In: *Journal of Research of the National Bureau of Standards, Section D: Radio Propagation* 67 (1963), pp. 303–323 (cit. on p. 91).

[30] E. A. Novikov. "Functionals and the random-force method in turbulence". In: *Journal of Experimental and Theoretical Physics (JETP)* 47 (1964). [Sov. Phys. JETP **20**, 1990 (1965)], p. 1919 (cit. on p. 91).

[31] P. Barthelemy, J. Bertolotti, and D. S. Wiersma. "A Lévy flight for light". In: *Nature* 453 (2008), pp. 495 –498. DOI: https://doi.org/10.1038/nature06948 (cit. on p. 92).

[32] H. C. Fogedby. "Langevin equations for continuous time Lévy flights". In: *Physical Review E* 50 (2 1994), pp. 1657–1660 (cit. on pp. 92, 93).

[33] J. P. Bouchaud and A. Georges. "Anomalous Diffusion in Disordered Media: Statistical Mechanisms, Models and Physical Applications". In: *Physics Reports* 195.4-5 (1990), pp. 127–293 (cit. on p. 92).

Chapter 4

CTRW on combs

Continuous time random walks (CTRW) have been discussed in Chapter 3 for various realisations of the diffusion limit, which leads to Brownian (normal) diffusion, subdiffusion and Lévy walks. In this chapter, we extend our discussion of the CTRW for the topologically constrained two dimensional case, known as a comb model [1]. Various realizations of anomalous and heterogeneous diffusion are considered with explanation of the geometry and mechanism of the phenomena. We shall also discuss three dimensional case of the comb known as a xyz-comb. Different generalizations of the comb model are considered in Refs. [2, 3, 4, 5, 6, 7, 8, 9].

We start our consideration from a general consideration of a subordination approach to anomalous diffusion, considered in Sec. 3.7.

4.1 CTRW and subordination

The subordination approach is a powerful technique for solving different fractional and generalized diffusion and Fokker-Planck equation. In what follows we will show how this approach works within the CTRW theory.

Let us return to Eqs. (3.138) and briefly repeat the main idea of the subordination function $h(t, u)$. We consider the coupled Langevin equations [10]

$$\begin{cases} \frac{d}{du}X(u) = r(u), \\ \frac{d}{du}T(u) = \tau(u). \end{cases} \tag{4.1}$$

From Eq. (3.142b), we obtain the subordination function as follows

$$h(u, t) = -\frac{\partial}{\partial u} \left\langle \Theta \left(t - T(u) \right) \right\rangle, \tag{4.2}$$

where $\Theta(z)$ is the Heaviside theta function. The Laplace transform then yields

$$\hat{h}(u,s) = -\frac{\partial}{\partial u}\frac{1}{s}\left\langle \int_0^\infty \delta\left(t - T(u)\right)e^{-st}\,dt \right\rangle = -\frac{\partial}{\partial u}\frac{1}{s}\langle e^{-s\,T(u)}\rangle. \quad (4.3)$$

Then taking into account the result in Eq. (3.141c) and performing the Laplace inversion, we immediately obtain the subordination function (3.143). Therefore, the PDF of the subordinated process can be given in terms of the subordination integral of a product of the parent process PDF $f(x,u)$ and the subordination function $h(u,t)$,

$$P(x,t) = \langle \delta(x - X(t)) \rangle = \int_0^\infty f(x,u)\,h(u,t)\,dt. \quad (4.4)$$

Let us consider how subordination "works" for solutions of the Fokker-Planck equations.

4.1.1 *From standard to generalized Fokker-Planck equation*

First we consider the standard Fokker-Planck equation

$$\frac{\partial}{\partial t}P(x,t) = L_{\mathrm{FP}}P(x,t), \quad (4.5)$$

with the Fokker-Planck operator

$$L_{\mathrm{FP}} \equiv \left[\frac{\partial}{\partial x}U'(x) + \mathcal{D}\frac{\partial^2}{\partial x^2}\right]. \quad (4.6)$$

After the Laplace transform, Eq. (4.5) reads

$$s\,\hat{P}(x,s) - \delta(x - x_0) = L_{\mathrm{FP}}\hat{P}(x,s). \quad (4.7)$$

Introducing the change of the variables $s \to s\hat{\gamma}(s)$, we obtain

$$s\hat{\gamma}(s)\,\hat{P}(x,s\hat{\gamma}(s)) - \delta(x - x_0) = L_{\mathrm{FP}}\hat{P}(x,s\hat{\gamma}(s)), \quad (4.8)$$

which can be rewritten as follows

$$s\underbrace{\left(\hat{\gamma}(s)\hat{P}(x,s\hat{\gamma}(s))\right)}_{=\hat{P}_{\mathrm{s}}(x,s)} - \delta(x - x_0) = \frac{1}{\hat{\gamma}(s)}L_{\mathrm{FP}}\underbrace{\left(\hat{\gamma}(s)\hat{P}(x,s\hat{\gamma}(s))\right)}_{=\hat{P}_{\mathrm{s}}(x,s)}. \quad (4.9)$$

Introducing a new function

$$\hat{P}_{\mathrm{s}}(x,s) = \hat{\gamma}(s)\hat{P}(x,s\hat{\gamma}(s)), \quad (4.10)$$

we arrive at the following equation for $\hat{P}_{\mathrm{s}}(x,s)$,

$$\hat{\gamma}(s)\left[s\hat{P}_{\mathrm{s}}(x,s) - \delta(x - x_0)\right] = L_{\mathrm{FP}}\hat{P}_{\mathrm{s}}(x,s). \quad (4.11)$$

The inverse Laplace transform then yields the following generalized Fokker-Planck equation

$$\int_0^t \gamma(t-t')\frac{\partial}{\partial t'}P_s(x,t')dt' = L_{FP}P_s(x,t) \qquad (4.12)$$

with the memory kernel $\gamma(t)$.

We note the $P_s(x,s)$ in Eq. (4.10) can be presented in the subordination form as follows

$$\hat{P}_s(x,s) = \hat{\gamma}(s)\hat{P}(x,s\hat{\gamma}(s)) = \hat{\gamma}(s)\int_0^\infty P(x,u)e^{-us\hat{\gamma}(s)}du$$

$$= \int_0^\infty P(x,u)\hat{h}(u,s)du \qquad (4.13)$$

where

$$\hat{h}(u,s) = \hat{\gamma}(s)e^{-us\hat{\gamma}(s)} \qquad (4.14)$$

is the subordination function in the Laplace space, while the integral (4.13) is the subordination integral. By the inverse Laplace transform of Eq. (4.14), the subordination function reads

$$h(u,t) = \mathcal{L}^{-1}\left[\hat{\gamma}(s)e^{-us\hat{\gamma}(s)}\right]. \qquad (4.15)$$

Therefore, the knowledge of the subordination function and the solution of a standard Fokker-Plank equation makes it possible to obtain the PDF of the subordinated process, governed by the generalized Fokker-Planck equation (4.5), by means of the subordination integral

$$P_s(x,t) = \int_0^\infty P(x,u)h(u,t)du$$

where $P(x,u)$ is the solution of the standard Fokker-Planck equation (4.05).

Remark 4.1 (Equivalent formulation). If we present the kernel $\hat{\gamma}(s)$ in the form of the reciprocal function[1]

$$\hat{\gamma}(s) = \frac{1}{s\hat{\eta}(s)}, \qquad (4.16)$$

then Eq. (4.11) reads

$$s\hat{P}_s(x,s) - \delta(x-x_0) = \frac{1}{s\hat{\eta}(s)}L_{FP}\hat{P}_s(x,s). \qquad (4.17)$$

The inverse Laplace transform yields

$$\frac{\partial}{\partial t}P_s(x,t) = \frac{\partial}{\partial t}\int_0^t \eta(t-t')L_{FP}P_s(x,t')dt', \qquad (4.18)$$

which is an equivalent representation of the generalized Fokker-Planck equation (4.12).

---

[1]For functions which satisfy the property $\hat{M}(s)\hat{K}(s) = 1/s$, i.e., the convolution integral equals 1, $M \star K \equiv 1$, is said to form a pair of Sonine kernels [11, 12].

4.2 Fractional Fokker-Planck equation

Let us consider a particular case of the generalized Fokker-Planck equation by using a power-law memory kernel $\gamma(t) = \frac{t^{-\mu}}{\Gamma(1-\mu)}$. Then, one arrives at the FFPE

$$_{C}D_{0+}^{\mu}P_{s}(x,t) = L_{FP}P_{s}(x,t), \tag{4.19}$$

where $_{C}D_{a+}^{\mu}f(t)$ is the Caputo time fractional derivative (1.196). From Remark 4.1 one finds that $\eta(t) = \mathcal{L}^{-1}\left[\frac{1}{s\hat{\gamma}(s)}\right] = \frac{t^{\mu-1}}{\Gamma(\mu)}$, which means that the FFPE (4.19) can be written in the equivalent form

$$\frac{\partial}{\partial t}P_{s}(x,t) = {}_{RL}D_{0+}^{1-\mu}L_{FP}P_{s}(x,t), \tag{4.20}$$

where $_{RL}D_{a+}^{\mu}$ is the R-L time fractional derivative (1.194).

4.3 Diffusion on comb structures: Derivation of the FFPE

A comb model is a particular example of a non-Markovian motion, which takes place due to its specific geometry realization inside a two-dimensional structure. It consists of a backbone along the structure x axis and fingers (or branches) along the y direction, continuously spaced along the x coordinate, as shown in Fig. 4.1. This special geometry is due to the special properties of the components of the diffusion tensor which are $\mathcal{D}_{xx} = \mathcal{D}_{x}\,\delta(y-y_{0})$, and $\mathcal{D}_{yy} = \mathcal{D}_{y} \neq 0$, while $\mathcal{D}_{xy} = \mathcal{D}_{yx} = 0$. That is, the transport along the x direction is possible along the backbone only, for $y = y_{0}$. In sequel we choose the backbone at $y_{0} = 0$. This model has been extensively explored for understanding various realizations of non-Markovian random walks. This situation is reflected in Ref. [1] (and references therein). Besides the interesting realizations of anomalous transport we also learn here how the latter can be described by means of the Fox H-functions.

From the CTRW point of view, a particle moving along the backbone can be stacked in the fingers, so there is no movement along the backbone until the particle returns back to the backbone. Therefore, the fingers play the role of traps for the x motion. The returning probability of the Brownian particle from the finger to the backbone corresponds to the waiting time PDF for the particle moving along the backbone, so for Brownian motion it scales as $\psi(t) \sim t^{-3/2}$ (cf. Eq. (3.61)). From the CTRW theory we know that such waiting times leads to anomalous diffusion with the MSD given by $\langle x^{2}(t) \rangle \sim t^{1/2}$.

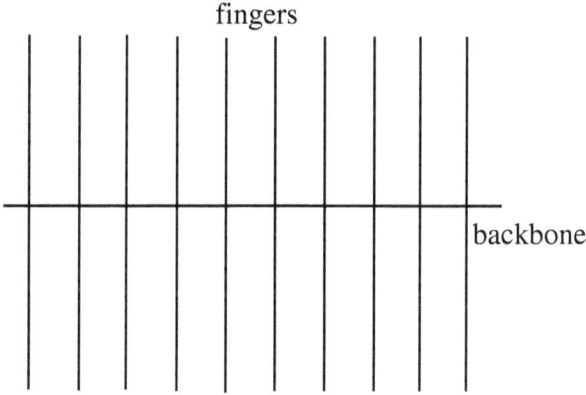

fingers

backbone

Fig. 4.1 Two dimensional comb structure. The backbone along the x direction has continuously distributed fingers (or branches) along y direction. Republished with permission of IOP Publishing, LTD, from J. Phys. A: Math. Theor. T. Sandev, A. Iomin and V. Méndez, 49(35), 355001 (2016).

4.3.1 *Fokker-Planck equation for comb structure*

Phenomenological formulation of the comb Fokker-Planck equation has been suggested in Ref. [13], and the equation reads

$$\frac{\partial}{\partial t} P(x, y, t) = \mathcal{D}_x \delta(y) \frac{\partial^2}{\partial x^2} P(x, y, t) + \mathcal{D}_y \frac{\partial^2}{\partial y^2} P(x, y, t), \qquad (4.21)$$

We should stress that the Dirac δ-function is incorporated inside the x component of the diffusion coefficient that insures non zero current along the backbone.

We look for the solution of the Fokker-Planck equation (4.21) for the initial condition $P(x, y, 0) = \delta(x)\delta(y)$. By the Laplace transform we have

$$s \hat{P}(x, y, s) - \delta(x)\delta(y) = \mathcal{D}_x \delta(y) \frac{\partial^2}{\partial x^2} \hat{P}(x, y, s) + \mathcal{D}_y \frac{\partial^2}{\partial x^2} \hat{P}(x, y, s). \quad (4.22)$$

The Fourier transform then yields

$$s \tilde{\hat{P}}(k_x, k_y, s) - 1 = -\mathcal{D}_x k_x^2 \tilde{\hat{P}}(k_x, y = 0, s) - \mathcal{D}_y k_y^2 \tilde{\hat{P}}(k_x, k_y, s), \qquad (4.23)$$

from where one obtains

$$\tilde{\hat{P}}(k_x, k_y, s) = \frac{1 - \mathcal{D}_x k_x^2 \tilde{\hat{P}}(k_x, y = 0, s)}{s + \mathcal{D}_y k_y^2}. \qquad (4.24)$$

Using the expression $\mathcal{F}^{-1}\left[\frac{2a}{a^2+k_y^2}\right] = e^{-a|y|}$, we obtain

$$\tilde{\hat{P}}(k_x, y, s) = \frac{s^{-1/2}e^{-\sqrt{s/\mathcal{D}_y}|y|}\left[1 - \mathcal{D}_x k_x^2 P(k_x, y = 0, s)\right]}{2\sqrt{\mathcal{D}_y}}. \tag{4.25}$$

This yields the explicit expression for $\tilde{\hat{P}}(k_x, y = 0, s)$. Substituting the obtained result in Eq. (4.24), we obtain

$$\tilde{\hat{P}}(k_x, k_y, s) = \frac{s^{1/2}}{\left(s + \mathcal{D}_y k_y^2\right)\left(s^{1/2} + \frac{\mathcal{D}_x}{2\sqrt{\mathcal{D}_y}}k_x^2\right)}. \tag{4.26}$$

This also yields

$$\tilde{\hat{P}}(k_x, k_y = 0, s) = \frac{s^{-1/2}}{s^{1/2} + \frac{\mathcal{D}_x}{2\sqrt{\mathcal{D}_y}}k_x^2} \tag{4.27}$$

and

$$\tilde{\hat{P}}(k_x = 0, k_y, s) = \frac{1}{s + \mathcal{D}_y k_y^2}. \tag{4.28}$$

As is known, diffusion along the backbone can be described by the marginal PDF

$$p_1(x, t) = \int_{-\infty}^{\infty} P(x, y, t)\, dy, \tag{4.29}$$

which after the Fourier-Laplace transform reads

$$p_1(k_x, s) = \int_{-\infty}^{\infty} P(k_x, y, s)\, dy. \tag{4.30}$$

By integration of the FPE (4.21) with respect to y, one finds

$$\tilde{\hat{p}}_1(k_x, s) = \tilde{\hat{P}}(k_x, k_y = 0, s) = \frac{s^{-1/2}}{s^{1/2} + \frac{\mathcal{D}_x}{2\sqrt{\mathcal{D}_y}}k_x^2}. \tag{4.31}$$

It can be rewritten in the form

$$s^{1/2}\tilde{\hat{p}}_1(k_x, s) - s^{-1/2} = -\frac{\mathcal{D}_x}{2\sqrt{\mathcal{D}_y}}k_x^2\,\tilde{\hat{p}}_1(k_x, s), \tag{4.32}$$

which by the inverse Fourier and Laplace transformations yields time fractional diffusion equation

$$_C D_{0+}^{1/2} p_1(x, t) = \frac{\mathcal{D}_x}{2\sqrt{\mathcal{D}_y}}\frac{\partial^2}{\partial x^2}p_1(x, t), \tag{4.33}$$

where $_C D_{0+}^{1/2} f(t)$ is the Caputo fractional derivative. The solution of Eq. (4.33) can be found first by inverse Laplace transform of Eq. (4.31), which reads

$$\tilde{p}_1(k_x, t) = \mathcal{L}^{-1} \left[\frac{s^{-1/2}}{s^{1/2} + \frac{\mathcal{D}_x}{2\sqrt{\mathcal{D}_y}} k_x^2} \right] = E_{1/2} \left(-\frac{\mathcal{D}_x}{2\sqrt{\mathcal{D}_y}} k_x^2 t^{1/2} \right), \qquad (4.34)$$

and then by inverse Fourier transform, which finally yields

$$p_1(x, t) = \mathcal{F}^{-1} \left[E_{1/2} \left(-\frac{\mathcal{D}_x}{2\sqrt{\mathcal{D}_y}} k_x^2 t^{1/2} \right) \right]$$

$$= \frac{1}{2|x|} H_{1,1}^{1,0} \left[\frac{|x|}{\left(\frac{\mathcal{D}_x}{2\sqrt{\mathcal{D}_y}} t^{1/2} \right)^{1/2}} \middle| \begin{matrix} (1, \frac{1}{4}) \\ (1, 1) \end{matrix} \right]. \qquad (4.35)$$

The MSD is calculated straightforwardly from Eq. (4.30), which yields

$$\langle x^2(t) \rangle = \mathcal{L}^{-1} \left[-\frac{\partial^2}{\partial k_x^2} \hat{\tilde{p}}_1(k_x, s) \right] \bigg|_{k_x=0} = 2 \left(\frac{\mathcal{D}_x}{2\sqrt{\mathcal{D}_y}} \right) \frac{t^{1/2}}{\Gamma(3/2)}. \qquad (4.36)$$

Diffusion in fingers is described by the marginal PDF

$$p_2(y, t) = \int_{-\infty}^{\infty} P(x, y, t) \, dx.$$

By integration of the Fokker-Planck equation (4.21) with respect to x, we find that the PDF along the fingers satisfies the standard diffusion equation

$$\frac{\partial}{\partial t} p_2(y, t) = \mathcal{D}_y \frac{\partial^2}{\partial y^2} p_2(y, t) \qquad (4.37)$$

for Brownian motion with Gaussian PDF, and the corresponding MSD reads

$$\langle y^2(t) \rangle = 2\mathcal{D}_y t. \qquad (4.38)$$

Remark 4.2 (Alternative solution). An alternative approach to the FPE (4.21) can be presented as well. By the Laplace transformation, it becomes

$$s \, P(x, y, s) - \delta(x)\delta(y) = \mathcal{D}_x \delta(y) \frac{\partial^2}{\partial x^2} P(x, y, s) + \mathcal{D}_y \frac{\partial^2}{\partial y^2} P(x, y, s). \qquad (4.39)$$

Then we are looking for the solution in the form

$$P(x, y, s) = g(x, s) e^{-r(x,s)|y|}, \qquad (4.40)$$

from where the marginal PDF is

$$\hat{p}_1(x,s) = \int_{-\infty}^{\infty} \hat{P}(x,y,s)\,dy = \frac{2g(x,s)}{r(x,s)}. \tag{4.41}$$

Using the formula $\frac{d}{dy}|y| = 2\theta(y) - 1$, from Eq. (4.39) one has

$$\frac{\partial}{\partial y}\hat{P}(x,y,s) = g(x,s)e^{-r(x,s)|y|}(-r(x,s))[2\theta(y) - 1]. \tag{4.42}$$

Accounting that $\frac{d}{dy}\theta(y) = \delta(y)$, it follows

$$\frac{\partial^2}{\partial y^2}\hat{P}(x,y,s) = -r(x,s)g(x,s)\left\{2\delta(y) - r(x,s)[2\theta(y) - 1]^2\right\}e^{-r(x,s)|y|}. \tag{4.43}$$

Next we use $f(y)\delta(y) = f(0)\delta(y)$, from where the equation transforms to

$$s\,g(x,s)e^{-r(x,s)|y|} - \delta(x)\underline{\delta(y)} = \mathcal{D}_x\underline{\delta(y)}\frac{\partial^2}{\partial x^2}g(x,s) - 2\mathcal{D}_y\underline{\delta(y)}r(x,s)g(x,s)$$
$$+ \mathcal{D}_y r^2(x,s)\underline{g(x,s)e^{-r(x,s)|y|}}. \tag{4.44}$$

Thus, we have a system of two equations:

$$s = \mathcal{D}_y r^2(x,s) \rightarrow r(x,s) = \sqrt{\frac{s}{\mathcal{D}_y}} \tag{4.45}$$

and

$$-\delta(x) = \mathcal{D}_x\frac{\partial^2}{\partial x^2}g(x,s) - 2\mathcal{D}_y r(x,s)g(x,s). \tag{4.46}$$

From Eq. (4.41) we find

$$\hat{p}_1(x,s) = 2\sqrt{\frac{\mathcal{D}_y}{s}}g(x,s), \tag{4.47}$$

and thus

$$-\delta(x) = \mathcal{D}_x\frac{\partial^2}{\partial x^2}g(x,s) - 2\sqrt{\mathcal{D}_y}s^{1/2}g(x,s). \tag{4.48}$$

From Eqs. (4.47) and (4.48) we obtain

$$s^{1/2}\hat{p}_1(x,s) - s^{-1/2}\delta(x) = \frac{\mathcal{D}_x}{2\sqrt{\mathcal{D}_y}}\frac{\partial^2}{\partial x^2}\hat{p}_1(x,s), \tag{4.49}$$

then the inverse Laplace transform yields the following time fractional diffusion equation

$$_C D_{0+}^{1/2}p_1(x,t) = \frac{\mathcal{D}_x}{2\sqrt{\mathcal{D}_y}}\frac{\partial^2}{\partial x^2}p_1(x,t), \tag{4.50}$$

which is the same as Eq. (4.33). Its solution is given in terms of the Fox H-function

$$p_1(x,t) = \frac{1}{2|x|} H^{1,0}_{1,1} \left[\frac{|x|}{\left(\frac{D_x}{2\sqrt{D_y}}t^{1/2}\right)^{1/2}} \middle| \begin{matrix}(1,\frac{1}{4}) \\ (1,1)\end{matrix}\right]. \qquad (4.51)$$

Its asymptotic behavior for

$$|x| / \left(\frac{D_x}{2\sqrt{D_y}}t^{1/2}\right)^{1/2} \gg 1$$

has the stretched exponential form,

$$p_1(x,t) \sim \frac{4^{1/3}}{\sqrt{6\pi}} \frac{|x|^{-1/3}}{\left(\frac{D_x}{2\sqrt{D_y}}t^{1/2}\right)^{1/3}} \times \exp\left(-\frac{3}{4^{4/3}}\frac{|x|^{4/3}}{\left(\frac{D_x}{2\sqrt{D_y}}t^{1/2}\right)^{2/3}}\right).$$

4.3.2 *Diffusion on 3D comb*

Let us discuss anomalous diffusion in the xyz-comb considered in Refs. [14, 15], see Fig. 4.2. The corresponding FPE reads

$$\frac{\partial}{\partial t}P(x,y,z,t) = D_x\delta(y)\delta(z)\frac{\partial^2}{\partial x^2}P(x,y,z,t) + D_y\delta(z)\frac{\partial^2}{\partial y^2}P(x,y,z,t)$$

$$+ D_z\frac{\partial^2}{\partial z^2}P(x,y,z,t), \qquad (4.52)$$

where $D_x\delta(y)\delta(z)$, $D_y\delta(z)$, and D_z are the diffusion coefficients. We assume the initial condition

$$P(x,y,z,t=0) = \delta(x-x')\delta(y-y')\delta(z-z'), \qquad (4.53)$$

and zero boundary conditions at infinity, $P(x,y,z,t) = 0$ and $\frac{\partial}{\partial q}P(x,y,z,t) = 0$, $q = \{x,y,z\}$, when $q = \pm\infty$. We notice that the xyz-comb is constructed by geometric constraints of the diffusive motion by means of the Dirac δ-functions in complete analogy with the 2D comb. That is, the motion along the x-direction occurs only at $y = z = 0$ that is dictated by the x and y components of the diffusion coefficient.

In this section, we concern with the transport properties along the x, y and z axes in the framework of corresponding PDFs. To this end, we

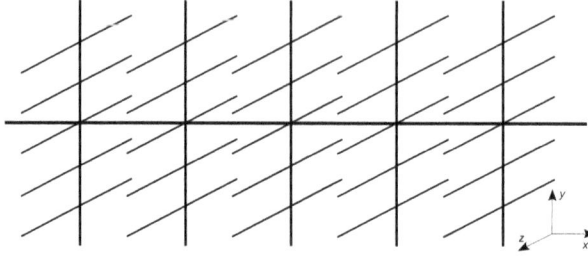

Fig. 4.2   Three dimensional comb structure. The backbone along the x direction has continuously distributed fingers along the y direction. The fingers also has perpendicular fingers continuously distributed along the z direction. Republished with permission of IOP Publishing, LTD, from J. Stat. Mech., E. K. Lenzi, T. Sandev, H. V. Ribeiro, P. Jovanovski, A. Iomin and L. Kocarev, 2020(5), 053203 (2020).

introduce the marginal PDFs $p_1(x,t)$, $p_2(y,t)$, $p_3(z,t)$ and $p_{12}(x,y,t)$, which are defined as follows

$$p_1(x,t) = \int_{-\infty}^{\infty} \int_{-\infty}^{\infty} P(x,y,z,t)\, dz\, dy, \tag{4.54}$$

$$p_2(y,t) = \int_{-\infty}^{\infty} \int_{-\infty}^{\infty} P(x,y,z,t)\, dz\, dx, \tag{4.55}$$

$$p_3(z,t) = \int_{-\infty}^{\infty} \int_{-\infty}^{\infty} P(x,y,z,t)\, dy\, dx, \tag{4.56}$$

$$p_{12}(x,y,t) = \int_{-\infty}^{\infty} P(x,y,z,t)\, dz. \tag{4.57}$$

From Eq. (4.57), it follows that $\tilde{\hat{p}}_{12}(k_x, k_y, s) = \tilde{\hat{P}}_z(k_x, k_y, k_z = 0, s)$. In the framework of such consideration the z-direction is considered as an auxiliary subspace, needed to describe the anomalous diffusive behavior along the y fingers by means of the Markovian process. Thus, by integration of PDF $P(x,y,z,t)$ with respect to z variable one analyzes the diffusive motion on the two dimensional (xy) comb, which is described by the corresponding marginal PDF $p_{12}(x,y,t)$. Further projection of the two dimensional comb dynamics on the one dimensional configuration space, makes it possible to consider diffusion of particles along the main backbone in the framework of the marginal PDF $p_1(x,t)$.

Employing the Fourier transform with respect to z and the Laplace transform with respect to t in Eq. (4.52), one obtains

$$s\bar{\hat{P}}_z(x, y, k_z, s) - \bar{P}_z(x, y, k_z, t = 0) = D_x \delta(y)\frac{\partial^2}{\partial x^2}\hat{P}(x, y, z = 0, s)$$

$$+ D_y \frac{\partial^2}{\partial y^2}\hat{P}(x, y, z = 0, s) - D_z k_z^2 \bar{\hat{P}}_z(x, y, k_z, s). \quad (4.58)$$

For $k_z = 0$, it follows

$$s\bar{\hat{P}}_z(x, y, k_z = 0, s) - \bar{P}_z(x, y, k_z = 0, t = 0) = D_x \delta(y)\frac{\partial^2}{\partial x^2}\hat{P}(x, y, z = 0, s)$$

$$+ D_y \frac{\partial^2}{\partial y^2}\hat{P}(x, y, z = 0, s).$$

$$(4.59)$$

Therefore, we can write

$$\bar{\hat{P}}_z(x, y, k_z, s) = \frac{1}{s + D_z k_z^2}\left[\bar{P}_z(x, y, k_z = 0, t = 0)\right.$$

$$+ D_x \delta(y)\frac{\partial^2}{\partial x^2}\hat{P}(x, y, z = 0, s)$$

$$\left. + D_y \frac{\partial^2}{\partial y^2}\hat{P}(x, y, z = 0, s)\right]. \quad (4.60)$$

Performing the inverse Fourier transform with respect to k_z, we find

$$\hat{P}(x, y, z, s) = \frac{1}{2\sqrt{D_z}} s^{-1/2} e^{-\sqrt{\frac{s}{D_z}}|z|}\left[\bar{P}_z(x, y, k_z = 0, t = 0)\right.$$

$$\left. + D_x \delta(y)\frac{\partial^2}{\partial x^2}\hat{P}(x, y, z = 0, s) + D_y \frac{\partial^2}{\partial y^2}\hat{P}(x, y, z = 0, s)\right],$$

$$(4.61)$$

and, thus,

$$\hat{P}(x, y, z = 0, s) = \frac{1}{2\sqrt{D_z}} s^{1/2} \bar{\hat{P}}_z(x, y, k_z = 0, s). \quad (4.62)$$

On the other hand, by integrating Eq. (4.52) with respect to z, we find

$$\hat{p}_{12}(x, y, s) = \bar{\hat{P}}_z(x, y, k_z = 0, s), \quad (4.63)$$

and the equation for $\hat{p}_{12}(x, y, s)$ becomes

$$s^{1/2}\hat{p}_{12}(x, y, s) - s^{-1/2} p_{12}(x, y, t = 0)$$

$$= \frac{D_y}{2\sqrt{D_z}}\left[\delta(y)\frac{D_x}{D_y}\frac{\partial^2}{\partial x^2} + \frac{\partial^2}{\partial y^2}\right]\hat{p}_{12}(x, y, s). \quad (4.64)$$

The inverse Laplace transform yields the time fractional diffusion equation on the two dimensional xy comb [16]

$$_{\mathrm{c}}D_{0+}^{1/2}p_{12}(x,y,t) = \frac{\mathcal{D}_y}{2\sqrt{\mathcal{D}_z}}\left[\delta(y)\frac{\mathcal{D}_x}{\mathcal{D}_y}\frac{\partial^2}{\partial x^2} + \frac{\partial^2}{\partial y^2}\right]p_{12}(x,y,t), \qquad (4.65)$$

where $_{\mathrm{c}}D_{0+}^{\mu}f(t)$ is the Caputo fractional derivative of order $0 < \mu < 1$. The time fractional diffusion equation for the 2D comb naturally appears by projection of the 3D comb motion on the xy-plane in the (x, y, k_z, s) space. This projection corresponds to a coarse-graining procedure, when the information about the z-motion is lost. This leads to a trap picture with the waiting time distribution, reflected in the Caputo fractional derivative in Eq. (4.65).

We can also solve Eq. (4.65) by using the same approach. For the marginal PDF along the x-direction, $p_1(x,t) = \int_{-\infty}^{\infty} p_{12}(x,y,t)\,dy$, we obtain the following time fractional diffusion equation with the Caputo fractional derivative of the order of $1/4$, that is,

$$_{\mathrm{c}}D_{0+}^{1/4}p_1(x,t) = \frac{\mathcal{D}_x}{2\sqrt{2\mathcal{D}_y\sqrt{\mathcal{D}_z}}}\frac{\partial^2}{\partial x^2}p_1(x,t). \qquad (4.66)$$

The solution of the Eq. (4.66) can be obtained in terms of the Fox H-function (2.17),

$$p_1(x,t) = \frac{1}{2|x|}H_{1,1}^{1,0}\left[\frac{|x|}{\left(\frac{\mathcal{D}_x}{2\sqrt{2\mathcal{D}_y\sqrt{\mathcal{D}_z}}}t^{1/4}\right)^{1/2}}\middle| \begin{matrix}(1,\frac{1}{8})\\(1,1)\end{matrix}\right]$$

$$= \frac{1}{2}\sum_{k=0}^{\infty}\frac{(-1)^k}{k!\,\Gamma(1-(k+1)/8)}\frac{|x|^k}{\left(\frac{\mathcal{D}_x}{2\sqrt{2\mathcal{D}_y\sqrt{\mathcal{D}_z}}}t^{1/4}\right)^{(k+1)/2}}. \qquad (4.67)$$

When

$$|x|\Bigg/\left(\frac{\mathcal{D}_x}{2\sqrt{2\mathcal{D}_y\sqrt{\mathcal{D}_z}}}t^{1/4}\right)^{1/2} \gg 1$$

the asymptotic behavior of the solution (4.67) reduces to a stretched exponential function as follows

$$p_1(x,t) \sim \frac{8^{3/7}}{\sqrt{7\pi}}\frac{|x|^{-3/7}}{\left(\frac{\mathcal{D}_x}{2\sqrt{2\mathcal{D}_y\sqrt{\mathcal{D}_z}}}t^{1/4}\right)^{2/7}}\times\exp\left(-\frac{7}{8^{8/7}}\frac{|x|^{8/7}}{\left(\frac{\mathcal{D}_x}{2\sqrt{2\mathcal{D}_y\sqrt{\mathcal{D}_z}}}t^{1/4}\right)^{4/7}}\right).$$

From the solution (4.67), we calculate the MSD along the main backbone (x-direction)

$$\langle x^2(t) \rangle = \int_{-\infty}^{\infty} x^2 \, p_1(x,t) \, dx = 2 \left(\frac{\mathcal{D}_x}{2\sqrt{2\mathcal{D}_y \sqrt{\mathcal{D}_z}}} \right) \frac{t^{1/4}}{\Gamma(5/4)}, \qquad (4.68)$$

which corresponds to subdiffusion (MSD $\sim t^{1/4}$). We pay attention on the fact that the transport exponent is twice as less as in subdiffusion in the 2D comb. This reduction of the transport exponent results from the 3D comb geometry, where the transport inside the y fingers is subdiffusive as well. To show this we consider the corresponding marginal PDF $p_2(y,t)$.

The marginal PDF along the y-direction, $p_2(y,t) = \int_{-\infty}^{\infty} dx \, p_{12}(x,y,t)$, satisfies the time fractional diffusion equation with the Caputo fractional derivative of the order $1/2$,

$$_C D_{0+}^{1/2} p_2(y,t) = \frac{\mathcal{D}_y}{2\sqrt{\mathcal{D}_z}} \frac{\partial^2}{\partial y^2} p_2(y,t) \qquad (4.69)$$

and the solution is

$$p_2(y,t) = \frac{1}{2|y|} H_{1,1}^{1,0} \left[\frac{|y|}{\left(\frac{\mathcal{D}_y}{2\sqrt{\mathcal{D}_z}} t^{1/2} \right)^{1/2}} \left| \begin{matrix} (1,\frac{1}{4}) \\ (1,1) \end{matrix} \right. \right]$$

$$= \frac{1}{2} \sum_{k=0}^{\infty} \frac{(-1)^k}{k! \, \Gamma(1-(k+1)/4)} \frac{|y|^k}{\left(\frac{\mathcal{D}_y}{2\sqrt{\mathcal{D}_z}} t^{1/2} \right)^{(k+1)/2}}. \qquad (4.70)$$

The asymptotic behavior of the equation for

$$|y| / \left(\frac{\mathcal{D}_y}{2\sqrt{\mathcal{D}_z}} t^{1/2} \right)^{1/2} \gg 1$$

is also a stretched exponential form, that is,

$$p_2(y,t) \sim \frac{4^{1/3}}{\sqrt{6\pi}} \frac{|y|^{-1/3}}{\left(\frac{\mathcal{D}_y}{2\sqrt{\mathcal{D}_z}} t^{1/2} \right)^{1/3}} \times \exp \left(-\frac{3}{4^{4/3}} \frac{|y|^{4/3}}{\left(\frac{\mathcal{D}_y}{2\sqrt{\mathcal{D}_z}} t^{1/2} \right)^{2/3}} \right).$$

Correspondingly, the MSD along the y-direction is

$$\langle y^2(t) \rangle = \int_{-\infty}^{\infty} y^2 \, p_2(y,t) \, dy = \frac{\mathcal{D}_y}{2\sqrt{\mathcal{D}_z}} \frac{t^{1/2}}{\Gamma(3/2)}, \qquad (4.71)$$

and it also presents anomalous subdiffusion (MSD $\sim t^{1/2}$). This anomalous dynamics, as before, appears as a result of the trapping events of a particle in the z-fingers.

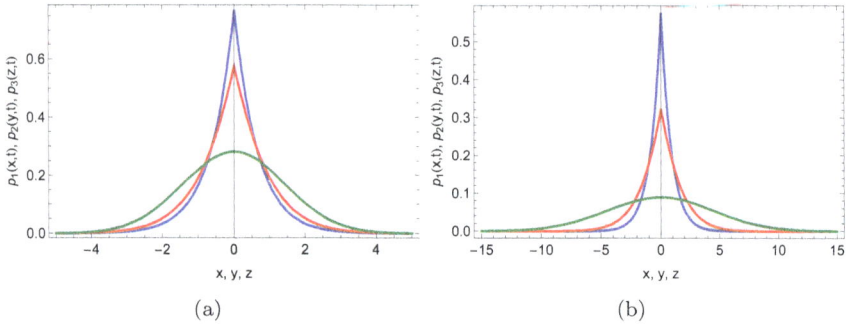

(a)   (b)

Fig. 4.3 Graphical representation of marginal PDFs, Eq. (4.66) (blue line) and Eq. (4.69) (red line) for $\mathcal{D}_x = \mathcal{D}_y = \mathcal{D}_z = 1$ and (a) $t = 1$, (b) $t = 10$. Green lines correspond to Brownian diffusion along the z-axis, according to Eq. (4.72). Republished with permission of IOP Publishing, LTD, from J. Stat. Mech., E. K. Lenzi, T. Sandev, H. V. Ribeiro, P. Jovanovski, A. Iomin and L. Kocarev, 2020(5), 053203 (2020).

For the PDF $p_3(z,t) = \int_{-\infty}^{\infty} \int_{-\infty}^{\infty} P(x,y,t)\,dxdy$, we have the standard diffusion equation

$$\frac{\partial}{\partial t} p_3(z,t) = \mathcal{D}_z \frac{\partial^2}{\partial z^2} p_3(z,t) \tag{4.72}$$

with the Gaussian PDF,

$$p_3(z,t) = \frac{1}{\sqrt{4\pi \mathcal{D}_z t}} e^{-\frac{z^2}{4\mathcal{D}_z t}} \tag{4.73}$$

and normal diffusion along the z-direction, that is,

$$\langle z^2(t) \rangle = \int_{-\infty}^{\infty} z^2\, p_3(z,t)\,dz = 2\mathcal{D}_z t. \tag{4.74}$$

In Fig. 4.3 a graphical representation of the marginal PDFs is shown. The cusps at the origin of the solutions (4.66) — blue line, and (4.69) — red line, which correspond to the slowly decaying initial condition, are clearly observed in Fig. 4.3. This effect of the cusps results from the scale-free waiting time PDFs, which have diverging characteristic time scales. In other words it is due to the power law waiting time PDFs.

4.3.3   *From drift-diffusion to FFPE*

4.3.3.1   *One dimensional Brownian motion with drift*

In ensuing chapters (Chapters 6 and 7) we shall consider resetting and search problems in the framework of an one dimensional diffusion process with a drift. We show that this "simple" addition of a drift term in corresponding equations leads to new physical effects, which are also based on well known results on the one dimensional diffusion-advection equation.

Therefore, to set the stage for the clear presentation of the analysis, we are first offering a short overview on the results related to the one dimensional diffusion-advection equation. The corresponding Fokker-Planck equation with a constant velocity V reads

$$\frac{\partial}{\partial t} P_0(x,t) = \left[\mathcal{D} \frac{\partial^2}{\partial x^2} - V \frac{\partial}{\partial x} \right] P_0(x,t). \tag{4.75}$$

The initial condition $P_0(x, t = 0) = \delta(x - x_0)$, and vanishing boundary conditions at infinity, $P_0(\pm\infty, t) = 0$ and $\frac{\partial}{\partial x} P_0(\pm\infty, t) = 0$ are imposed. In Laplace space it reads

$$sP_0(x,s) - \delta(x - x_0) = \left[\mathcal{D} \frac{\partial^2}{\partial x^2} - V \frac{\partial}{\partial x} \right] P_0(x,s). \tag{4.76}$$

The solution of Eq. (4.75) is

$$P_0(x,t) = \frac{1}{\sqrt{4\pi \mathcal{D}t}} e^{-\frac{(x-x_0-Vt)^2}{4\mathcal{D}t}}, \tag{4.77}$$

and in the sequel we shall need its Laplace image, which is

$$P_0(x,s) = \frac{1}{2\mathcal{D}} \frac{1}{\sqrt{\frac{s}{\mathcal{D}} + \frac{V^2}{4\mathcal{D}^2}}} e^{\frac{V}{2\mathcal{D}}(x-x_0) - \sqrt{\frac{s}{\mathcal{D}} + \frac{V^2}{4\mathcal{D}^2}}|x-x_0|}. \tag{4.78}$$

Note that the PDF (4.77) is normalized

$$\langle x^0(t) \rangle_0 = \int_{-\infty}^{\infty} P_0(x,t)\, dx = \frac{1}{\sqrt{4\pi \mathcal{D}t}} \int_{-\infty}^{\infty} e^{-\frac{(x-x_0-Vt)^2}{4\mathcal{D}t}}\, dx$$

$$= \frac{2}{\sqrt{4\pi \mathcal{D}t}} \int_0^{\infty} e^{-\frac{y^2}{4\mathcal{D}t}}\, dy = 1, \tag{4.79}$$

and respectively

$$\langle x^0(s) \rangle_0 = \int_{-\infty}^{\infty} P_0(x,s)\, dx = \frac{1}{s}. \tag{4.80}$$

The MSD can be easily find from the expression

$$\langle x^2(s) \rangle_0 = \int_{-\infty}^{\infty} x^2 P_0(x,s)\, dx = \frac{x_0^2}{s} + \frac{2(\mathcal{D} + x_0 V)}{s^2} + \frac{2V^2}{s^3} \tag{4.81}$$

that yields

$$\langle x^2(t) \rangle_0 = \int_{-\infty}^{\infty} x^2 P_0(x,t)\, dx = 2\mathcal{D}t + (x_0 + Vt)^2. \tag{4.82}$$

Correspondingly, the short time diffusive behavior ($\langle x^2(t) \rangle_0 \sim t$) turns to the ballistic motion in the long time limit $\langle x^2(t) \rangle_0 \sim t^2$.

4.3.3.2 *Brownian motion with drift on 2D comb*

Now we consider Brownian motion on a two dimensional comb in presence of a drift along the backbone. The corresponding Fokker-Planck equation reads

$$\frac{\partial}{\partial t} P(x, y, t) = \delta(y) \left[-v \frac{\partial}{\partial x} + \mathcal{D}_x \frac{\partial^2}{\partial x^2} \right] P(x, y, t) + \mathcal{D}_y \frac{\partial^2}{\partial y^2} P(x, y, t).$$
(4.83)

The initial condition is $P(x, t, t = 0) = \delta(x)\delta(y)$ and zero boundary conditions are chosen at infinity. By the Laplace-Fourier transformation the equation transforms to

$$sP(k_x, k_y, s) - 1 = - \left(\imath v k_x + \mathcal{D}_x k_x^2 \right) P(k_x, y = 0, s) - \mathcal{D}_y k_y^2 P(k_x, k_y, s).$$
(4.84)

We solve it in the same way as the FPE for the comb without the drift term. The PDF in Fourier-Laplace space is

$$P(k_x, k_y, s) = \frac{s^{1/2}}{\left(s + \mathcal{D}_y k_y^2 \right) \left(s^{1/2} + \left[\imath \frac{v}{2\sqrt{\mathcal{D}_y}} k_x + \frac{\mathcal{D}_x}{2\sqrt{\mathcal{D}_y}} k_x^2 \right] \right)}.$$
(4.85)

Therefore, diffusion along the backbone can be described by the marginal PDF $p_1(x, t) = \int_{-\infty}^{\infty} P(x, y, t)\, dy$, which reads

$$p_1(k_x, s) = P(k_x, k_y = 0, s) = \frac{s^{-1/2}}{s^{1/2} + \left[\imath \frac{v}{2\sqrt{\mathcal{D}_y}} k_x + \frac{\mathcal{D}_x}{2\sqrt{\mathcal{D}_y}} k_x^2 \right]}.$$
(4.86)

The algebraic equation corresponds to the time fractional Fokker-Planck equation for the marginal PDF along the backbone [17]

$$_C D_{0+}^{1/2} p_1(x, t) = \left[-\frac{v}{2\sqrt{\mathcal{D}_y}} \frac{\partial}{\partial x} + \frac{\mathcal{D}_x}{2\sqrt{\mathcal{D}_y}} \frac{\partial^2}{\partial x^2} \right] p_1(x, t).$$
(4.87)

For the MSD, we obtain

$$\langle x^2(t) \rangle = x_0^2 + \left(\frac{\mathcal{D}_x}{\sqrt{\mathcal{D}_y}} + \frac{x_0 v}{\sqrt{\mathcal{D}_y}} \right) \frac{t^{1/2}}{\Gamma(3/2)} + \frac{v^2}{2\mathcal{D}_y} t.$$
(4.88)

Therefore, in the long time limit, it scales as $\langle x^2(t) \rangle \sim \frac{v^2}{2\mathcal{D}_y} t$.

It is worth noting that the FPE (4.83) can be solved by presenting the PDF in the form $P(x, y, s) = g(x, s) \exp\left(-\sqrt{\frac{s}{\mathcal{D}_y}}|y|\right)$, as it was done for the case without drift.

Remark 4.3 (Solution with subordination approach[2]). The FPE (4.83) can also be solved by using the subordination approach. Starting from the Fokker-Planck equation with a constant drift in the one dimension

$$\frac{\partial}{\partial t} P(x, t) = \left[-V \frac{\partial}{\partial x} + \mathcal{D} \frac{\partial^2}{\partial x^2}\right] P(x, t), \tag{4.89}$$

and performing the Laplace transform, we have

$$s \hat{P}(x, s) - \delta(x - x_0) = \left[-V \frac{\partial}{\partial x} + \mathcal{D} \frac{\partial^2}{\partial x^2}\right] \hat{P}(x, s). \tag{4.90}$$

By exchanging $s \to s^{1/2}$, we obtain

$$s^{1/2} \hat{P}(x, s^{1/2}) - \delta(x - x_0) = \left[-V \frac{\partial}{\partial x} + \mathcal{D} \frac{\partial^2}{\partial x^2}\right] \hat{P}(x, s^{1/2}) \tag{4.91}$$

i.e.,

$$s \underbrace{\left(s^{-1/2} \hat{P}(x, s^{1/2})\right)}_{=\hat{p}_1(x,s)} - \delta(x - x_0) = s^{1/2} \left[-V \frac{\partial}{\partial x} + \mathcal{D} \frac{\partial^2}{\partial x^2}\right] \underbrace{\left(s^{-1/2} \hat{P}(x, s^{1/2})\right)}_{=\hat{p}_1(x,s)}. \tag{4.92}$$

If we define $\hat{p}_1(x, s) = \left(s^{-1/2} \hat{P}(x, s^{1/2})\right)$, $V = \frac{v}{2\sqrt{\mathcal{D}_y}}$ and $\mathcal{D} = \frac{\mathcal{D}_x}{2\sqrt{\mathcal{D}_y}}$, the equation (4.92) reduces to

$$s^{1/2} \hat{p}_1(x, s) - s^{-1/2} \delta(x - x_0) = \left[-V \frac{\partial}{\partial x} + \mathcal{D} \frac{\partial^2}{\partial x^2}\right] \hat{p}_1(x, s) \tag{4.93}$$

which by the inverse Laplace transform becomes

$$_C D_{0+}^{1/2} p_1(x, t) = \left[-V \frac{\partial}{\partial x} + \mathcal{D} \frac{\partial^2}{\partial x^2}\right] p_1(x, t). \tag{4.94}$$

For the marginal PDF

$$\hat{p}_1(x, s) = s^{-1/2} \hat{P}(x, s^{1/2}) = s^{-1/2} \int_0^\infty \hat{P}(x, u) e^{-us^{1/2}} du, \tag{4.95}$$

---

[2]We follow arguments of Refs. [18, 19, 20].

we arrive at the subordination integral

$$\hat{p}_1(x,s) = \int_0^\infty P(x,u)\hat{h}(u,s)\,du \tag{4.96}$$

with the subordination function

$$\hat{h}(u,s) = s^{-1/2}e^{-us^{1/2}} \tag{4.97}$$

in Laplace space. By the inverse Laplace transform it becomes

$$h(u,t) = \mathcal{L}^{-1}\left[s^{-1/2}e^{-us^{1/2}}\right] = \frac{1}{\sqrt{\pi t}}\exp\left(-u^2/4t\right), \tag{4.98}$$

and the subordination integral has the form [19]

$$p_1(x,t) = \int_0^\infty P(x,u)h(u,t)\,du, \tag{4.99}$$

where

$$P(x,u) = \frac{1}{\sqrt{4\pi Du}}\exp\left(-\frac{(x-x_0-Vu)^2}{4Du}\right)$$

is the Galilei-shifted Gaussian PDF.

We note that from Eqs. (4.95) and (4.78) it follows that the marginal PDF in Laplace space reads

$$\hat{p}_1(x,s) = s^{-1/2}\hat{P}(x,s^{1/2}) = \frac{s^{-1/2}}{2D\sqrt{\frac{s^{1/2}}{D} + \frac{V^2}{4D^2}}}e^{\frac{V}{2D}(x-x_0) - \sqrt{\frac{s^{1/2}}{D} + \frac{V^2}{4D^2}}|x-x_0|}.$$

$$\tag{4.100}$$

This solution can be used for graphical representation of the PDF $p_1(x,t)$ by using the numerical inverse Laplace transform, see Example E.7 in Appendix E.1.

References

[1] A. Iomin, V. Mèndez, and W. Horsthemke. *Fractional Dynamics in Comb-like Structures*. Singapore: World Scientific, 2018 (cit. on pp. 99, 102).

[2] L. Liu, L. Zheng, and F. Liu. "Temporal anomalous diffusion and drift of particles in a comb backbone with fractional Cattaneo-Christov flux". In: *Journal of Statistical Mechanics: Theory and Experiment* 2017.4 (2017), p. 043208 (cit. on p. 99).

[3] L. Liu et al. "Comb model for the anomalous diffusion with dual-phase-lag constitutive relation". In: *Communications in Nonlinear Science and Numerical Simulation* 63 (2018), pp. 135–144 (cit. on p. 99).

[4] A. Iomin, V. Méndez, and W. Horsthemke. "Comb model: Non-Markovian versus Markovian". In: *Fractal and Fractional* 3.4 (2019), p. 54 (cit. on p. 99).

[5] K. Suleiman et al. "Anomalous diffusion on Archimedean spiral structure with Cattaneo flux model". In: *Journal of Molecular Liquids* 319 (2020), p. 114256 (cit. on p. 99).

[6] A. A. Tateishi et al. "Quenched and annealed disorder mechanisms in comb models with fractional operators". In: *Physical Review E* 101.2 (2020), p. 022135 (cit. on p. 99).

[7] K. Suleiman et al. "The Effect of Geometry on the Diffusion: Branched Archimedean spiral". In: *International Communications in Heat and Mass Transfer* 117 (2020), p. 104733 (cit. on p. 99).

[8] Z. Wang and L. Zheng. "Anomalous diffusion in inclined comb-branch structure". In: *Physica A* 549 (2020), p. 123889 (cit. on p. 99).

[9] Z. Wang, P. Lin, and E. Wang. "Modeling multiple anomalous diffusion behaviors on comb-like structures". In: *Chaos, Solitons & Fractals* 148 (2021), p. 111009 (cit. on p. 99).

[10] H. C. Fogedby. "Langevin equations for continuous time Lévy flights". In: *Physical Review E* 50 (2 1994), pp. 1657–1660 (cit. on p. 99).

[11] N. Sonine. "Sur la généralisation d'une formule d'Abel". In: *Acta Mathematica* 4.1 (1884), pp. 171–176 (cit. on p. 101).

[12] A. N. Kochubei. "General fractional calculus, evolution equations, and renewal processes". In: *Integral Equations and Operator Theory* 71.4 (2011), pp. 583–600 (cit. on p. 101).

[13] V. E. Arkhincheev and E. M. Baskin. "Anomalous diffusion and drift in a comb model of percolation clusters". In: *Journal of Experimental and Theoretical Physics (JETP)* 73.1 (1991), pp. 161–300 (cit. on p. 103).

[14] V. Méndez and A. Iomin. "Comb-like models for transport along spiny dendrites". In: *Chaos, Solitons & Fractals* 53 (2013), pp. 46–51 (cit. on p. 107).

[15] E. K. Lenzi et al. "Anomalous diffusion and random search in xyz-comb: exact results". In: *Journal of Statistical Mechanics: Theory and Experiment* 2020.5 (2020), p. 053203 (cit. on p. 107).

[16]   T. Sandev et al. "Comb model with slow and ultraslow diffusion". In: *Mathematical Modelling of Natural Phenomena* 11.3 (2016), pp. 18–33 (cit. on p. 110).

[17]   T. Sandev and Z. Tomovski. *Fractional Equations and Models*. Springer Nature, 2019 (cit. on p. 114).

[18]   E. Barkai, R. Metzler, and J. Klafter. "From continuous time random walks to the fractional Fokker-Planck equation". In: *Physical Review E* 61.1 (2000), p. 132 (cit. on p. 115).

[19]   E. Barkai. "Fractional Fokker-Planck equation, solution, and application". In: *Physical Review E* 63.4 (2001), p. 046118 (cit. on pp. 115, 116).

[20]   R. Metzler, E. Barkai, and J. Klafter. "Anomalous diffusion and relaxation close to thermal equilibrium: A fractional Fokker-Planck equation approach". In: *Physical Review Letters* 82.18 (1999), p. 3563 (cit. on p. 115).

Chapter 5

Heterogeneous diffusion processes

In this chapter, we consider a heterogeneous diffusion process, one dimensional geometric Brownian motion, and turbulent diffusion in the two dimensional comb geometry. The macroscopic description of these diffusion processes are discussed in the framework of the Fokker-Planck equations with inhomogeneous transport-diffusion coefficients, which are functions of the walker positions in the configuration space. This complicated behavior of the transport coefficients results from a multiplicative white noise in a microscopic Langevin equation, which in its turn results from the inhomogeneous environment. As is well known, in the case of multiplicative white noise, a functional integral, as a solution of the Langevin equation, should be amended with an appropriate interpretation for the integral of the noise term [1, 2], see also Appendix B.

Such diffusion models with position-dependent diffusion coefficient are used to describe transport processes on random fractals [3, 4], and in heterogeneous media [5, 6, 7, 8, 9, 10, 11, 12, 13, 14, 15], and could be applied in various studies on biological cells [16, 17] and porous media [18, 19].

5.1   Heterogeneous diffusion

Heterogeneous diffusion can be described by the following Langevin equation with position dependent diffusion coefficient, see Ref. [15],

$$\dot{x}(t) = \sqrt{2\,\mathcal{D}(x)}\,\xi(t), \tag{5.1}$$

where $\xi(t)$ is a white noise of zero mean.

The corresponding Fokker-Planck equation reads

$$\frac{\partial}{\partial t}P(x,t) = \frac{\partial}{\partial x}\left\{\mathcal{D}(x)^{1-A/2}\frac{\partial}{\partial x}\left[\mathcal{D}(x)^{A/2}P(x,t)\right]\right\}, \tag{5.2}$$

where $P(x,t)$ is the PDF and $\mathcal{D}(x)$, is the position-dependent diffusion coefficient of the power-law form

$$\mathcal{D}(x) = \mathcal{D}_\alpha |x|^\alpha, \quad \alpha \le 2. \tag{5.3}$$

The case with $\alpha = 2$ corresponds to the geometric Brownian motion and will be considered separately in sequel sections. For different values $A = (2,1,0)$, Eq. (5.3) results from different interpretations of corresponding stochastic Langevin equation. In particular, $A = 0$ corresponds to a so called isothermal Klimontovich-Hänggi interpretation, while $A = 1$ and $A = 2$ are relevant for the most popular Stratonovich and Itô interpretations, respectively. For latter cases, see also discussion in Appendix B.

Let us show how the general FPE (5.2) results from the general form of the stochastic differential equation

$$\delta X(t) = X(t + \tau) - X(t) = f[X(t)]\tau + g[X(t + \eta\tau)]\delta W, \tag{5.4}$$

where τ is a small time, $\eta \in [0, 1]$, and $\delta W = W(t + \tau) - W(t)$. We take the Wiener process as a zero-mean Gaussian process with stationary independent increments,

$$\langle W(t) \rangle = 0, \quad \langle W(t)W(t') \rangle = \min(t, t'). \tag{5.5}$$

The choice[1] of $\eta = 0, 1/2, 1$ corresponds to the Itô, Stratonovich and isothermal Klimontovich-Hänggi interpretation of the stochastic differential equation (5.1), where $g(x) = \sqrt{2\mathcal{D}(x)}$.

Let us obtain a general solution to Eq. (5.2), with the power-law diffusion coefficient (5.3). The Laplace transform of Eq. (5.2) yields

$$s\hat{P}(x,s) - \delta(x) = \mathcal{D}_\alpha \frac{\partial}{\partial x} \left\{ |x|^{\frac{(2-A)\alpha}{2}} \frac{\partial}{\partial x} \left[|x|^{\frac{A\alpha}{2}} \hat{P}(x,s) \right] \right\}. \tag{5.6}$$

Performing differentiation with respect to x, one finds

$$s\hat{P}(x,s) - \delta(x) = \mathcal{D}_\alpha \left[A\alpha\delta(x)|x|^{\alpha-1}\hat{P}(x,s) + \frac{A(\alpha-1)\alpha}{2}|x|^{\alpha-2}\hat{P}(x,s) \right.$$
$$\left. + (2\theta(x) - 1)\frac{(A+2)\alpha}{2}|x|^{\alpha-1}\frac{\partial}{\partial x}\hat{P}(x,s) + |x|^\alpha \frac{\partial^2}{\partial x^2}\hat{P}(x,s) \right]. \tag{5.7}$$

Taking into account that the Fokker-Planck equation is symmetrical with respect to inversion $x \to -x$, we can consider the solution for the non-negative x, when $x = |x|$ and then extend it symmetrically for the entire

---

[1]Please do not confuse here the interpretation parameter η with the memory kernel $\eta(t)$.

x axis. Therefore, using the variable change $\hat{P}(|x|, s) = \mathcal{C}(s)\hat{f}(|x|, s) = \mathcal{C}(s)\hat{f}(y, s)$, where $\mathcal{C}(s)$ is a function of s, we transform Eq. (5.7) to

$$s\hat{f}(y, s) - \frac{\delta(x)}{\mathcal{C}(s)} = \mathcal{D}_\alpha \frac{A(\alpha - 1)\alpha}{2} y^{\alpha - 2}\hat{f}(y, s) + \mathcal{D}_\alpha A\alpha y^{\alpha - 1}\delta(x)\hat{f}(y, s)$$

$$+ \mathcal{D}_\alpha \frac{(A + 2)\alpha}{2} y^{\alpha - 1}\frac{\partial}{\partial y}\hat{f}(y, s) + 2\mathcal{D}_\alpha y^\alpha \delta(x)\frac{\partial}{\partial y}\hat{f}(y, s)$$

$$+ \mathcal{D}_\alpha y^\alpha \frac{\partial^2}{\partial y^2}\hat{f}(y, s). \tag{5.8}$$

Separating terms with $\delta(x)$, we obtain two independent equations

$$\frac{\partial^2}{\partial y^2}\hat{f}(y, s) + \frac{(A + 2)\alpha/2}{y}\frac{\partial}{\partial y}\hat{f}(y, s) + \left[-\frac{s/\mathcal{D}_\alpha}{y^\alpha} + \frac{A(\alpha - 1)\alpha/2}{y^2}\right]\hat{f}(y, s) = 0, \tag{5.9}$$

$$-1 = \mathcal{C}(s)\mathcal{D}_\alpha \left[A\alpha y^{\alpha - 1}\hat{f}(y, s) + 2y^\alpha \frac{\partial}{\partial y}\hat{f}(y, s)\right]\Big|_{y=0}. \tag{5.10}$$

Equation (5.9) is the Lommel-type differential equation

$$z''(y) + \frac{1 - 2\bar{\beta}}{y}z'(y) + \left[\left(a\bar{\alpha}y^{\bar{\alpha}-1}\right)^2 + \frac{\bar{\beta}^2 - \nu^2\bar{\alpha}^2}{y^2}\right]z(y) = 0, \tag{5.11}$$

where a, ν, $\bar{\alpha}$ and $\bar{\beta}$ are parameters, while primes for z denote derivatives with respect to y. The solution of Eq. (5.11) is

$$z(y) = y^{\bar{\beta}}Z_\nu \left(\imath a y^{\bar{\alpha}}\right),$$

where $Z_\nu(y) = C_1 J_\nu(y) + C_2 Y_\nu(y)$ is the Bessel function. The boundary conditions at infinity are equal to zero. Therefore, the solution reads

$$z(y) = y^{\bar{\beta}}K_\nu \left(a y^{\bar{\alpha}}\right),$$

where $K_\nu(y)$ is the modified Bessel function (of the third kind). Here we also find the relations

$$a = \frac{2}{2 - \alpha}\sqrt{\frac{s}{\mathcal{D}_\alpha}}, \quad \bar{\alpha} = \frac{2 - \alpha}{2}, \quad \bar{\beta} = \frac{2 - (A + 2)\alpha}{4},$$

$$\nu = \frac{[2 - (A + 2)\alpha]^2 - 8A(\alpha - 1)\alpha}{8(2 - \alpha)}. \tag{5.12}$$

Inserting the obtained solution $\hat{f}(y, s)$ in Eq. (5.10), we obtain $\mathcal{C}(s)$.

5.1.1 *Stratonovich interpretation*

For $A = 1$, the solution to Eq. (5.9) is

$$\hat{f}(y, s) = y^{\frac{2-3\alpha}{4}} Z_{\frac{1}{2}}\left(\imath \frac{2}{2-\alpha}\sqrt{\frac{s}{\mathcal{D}_\alpha}} y^{\frac{2-\alpha}{2}}\right)$$

$$= y^{\frac{2-3\alpha}{4}} K_{\frac{1}{2}}\left(\frac{2}{2-\alpha}\sqrt{\frac{s}{\mathcal{D}_\alpha}} y^{\frac{2-\alpha}{2}}\right). \tag{5.13}$$

From Eqs. (5.13) and (5.10) we find the constant $\mathcal{C}(s)$ which is $\mathcal{C}(s) = \mathcal{D}_\alpha^{-3/4}\frac{s^{-1/4}}{\sqrt{(2-\alpha)\pi}}$. The PDF in Laplace space reads

$$\hat{P}(x, s) = \mathcal{D}_\alpha^{-3/4}\frac{s^{-1/4}}{\sqrt{(2-\alpha)\pi}}|x|^{\frac{2-3\alpha}{4}} K_{\frac{1}{2}}\left(\frac{2}{2-\alpha}\sqrt{\frac{s}{\mathcal{D}_\alpha}}|x|^{\frac{2-\alpha}{2}}\right)$$

$$= \frac{|x|^{-\alpha/2}}{\sqrt{4\mathcal{D}_\alpha}}s^{-1/2}\exp\left(-\frac{2}{2-\alpha}\frac{s^{1/2}}{\sqrt{\mathcal{D}_\alpha}}|x|^{(2-\alpha)/2}\right), \tag{5.14}$$

where we use $K_{1/2}(x) = \sqrt{\pi/(2x)}\,e^{-x}$. Performing the inverse Laplace transform, we find the PDF as follows (see also Ref. [7])

$$P(x, t) = \frac{|x|^{-\alpha/2}}{\sqrt{4\pi\mathcal{D}_0 t}}\exp\left(-\frac{|x|^{2-\alpha}}{(2-\alpha)^2\mathcal{D}_0 t}\right). \tag{5.15}$$

The PDF is normalized, and the MSD is

$$\langle x^2(t)\rangle = \mathcal{L}^{-1}\left[\int_{-\infty}^\infty x^2\,\hat{P}(x, s)\,dx\right]$$

$$= \frac{(2-\alpha)^{\frac{4}{2-\alpha}}\Gamma\left(\frac{4-\alpha}{2-\alpha}\right)\Gamma\left(\frac{6-\alpha}{2(2-\alpha)}\right)}{\sqrt{\pi}}\mathcal{D}_\alpha^{\frac{2}{2-\alpha}}\mathcal{L}^{-1}\left[s^{-1-\frac{2}{2-\alpha}}\right]$$

$$= \frac{(2-\alpha)^{\frac{4}{2-\alpha}}\Gamma\left(\frac{4-\alpha}{2-\alpha}\right)\Gamma\left(\frac{6-\alpha}{2(2-\alpha)}\right)}{\sqrt{\pi}}\mathcal{D}_\alpha^{\frac{2}{2-\alpha}}\frac{t^{\frac{2}{2-\alpha}}}{\Gamma\left(1+\frac{2}{2-\alpha}\right)}. \tag{5.16}$$

5.1.2 *Isothermal Klimontovich-Hänggi interpretation*

For $A = 0$ we consider the Klimontovich-Hänggi interpretation of the heterogeneous diffusion equation, with the solution

$$\hat{f}(y, s) = y^{\frac{1-\alpha}{2}} Z_{\frac{1-\alpha}{2-\alpha}}\left(\imath \frac{2}{2-\alpha}\sqrt{\frac{s}{\mathcal{D}_\alpha}} y^{\frac{2-\alpha}{2}}\right)$$

$$= y^{\frac{1-\alpha}{2}} K_{\frac{1-\alpha}{2-\alpha}}\left(\frac{2}{2-\alpha}\sqrt{\frac{s}{\mathcal{D}_\alpha}} y^{\frac{2-\alpha}{2}}\right). \tag{5.17}$$

From Eq. (5.10) and series representation of $K_\nu(y)$, we have

$$\mathcal{C}(s) = \mathcal{D}_\alpha^{-(3-\alpha)/[2(2-\alpha)]} \frac{s^{-1+(3-\alpha)/[2(2-\alpha)]}}{\Gamma\left(\frac{1}{2-\alpha}\right)(2-\alpha)^{1/(2-\alpha)}}.$$

The PDF in Laplace space reads

$$\hat{P}(x,s) = \mathcal{D}_\alpha^{-(3-\alpha)/[2(2-\alpha)]} \frac{s^{-(1-\alpha)/[2(2-\alpha)]}}{\Gamma\left(\frac{1}{2-\alpha}\right)(2-\alpha)^{1/(2-\alpha)}}$$

$$\times |x|^{\frac{1-\alpha}{2}} K_{\frac{1-\alpha}{2-\alpha}}\left(\frac{2}{2-\alpha}\sqrt{\frac{s}{\mathcal{D}_\alpha}}|x|^{\frac{2-\alpha}{2}}\right). \tag{5.18}$$

The inverse Laplace transform finally yields (see also Ref. [8])

$$P(x,t) = \frac{1}{2\Gamma\left(\frac{1}{2-\alpha}\right)(2-\alpha)^{\frac{\alpha}{2-\alpha}}[\mathcal{D}_\alpha t]^{1/(2-\alpha)}} \exp\left(-\frac{|x|^{2-\alpha}}{(2-\alpha)^2\mathcal{D}_0 t}\right). \tag{5.19}$$

Note that the PDF is also normalized to 1, while the MSD is

$$\langle x^2(t)\rangle = 2(2-\alpha)^{\frac{2+\alpha}{2-\alpha}} \frac{\Gamma\left(\frac{2}{2-\alpha}\right)\Gamma\left(\frac{3}{2-\alpha}\right)}{\Gamma\left(\frac{1}{2-\alpha}\right)} \mathcal{D}_\alpha^{\frac{2}{2-\alpha}} \frac{t^{\frac{2}{2-\alpha}}}{\Gamma\left(1+\frac{2}{2-\alpha}\right)}. \tag{5.20}$$

5.1.3 *Itô interpretation*

The Itô interpretation is obtained for $A = 2$. The corresponding solution reads

$$\hat{f}(y,s) = y^{\frac{1-2\alpha}{2}} Z_{\frac{1}{2-\alpha}}\left(\imath\frac{2}{2-\alpha}\sqrt{\frac{s}{\mathcal{D}_\alpha}}y^{\frac{2-\alpha}{2}}\right)$$

$$= y^{\frac{1-2\alpha}{2}} K_{\frac{1}{2-\alpha}}\left(\frac{2}{2-\alpha}\sqrt{\frac{s}{\mathcal{D}_\alpha}}y^{\frac{2-\alpha}{2}}\right), \tag{5.21}$$

while the coefficient $\mathcal{C}(s)$ is

$$\mathcal{C}(s) = \mathcal{D}_\alpha^{-(3-2\alpha)/[2(2-\alpha)]} \frac{s^{-1/[2(2-\alpha)]}}{\Gamma\left(\frac{1-\alpha}{2-\alpha}\right)(2-\alpha)^{(1-\alpha)/(2-\alpha)}}.$$

The PDF in Laplace space becomes

$$\hat{P}(x,s) = \mathcal{D}_\alpha^{-(3-2\alpha)/[2(2-\alpha)]} \frac{s^{-1/[2(2-\alpha)]}}{\Gamma\left(\frac{1-\alpha}{2-\alpha}\right)(2-\alpha)^{(1-\alpha)/(2-\alpha)}}$$

$$\times |x|^{\frac{1-2\alpha}{2}} K_{\frac{1}{2-\alpha}}\left(\frac{2}{2-\alpha}\sqrt{\frac{s}{\mathcal{D}_\alpha}}|x|^{\frac{2-\alpha}{2}}\right), \tag{5.22}$$

which also yields the normalized PDF as follows

$$P(x,t) = \frac{(2-\alpha)^{\frac{\alpha}{2-\alpha}} |x|^{-\alpha}}{2\,\Gamma\left(\frac{1-\alpha}{2-\alpha}\right) [\mathcal{D}_\alpha t]^{(1-\alpha)/(2-\alpha)}} \exp\left(-\frac{|x|^{2-\alpha}}{(2-\alpha)^2 \mathcal{D}_0 t}\right). \tag{5.23}$$

For the MSD we have

$$\langle x^2(t) \rangle = (2-\alpha)^{\frac{4}{2-\alpha}} \frac{\Gamma\left(\frac{4-\alpha}{2-\alpha}\right) \Gamma\left(\frac{3-\alpha}{2-\alpha}\right)}{\Gamma\left(\frac{1-\alpha}{2-\alpha}\right)} \mathcal{D}_\alpha^{\frac{2}{2-\alpha}} \frac{t^{\frac{2}{2-\alpha}}}{\Gamma\left(1+\frac{2}{2-\alpha}\right)}. \tag{5.24}$$

5.2 Turbulent diffusion and geometric Brownian motion

5.2.1 *Geometric Brownian motion and turbulent diffusion*

It is well known that at turbulent diffusion a contaminant spreads very fast, for example, in case of Richardson diffusion, the MSD behaves as t^3 [20, 21]. Dating back to work by Kolmogorov and Obukhov, it suggests this turbulent acceleration by means of a Gaussian delta correlated noise ξ, added to the dynamical system [22], $\ddot{x} + \xi(t) = 0$. This turbulent spread, however, can be essentially increased due to a multiplicative noise [23, 24], where the MSD grows exponentially with time.

As an example, a geometric Brownian motion (GBM) is described by means of the Langevin equation [23, 25]

$$dx(t) = \mu\,x(t)\,dt + \sigma\,x(t)\,dB(t), \quad x_0 = x(0), \tag{5.25}$$

where $x(t)$ is the particle position, μ is the drift, $\sigma > 0$ is the volatility, and $B(t)$ is the standard Brownian motion. This stochastic process corresponds to the well-known Black-Scholes model where $x(t) > 0$ is an asset price,

$$x(t) = x_0\, e^{\sigma B(t) + \mu t}, \quad x_0 = x(0) > 0. \tag{5.26}$$

Depending on the interpretation of Eq. (5.25), the corresponding Fokker-Planck equation can have different forms.

5.2.1.1 *Itô interpretation of GBM*

In finance math literature the Itô convention is the standard interpretation. When the dynamics of the asset price follows a GBM, then a risk-neutral distribution (probability distribution which takes into account the risk of future price fluctuations) can be easily found by solving the corresponding Fokker-Planck equation

$$\frac{\partial}{\partial t} f(x,t) = -\mu \frac{\partial}{\partial x} x f(x,t) + \frac{\sigma^2}{2} \frac{\partial^2}{\partial x^2} x^2 f(x,t), \tag{5.27}$$

with initial condition $f(x, t=0) = \delta(x - x_0)$.

By performing the Laplace transform with respect to t and the Mellin transform with respect to x, we obtain

$$\hat{\bar{f}}(q,s) = \frac{x_0^{q-1}}{s - \left[\frac{\sigma^2}{2}(q-1)(q-2) + \mu(q-1)\right]}, \tag{5.28}$$

where we use also the Mellin transform of the Dirac delta-function in Eq. (1.66). Then the inverse Laplace transform yields

$$\bar{f}(q,t) = x_0^{q-1} \times \exp\left(\frac{\sigma^2}{2}\left[q + \frac{1}{2}\left(\frac{2\mu}{\sigma^2} - 3\right)\right]^2 t - \frac{\left(\mu - \frac{\sigma^2}{2}\right)^2}{2\sigma^2}t\right), \tag{5.29}$$

where we use the Laplace transform of the exponential function in Eq. (1.35). Applying the inverse Mellin transform and looking for the solution in the form of the convolution integral of two functions, $\mathcal{M}\{h(x)\}(q) = \bar{H}(q)$ and $\mathcal{M}\{g(x)\}(q) = \bar{G}(q)$, see Eq. (1.61), we obtain the solution of the FPE (5.27) for the GBM

$$f(x,t) = \int_0^\infty \delta(r-x_0) \times \frac{\exp\left(-\frac{\left[\log\frac{x}{r} - \left(\mu - \frac{\sigma^2}{2}\right)t\right]^2}{2\sigma^2 t}\right)}{(x/r)\sqrt{2\pi\sigma^2 t}}\frac{dr}{r}$$

$$= \frac{1}{x\sqrt{2\pi\sigma^2 t}} \times \exp\left(-\frac{\left[(\log x - \log x_0) - \left(\mu - \frac{\sigma^2}{2}\right)t\right]^2}{2\sigma^2 t}\right). \tag{5.30}$$

Here we use Eq. (1.66), i.e., $\mathcal{M}^{-1}\left\{x_0^{q-1}\right\} = \delta(x - x_0)$, and

$$\mathcal{M}^{-1}\left\{\exp\left(\frac{\sigma^2}{2}\left[q + \frac{1}{2}\left(\frac{2\mu}{\sigma^2} - 3\right)\right]^2 t - \frac{\left(\mu - \frac{\sigma^2}{2}\right)^2}{2\sigma^2}t\right)\right\}$$

$$= \frac{1}{x\sqrt{2\pi\sigma^2 t}} \times \exp\left(-\frac{\left[\log x - \left(\mu - \frac{\sigma^2}{2}\right)t\right]^2}{2\sigma^2 t}\right).$$

We also used the properties of the inverse Mellin transform, according to Eqs. (1.60) and (1.67). Eventually, the obtained result (5.30) corresponds to the log-normal distribution.

The GBM process is also implemented in the Wolfram Language by `GeometricBrownianMotionProcess` $[\mu, \sigma, x_0]$, while the log-normal distribution by `LogNormalDistribution`$[\mu, \sigma]$.

5.2.1.2 *Stratonovich interpretation of GBM*

For the Stratonovich interpretation, the corresponding Fokker-Planck equation reads

$$\frac{\partial}{\partial t} f_S(x,t) = -\mu \left(\frac{\partial}{\partial x} x \right) f_S(x,t) + \frac{\sigma^2}{2} \left(\frac{\partial}{\partial x} x \right)^2 f_S(x,t). \qquad (5.31)$$

Setting $\mu = 0$ and $\sigma^2/2 = \mathcal{D}$, we arrive at the diffusion-advection equation as follows

$$\frac{\partial}{\partial t} P(x,t) = \mathcal{D} \left(\frac{\partial}{\partial x} x \right)^2 P(x,t). \qquad (5.32)$$

Here we find the solution of the FPE (5.31), with the initial condition $f_S(x, t = 0) = \delta(x - x_0)$ and the boundary conditions $f_S(x,t) = \partial_x f_S(x,t) = 0$ at $x = \infty$. Taking into account that Eq. (5.31) is symmetrical and considering $x > 0$, we introduce the new variable $y = \log x$ and the derivatives with respect to y

$$\frac{\partial}{\partial x} = e^{-y} \frac{\partial}{\partial y}, \quad \frac{\partial^2}{\partial x^2} = -e^{-2y} \frac{\partial}{\partial y} + e^{-2y} \frac{\partial^2}{\partial y^2}.$$

Then Eq. (5.31) reads

$$\frac{\partial}{\partial t} f_S(y,t) = \left(\frac{\sigma^2}{2} - \mu \right) f_S(y,t) + (\sigma^2 - \mu) \frac{\partial}{\partial y} f_S(y,t) + \frac{\sigma^2}{2} \frac{\partial^2}{\partial y^2} f_S(y,t). \qquad (5.33)$$

Looking for the solution in the form $f_S(y,t) = e^{-y} f(y,t)$, we obtain the Fokker-Planck equation for the function $f(y,t)$ as follows

$$\frac{\partial}{\partial t} f(y,t) = -\mu \frac{\partial}{\partial y} f(y,t) + \frac{\sigma^2}{2} \frac{\partial^2}{\partial y^2} f(y,t). \qquad (5.34)$$

From the initial condition $f(x, t = 0) = \delta(x - x_0)$, we have

$$f(y, t = 0) = f(x, t = 0) \frac{dx}{dy} = \delta(e^y - e^{y_0}) e^y = \frac{\delta(y - y_0)}{e^{y_0}} e^y = \delta(y - y_0), \qquad (5.35)$$

and thus, the solution of Eq. (5.34) for the initial condition (5.35) is the Galilei shifted Gaussian [26]

$$f(y,t) = \frac{1}{\sqrt{2\pi\sigma^2 t}} \times \exp\left(-\frac{[(y - y_0) - \mu t]^2}{2\sigma^2 t} \right), \qquad (5.36)$$

which yields

$$f_S(y,t) = \frac{e^{-y}}{\sqrt{2\pi\sigma^2 t}} \times \exp\left(-\frac{[(y-y_0)-\mu t]^2}{2\sigma^2 t}\right). \tag{5.37}$$

Eventually, the solution of Eq. (5.31) reads [27]

$$f_S(x,t) = \frac{1}{x\sqrt{2\pi\sigma^2 t}} \times \exp\left(-\frac{[(\log x - \log x_0)-\mu t]^2}{2\sigma^2 t}\right). \tag{5.38}$$

For $\mu = 0$, the solution (5.38) reduces to

$$f_S(x,t) = \frac{1}{x\sqrt{2\pi\sigma^2 t}} \times \exp\left(-\frac{\log^2 \frac{x}{x_0}}{2\sigma^2 t}\right). \tag{5.39}$$

Remark 5.1 (Alternative solution by the Mellin-Laplace transform). Note that the Mellin transform can be applied to obtain Eq. (5.39), as well. Therefore, performing the Mellin and the Laplace transforms in Eq. (5.31), we have

$$\bar{\tilde{f}}_S(q,s) = \frac{x_0^{q-1}}{s - \left[\frac{\sigma^2}{2}(q-1)^2 + \mu(q-1)\right]}. \tag{5.40}$$

Then the inverse Laplace transform yields

$$\tilde{f}_S(q,t) = x_0^{q-1} \times \exp\left(\frac{\sigma^2}{2}\left[q + \left(\frac{\mu^2}{\sigma^2} - 1\right)\right]^2 t - \frac{\mu^2}{2\sigma^2}t\right). \tag{5.41}$$

Applying the inverse Mellin transform and looking for the solution in the form of the convolution integral

$$\mathcal{M}^{-1}[f(q)\,g(q)] = \int_0^\infty f(r)\,g(x/r)\frac{dr}{r},$$

we obtain the solution as follows

$$f_S(x,t) = \int_0^\infty \delta(r - x_0) \times \frac{\exp\left(-\frac{[\log\frac{x}{r}-\mu t]^2}{2\sigma^2 t}\right)}{(x/r)\sqrt{2\pi\sigma^2 t}}\frac{dr}{r}$$

$$= \frac{1}{x\sqrt{2\pi\sigma^2 t}} \times \exp\left(-\frac{[(\log x - \log x_0)-\mu t]^2}{2\sigma^2 t}\right). \tag{5.42}$$

Here we use the property (1.60) and Eqs. (1.66) and (1.67) for the Mellin transform, as well as the expression

$$g(x) = \mathcal{M}^{-1}\left[\exp\left(\frac{\sigma^2}{2}\left[q + \left(\frac{\mu^2}{\sigma^2} - 1\right)\right]^2 t\right)\right]$$

$$= \frac{1}{x\sqrt{2\pi\sigma^2 t}} \times \exp\left(-\frac{[\log x - \mu t]^2}{2\sigma^2 t}\right).$$

5.2.1.3 *Klimontovich-Hänggi interpretation of GBM*

The corresponding Fokker-Planck equation for the Klimontovich-Hänggi (isothermal) interpretation is given by

$$\frac{\partial}{\partial t} f_{KH}(x,t) = -\mu \left(\frac{\partial}{\partial x} x \right) f_{KH}(x,t) + \frac{\sigma^2}{2} \left(\frac{\partial}{\partial x} x^2 \frac{\partial}{\partial x} \right) f_{KH}(x,t). \quad (5.43)$$

We consider the same initial and boundary conditions as for the other interpretations considered above.

For $x > 0$, let us introduce the new variable $y = \log x$ and the derivatives with respect to y. Thus, Eq. (5.31) becomes

$$\frac{\partial}{\partial t} f_{KH}(y,t) = -\mu f_{KH}(y,t) + \left(\frac{\sigma^2}{2} - \mu \right) f_{KH}(y,t) + \frac{\sigma^2}{2} \frac{\partial^2}{\partial y^2} f_{KH}(y,t). \quad (5.44)$$

By using $f_{KH}(y,t) = e^{-y} f(y,t)$, we obtain the Fokker-Planck equation

$$\frac{\partial}{\partial t} f(y,t) = - \left(\mu + \frac{\sigma^2}{2} \right) \frac{\partial}{\partial y} f(y,t) + \frac{\sigma^2}{2} \frac{\partial^2}{\partial y^2} f(y,t) \quad (5.45)$$

with the solution in the form of the Galilei shifted Gaussian [26]

$$f(y,t) = \frac{1}{\sqrt{2\pi\sigma^2 t}} \times \exp \left(-\frac{\left[(y - y_0) - \left(\mu + \frac{\sigma^2}{2} \right) t \right]^2}{2\sigma^2 t} \right). \quad (5.46)$$

Therefore, we have

$$f_{KH}(x,t) = \frac{1}{x\sqrt{2\pi\sigma^2 t}} \times \exp \left(-\frac{\left[(\log x - \log x_0) - \left(\mu + \frac{\sigma^2}{2} \right) t \right]^2}{2\sigma^2 t} \right). \quad (5.47)$$

5.2.2 *Turbulent diffusion due to comb geometry*

Another example where turbulent diffusion can be realized due to the multiplicative noise is the comb model, considered in Chapter 4. In this case, considering relative diffusion of a pair of particles, see Fig. 5.1, namely the distance between them, one finds that the relative diffusivity of two particles grows with the inter-particle distance. The mechanism of turbulence is due to the comb geometry, which plays essential role when two nearest particles move into different (orthogonal) directions that contributes to the exponential growth of the inter-particle distance.

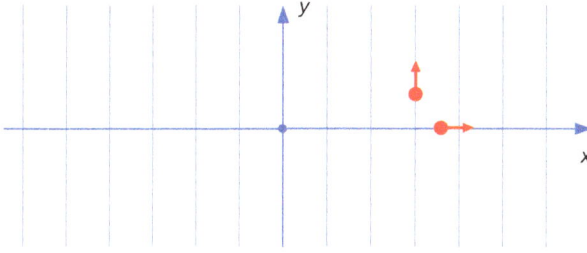

Fig. 5.1 Comb model with two tracers moving in different directions. Reprinted figure with permission from T. Sandev, A. Iomin and L. Kocarev, Phys. Rev. E, 102, 042109 (2020). Copyright (2020) by the American Physical Society.

To understand this process, let us consider a Langevin equation in a so-called Matheron - de Marsily form [28]

$$\dot{x}_1 = v(x_2)\, x_1 \quad \dot{x}_2 = \xi(t), \quad v(x_2) = v\, \delta(x_2). \tag{5.48}$$

Here $\xi(t)$ is a random Gaussian delta correlated process (white noise) $\langle \xi(t)\xi(t') \rangle = 2\mathcal{D}\, \delta(t - t')$, where $\mathcal{D}$ is a diffusion coefficient. Specific form of Eq. (5.48) corresponds exactly to the comb geometry, when a random exponential spread $\exp[\int v(B(\tau)\, d\tau]$ is possible along the x axis in complete analogy with Eq. (5.26), since $x_2 = B(t) = \int^t \xi(\tau)\, d\tau$ is the Brownian process (functional), taking place along the y axis. Here, the specific property of the comb is that the random exponential spread is possible only along the backbone at $y = 0$ in the form of randomly-inhomogeneous advection with the velocity parameter v, while normal Brownian diffusion in the y axis is homogeneous. Introducing a distribution function $P(x, y, t) = \langle \delta(x_1 - x)\delta(x_2 - y) \rangle$, we obtain the Fokker-Planck equation in the form of a 2D comb model [29]

$$\frac{\partial}{\partial t} P(x, y, t) = -v\, \delta(y)\, \frac{\partial}{\partial x}\{x\, P(x, y, t)\} + \mathcal{D}\, \frac{\partial^2}{\partial y^2} P(x, y, t). \tag{5.49}$$

This FPE conserves the probability flow, $\int P(x, y, t)\, dx\, dy = 1$.

It must be admitted that the diffusion coefficients in Eqs. (5.32) and (5.49) with the same nomenclature $\mathcal{D}$ are of different dimensions. Therefore, to avoid any contradictions in the nomenclature, we consider all equations in the dimensionless form when all parameters and variables are dimensionless.[2]

The solution of Eq. (5.49), for the initial condition $P(x, y, t = 0) = \delta(x - x_0)\delta(y)$, and $x > x_0 > 0$, can be obtained by the Laplace transform, which yields

$$s\hat{P}(x, y, s) - \delta(x - x_0) = -v\,\delta(y)\frac{\partial}{\partial x}\{x\,\hat{P}(x, y, s)\} + \mathcal{D}\frac{\partial^2}{\partial y^2}\hat{P}(x, y, s). \quad (5.50)$$

Looking for the solution to Eq. (5.50) in the form

$$\hat{P}(x, y, s) = g(x, s) \times e^{-\sqrt{\frac{s}{\mathcal{D}}}|y|}, \quad (5.51)$$

we obtain an equation for the backbone anomalous transport

$$v\,x\frac{\partial}{\partial x}g(x, s) = -\left(v + 2\sqrt{\mathcal{D}}\,s^{1/2}\right)g(x, s) + \delta(x - x_0). \quad (5.52)$$

The solution is

$$g(x, s) = \frac{\theta(x - x_0)}{v\,x} \times e^{-\frac{2\sqrt{\mathcal{D}}}{v}\sqrt{s}\,\log\frac{x}{x_0}}, \quad (5.53)$$

where $\theta(z)$ is the Heaviside theta function. By integration of Eq. (5.51) over y we obtain the marginal PDF,

$$\hat{p}_1(x, s) = \int_{-\infty}^{\infty} \hat{P}(x, y, s)\,dy = 2\sqrt{\frac{\mathcal{D}}{s}}\,g(x, s), \quad (5.54)$$

which reads

$$\hat{p}_1(x, s) = \frac{2\sqrt{\mathcal{D}}}{v}\frac{\theta(x - x_0)}{x} \times s^{-1/2}e^{-\frac{2\sqrt{\mathcal{D}}}{v}\sqrt{s}\,\log\frac{x}{x_0}}. \quad (5.55)$$

The inverse Laplace transform, yields the PDF $P(x, y, t)$ as follows

$$P(x, y, t) = \frac{\left(\frac{2\sqrt{\mathcal{D}}}{v}\log\frac{x}{x_0} + \frac{|y|}{\sqrt{\mathcal{D}}}\right)}{x\sqrt{4\pi\,v^2\,t^3}}\,\theta(x - x_0)\exp\left(-\frac{\left(\frac{2\sqrt{\mathcal{D}}}{v}\log\frac{x}{x_0} + \frac{|y|}{\sqrt{\mathcal{D}}}\right)^2}{4\,t}\right), \quad (5.56)$$

where $x > x_0 > 0$. Then the marginal PDF is

$$p_1(x, t) = \frac{2\,\theta(x - x_0)}{x\sqrt{4\pi\left(\frac{v}{2\sqrt{\mathcal{D}}}\right)^2 t}} \times \exp\left(-\frac{\log^2\frac{x}{x_0}}{4\left(\frac{v}{2\sqrt{\mathcal{D}}}\right)^2 t}\right). \quad (5.57)$$

---

[2]In particular, to carry out this dimensionless procedure for the comb model, the time and space scaling parameters can be introduced. Let $\bar{v}$ and $\bar{D}$ are unite velocity and diffusivity correspondingly, and their dimensionality is $[\bar{v}] = LT^{-1}$, $[\bar{D}] = L^2T^{-1}$. That is $[\bar{t}] = T$, $[\bar{x}] = [\bar{y}] = L$, are dimension variables as well. Then $\bar{D}/\bar{v}^2$ and $\bar{D}/\bar{v}$ are the time and space scaling parameters. This yields the dimensionless procedure as follows $x \cdot [\bar{v}/\bar{D}] \to x$, $y \cdot [\bar{v}/\bar{D}] \to y$, $t \cdot [\bar{v}^2/\bar{D}] \to t$, $v/\bar{v} \to v$, and $D/\bar{D} \to D$.

The solutions (5.56) and (5.57) can be symmetrically extended on the entire x axis, taking modulus $|x|$. Note that $|x| > x_0$, because a tracer cannot move upstream.

The marginal PDF is the solution of a fractional Fokker-Planck equation. Substituting Eq. (5.54) in Eq. (5.52) and performing the inverse Laplace transform, one obtains

$$_C D_{0+}^{1/2} p_1(x, t) = -\frac{v}{2\sqrt{\mathcal{D}}} \frac{\partial}{\partial x} \{ x \, p_1(x, t) \}, \qquad (5.58)$$

where the initial condition is $p_1(x, t = 0) = \delta(x - x_0)$. Here we used the Laplace transform formula of the Caputo fractional derivative.

For the MSD, $\langle x^2(t) \rangle = \int x^2 \, p_1(x, t) \, dx$, we obtain the solution in the form of the one parameter M-L function,

$$\langle x^2(t) \rangle = x_0^2 \, E_{\frac{1}{2}} \left(\frac{v}{\sqrt{\mathcal{D}}} t^{1/2} \right). \qquad (5.59)$$

Its asymptotic behavior in the long time limit is the exponential growth, according to $\langle x^2(t) \rangle \sim 2 \, x_0^2 \, e^{4(\frac{v}{2\sqrt{\mathcal{D}}})^2 t}$, while the short time behavior is subdiffusive, $\langle x^2(t) - x_0^2 \rangle \sim 2 x_0^2 (\frac{v}{2\sqrt{\mathcal{D}}}) \frac{t^{1/2}}{\Gamma(3/2)}$. Note that this subdiffusion is due to advection, characterized by v, which differs from Eq. (4.33). However, diffusion in fingers is still described by Eqs. (4.37) and (4.38).

Remark 5.2 (Alternative solution by subordination). Let us first consider ordinary form of Eq. (5.58),

$$\frac{\partial}{\partial t} f(x, t) = -\nu \frac{\partial}{\partial x} \{ x \, f(x, t) \}. \qquad (5.60)$$

The Laplace transform yields

$$s \, \hat{f}(x, s) - \delta(x - x_0) = -\nu \frac{\partial}{\partial x} \left\{ x \, \hat{f}(x, s) \right\}, \qquad (5.61)$$

from where we find the solution in the Laplace domain, for $x_0 > 0$,

$$\hat{f}(x, s) = \frac{\theta(x - x_0)}{\nu \, x} \times e^{-\frac{s}{\nu} \log \frac{x}{x_0}}. \qquad (5.62)$$

The inverse Laplace transform gives the solution

$$f(x, t) = \frac{\theta(x - x_0)}{\nu \, x} \times \delta \left(t - \frac{1}{\nu} \log \frac{x}{x_0} \right). \qquad (5.63)$$

This result can also be obtained from the solution (5.38) in the limit $\frac{\sigma^2}{2} = \mathcal{D} \to 0$, by using the limit representation

$$\delta(z) = \lim_{\epsilon \to 0} \frac{1}{\sqrt{4\pi\epsilon}} \times \exp \left(-\frac{z^2}{4\epsilon} \right) \quad \text{and} \quad \delta(\alpha z) = \frac{1}{|\alpha|} \delta(z).$$

To apply a subordination approach, we present the FFPE (5.58) in Laplace space,

$$s^{1/2}\,\hat{p}_1(x,s) - s^{-1/2}\,\delta(x - x_0) = -\frac{v}{2\sqrt{\mathcal{D}}}\,\frac{\partial}{\partial x}\,\{x\,\hat{p}_1(x,s)\}. \qquad (5.64)$$

Rewriting Eq. (5.64) in the form

$$s\,\hat{p}_1(x,s) - \delta(x - x_0) = -\frac{v}{2\sqrt{\mathcal{D}}}\,s\left[s^{-1/2}\,\frac{\partial}{\partial x}\,\{x\,\hat{p}_1(x,s)\}\right], \qquad (5.65)$$

we obtain by the inverse Laplace transform

$$\frac{\partial}{\partial t}p_1(x,t) = -\frac{v}{2\sqrt{\mathcal{D}}}\,\frac{d}{dt}\int_0^t \frac{(t - t')^{-1/2}}{\Gamma(1/2)}\,\frac{\partial}{\partial x}\,\{x\,p_1(x,t')\}\,dt', \qquad (5.66)$$

which is an equivalent representation of Eq. (5.58).

Let us now use the subordination approach to find the solution of Eq. (5.66). We present the solution of Eq. (5.66) as follows

$$p_1(x,t) = \int_0^\infty f(x,u)\,h(u,t)\,du, \qquad (5.67)$$

where $f(x,u)$ satisfies Eq. (5.60). Note that it satisfies formally, since here $f(x,u)$ is a part of equation for $p_1(x,t)$ and the constant ν must be replaced by $v^2/2\sqrt{\mathcal{D}}$. This also relates to the solution (5.63). The function $h(u,t)$ is the PDF subordinating the process governed by Eq. (5.66) to the process governed by Eq. (5.60). The function $h(u,t)$ subordinates the processes from the time scale t (physical time) to the time scale u (operational time). In Laplace space, Eq. (5.67) reads

$$\hat{p}_1(x,s) = \int_0^\infty e^{-st}\,p_1(x,t)\,dt = \int_0^\infty f(x,u)\,\hat{h}(u,s)\,du, \qquad (5.68)$$

where $\hat{h}(u,s) = \mathcal{L}\,[h(u,t)]$. Accounting the relation between the subordinating function and its Laplace image

$$\hat{h}(u,s) = s^{-1/2}\,e^{-u\,s^{1/2}} \quad \rightarrow \quad h(u,t) = \frac{1}{\sqrt{\pi\,t}}\,e^{-\frac{u^2}{4\,t}}, \qquad (5.69)$$

we have

$$\hat{p}_1(x,s) = \int_0^\infty f(x,u)\,s^{-1/2}\,e^{-u\,s^{1/2}}\,du$$

$$= s^{-1/2}\int_0^\infty f(x,u)\,e^{-u\,s^{1/2}}\,du$$

$$= s^{-1/2}\,\hat{f}(x,s^{1/2}). \qquad (5.70)$$

Substituting Eq. (5.70) in Eq. (5.65), and performing the variable change $s \to s^{1/2}$, we obtain the equation for the image $\hat{p}_1(x, s)$ as follows

$$s\,\hat{p}_1(x, s) - p_1(x, 0) = s\left[-\frac{v}{2\sqrt{\mathcal{D}}}\,s^{-1/2}\,\frac{\partial}{\partial x}x\,\hat{p}_1(x, s)\right], \quad (5.71)$$

which is exactly the same as Eq. (5.65). Therefore, from Eqs. (5.63), (5.67) and (5.69), for the solution we find

$$p_1(x, t) = \int_0^\infty \frac{\theta(x - x_0)e^{-\frac{u^2}{4t}}}{\sqrt{\pi t}\left(\frac{v}{2\sqrt{\mathcal{D}}}\right)x}\,\delta\left(u - \frac{1}{\left(\frac{v}{2\sqrt{\mathcal{D}}}\right)}\log\frac{x}{x_0}\right)du$$

$$= \frac{2\,\theta(x - x_0)}{x\sqrt{4\pi\left(\frac{v}{2\sqrt{\mathcal{D}}}\right)^2 t}} \times \exp\left(-\frac{\log^2\frac{x}{x_0}}{4\left(\frac{v}{2\sqrt{\mathcal{D}}}\right)^2 t}\right). \quad (5.72)$$

5.2.3 *Solution of turbulent diffusion equation on comb*

Let us extend our consideration of the two dimensional comb turbulent diffusion and consider it in the framework of the comb model [29] as follows

$$\frac{\partial}{\partial t}P(x, y, t) = \mathcal{D}_x\left(\frac{\partial}{\partial x}x\right)^2 P(x, y, t) + \mathcal{D}_y\frac{\partial^2}{\partial y^2}P(x, y, t) \quad (5.73)$$

with the initial condition $P(x, y, t = 0) = \delta(x - x_0)\delta(y)$. By the Laplace transform, we obtain

$$s\hat{P}(x, y, s) - \delta(x - x_0)\delta(y) = \mathcal{D}_x\left(\frac{\partial}{\partial x}x\right)^2 \hat{P}(x, y, s) + \mathcal{D}_y\frac{\partial^2}{\partial y^2}\hat{P}(x, y, s). \quad (5.74)$$

Again, looking for the solution as follows

$$\hat{P}(x, y, s) = g(x, s) \times e^{-\sqrt{\frac{s}{\mathcal{D}_y}}|y|}, \quad (5.75)$$

one finds

$$\hat{p}_1(x, s) = 2\sqrt{\mathcal{D}_y} \times s^{-1/2}\,g(x, s), \quad (5.76)$$

which can be presented in the form of the marginal PDF as follows

$$\hat{P}(x, y, s) = \frac{s^{1/2}}{2\sqrt{\mathcal{D}_y}}\,\hat{p}_1(x, s) \times e^{-\sqrt{\frac{s}{\mathcal{D}_y}}|y|}.$$

The equation for the marginal PDF in Laplace space then reads

$$s^{1/2}\,\hat{p}_1(x, s) - s^{-1/2}\,\delta(x - x_0) = \frac{\mathcal{D}_x}{2\sqrt{\mathcal{D}_y}}\left(\frac{\partial}{\partial x}x\right)^2 \hat{p}_1(x, s). \quad (5.77)$$

The inverse Laplace transformation yields the time fractional turbulent diffusion equation of the form

$$\mathrm{c}D_{0+}^{1/2}p_1(x,t) = \mathcal{D}_1\left(\frac{\partial}{\partial x}x\right)^2 p_1(x,t). \tag{5.78}$$

Following the subordination approach, the solution of Eq. (5.78) is given by the subordination integral (5.68)

$$\hat{p}_1(x,s) = \int_0^\infty f(x,u)\,\hat{h}(u,s)\,du,$$

where $f(x,u)$ now the GBM solution (5.39) of the turbulent diffusion equation (5.32),

$$f(x,u) = \frac{1}{x\sqrt{4\pi\mathcal{D}_1 u}} \times \exp\left(-\frac{\log^2\frac{x}{x_0}}{4\mathcal{D}_1 u}\right), \tag{5.79}$$

and

$$\hat{h}(u,s) = s^{-1/2}\,e^{-u\,s^{1/2}}. \tag{5.80}$$

Substituting explicit expressions in Eqs. (5.79) and (5.80) into the subordination integral, we obtain the solution in Laplace space

$$\hat{p}_1(x,s) = \frac{1}{2\,x\,\sqrt{\mathcal{D}_1}}\,s^{-3/4} \times \exp\left(-\frac{\left|\log\frac{x}{x_0}\right|}{\sqrt{\mathcal{D}_1}}s^{1/4}\right)$$

$$= \frac{s^{-3/4}}{x\,\sqrt{\mathcal{D}_1}}H_{0,1}^{1,0}\left[\frac{\log^2\frac{x}{x_0}}{\mathcal{D}_1}s^{1/2}\,\middle|\,\begin{matrix}-\\(0,2)\end{matrix}\right], \tag{5.81}$$

where $H_{p,q}^{m,n}(z)$ is the Fox H-function (2.17). In Eq. (5.81) we use properties (2.42) and (2.25) of the Fox H-function.

From the inverse Laplace transform for the marginal PDF, we finally obtain

$$p_1(x,t) = \frac{1}{x\sqrt{\mathcal{D}_1\,t^{1/2}}}H_{1,1}^{1,0}\left[\frac{\log^2\left|\frac{x}{x_0}\right|}{\mathcal{D}_1\,t^{1/2}}\,\middle|\,\begin{matrix}(3/4,1/2)\\(0,2)\end{matrix}\right]. \tag{5.82}$$

5.2.4   *Generalized geometric Brownian motion*

Let us introduce a generalized GBM (gGBM) by means of the subordination integral

$$P(x,t) = \int_0^\infty f(x,u)h(u,t)\,du, \tag{5.83}$$

where $f(x, u)$ satisfies the FPE (5.31) for the standard GBM (in the Stratonovich interpretation). Consequently, we subordinate the processes in the physical time scale t to the GBM in the operational time scale u. The function $h(u, t)$ is the PDF subordinating the gGBM $x(t)$ to the standard GBM. In Laplace space, Eq. (5.83) reads

$$\hat{P}(x, s) = \mathcal{L}\{P(x, t)\} = \int_0^\infty e^{-st} P(x, t)\, dt = \int_0^\infty f(x, u)\hat{h}(u, s)\, du,$$
(5.84)

where $\hat{h}(u, s) = \mathcal{L}\{h(u, t)\}$. Considering

$$\hat{h}(u, s) = \frac{\hat{\Psi}(s)}{s} e^{-u\hat{\Psi}(s)} = \frac{1}{s\hat{\eta}(s)} e^{-\frac{u}{\hat{\eta}(s)}},$$
(5.85)

we have

$$\hat{P}(x, s) = \frac{1}{s\hat{\eta}(s)} \int_0^\infty f(x, u) e^{-\frac{u}{\hat{\eta}(s)}}\, du = \frac{1}{s\hat{\eta}(s)} \hat{f}\left(x, \frac{1}{\hat{\eta}(s)}\right).$$
(5.86)

By the Laplace transform of the FPE (5.31) for the GBM, and using relation (5.86), one finds that the PDF $P(x, s)$ satisfies

$$s\hat{P}(x, s) - P(x, 0) = s\,\hat{\eta}(s)\left[-\mu\frac{\partial}{\partial x} x\hat{P}(x, s) + \frac{\sigma^2}{2}\frac{\partial^2}{\partial x^2} x^2\hat{P}(x, s)\right].$$
(5.87)

After the inverse Laplace transform we arrive at the generalised Fokker-Planck equation [30, 31, 32]

$$\frac{\partial}{\partial t} P(x, t) = \frac{\partial}{\partial t} \int_0^t \eta(t - t')\left[-\mu\frac{\partial}{\partial x} x P(x, t') + \frac{\sigma^2}{2}\frac{\partial^2}{\partial x^2} x^2 P(x, t')\right] dt',$$
(5.88)

where $\eta(t)$ is a so-called memory kernel. One observes that for $\eta(t) = 1$ we arrive at the FPE (5.27) for the GBM.

From Eqs. (5.84) and (5.85), one finds the PDF in Laplace space [31, 32],

$$\hat{P}(x, s) = \int_0^\infty \frac{1}{x\sqrt{2\pi\sigma^2 u}} \times \exp\left(-\frac{\left[\log x - \log x_0 - \bar{\mu}u\right]^2}{2\sigma^2 u}\right) \frac{e^{-\frac{u}{\hat{\eta}(s)}}}{s\hat{\eta}(s)}\, du$$

$$= \frac{1/[s\hat{\eta}(s)]}{x\sqrt{\bar{\mu}^2 + \frac{2\sigma^2}{\hat{\eta}(s)}}} \begin{cases} \exp\left(-\frac{\log x - \log x_0}{\sigma^2}\left[\sqrt{\bar{\mu}^2 + \frac{2\sigma^2}{\hat{\eta}(s)}} - \bar{\mu}\right]\right), & x > x_0, \\ 1, & x = x_0, \\ \exp\left(\frac{\log x - \log x_0}{\sigma^2}\left[\sqrt{\bar{\mu}^2 + \frac{2\sigma^2}{\hat{\eta}(s)}} + \bar{\mu}\right]\right), & x < x_0. \end{cases}$$
(5.89)

Remark 5.3. We note that Eq. (5.87) can be written in an equivalent form as [32]

$$\int_0^t \gamma(t-t')\frac{\partial}{\partial t'}P(x,t')\,dt' = -\mu\frac{\partial}{\partial x}xP(x,t) + \frac{\sigma^2}{2}\frac{\partial^2}{\partial x^2}x^2 P(x,t), \quad (5.90)$$

where the memory kernel $\gamma(t)$ is connected to $\eta(t)$ in Laplace space as $\hat{\gamma}(s) = 1/[s\hat{\eta}(s)]$, see Eq. (4.12) and Ref. [33].

5.2.4.1 Calculation of moments of generalized GBM

Let us estimate the n-th moment by means of Eq. (5.88). Multiplying both sides of Eq. (5.88) by x^n, we have

$$\frac{\partial}{\partial t}\langle x^n(t)\rangle = \left[\frac{\sigma^2}{2}n(n-1)+\mu n\right]\mu\frac{d}{dt}\int_0^t \eta(t-t')\langle x^n(t')\rangle\,dt'. \quad (5.91)$$

In Laplace space, it reads

$$\langle \hat{x}^n(s)\rangle = x_0^n\,\frac{s^{-1}}{1-\hat{\eta}(s)\left[\frac{\sigma^2}{2}n(n-1)+\mu n\right]}. \quad (5.92)$$

For $n=0$ we obtain the normalization condition

$$\langle x^0(t)\rangle = \int_{-\infty}^{\infty} f(x,t)\,dx = 1.$$

For $n=1$, Eq. (5.91) reduces to equation for the mean value

$$\frac{\partial}{\partial t}\langle x(t)\rangle = \mu\frac{d}{dt}\int_0^t \eta(t-t')\,\langle x(t')\rangle\,dt', \quad (5.93)$$

whose Laplace image is

$$\langle \hat{x}(s)\rangle = x_0\,\frac{s^{-1}}{1-\mu\hat{\eta}(s)}. \quad (5.94)$$

In terms of the memory kernel $\gamma(t)$, Eq. (5.94) reads

$$\langle \hat{x}(s)\rangle = x_0\,\frac{\hat{\gamma}(s)}{s\hat{\gamma}(s)-\mu}. \quad (5.95)$$

We note that for the standard case with $\eta(t)=1$, that is $\gamma(t)=\delta(t)$, ($\hat{\eta}(s)=1/s$, i.e., $\hat{\gamma}(s)=1$) we recover the result for the GBM,

$$\langle x(t)\rangle = x_0\,\mathcal{L}^{-1}\left[\frac{1}{s-\mu}\right] = x_0\,e^{\mu t}. \quad (5.96)$$

For $n=2$ we obtain the equation for the second moment, or the MSD,

$$\frac{\partial}{\partial t}\langle x^2(t)\rangle = (\sigma^2+2\mu)\frac{d}{dt}\int_0^t \eta(t-t')\langle x^2(t')\rangle\,dt'. \quad (5.97)$$

The Laplace image is

$$\langle \hat{x}^2(s) \rangle = x_0^2 \frac{s^{-1}}{1 - (\sigma^2 + 2\mu)\hat{\eta}(s)}, \tag{5.98}$$

or

$$\langle \hat{x}^2(s) \rangle = x_0^2 \frac{\hat{\gamma}(s)}{s\hat{\gamma}(s) - (\sigma^2 + 2\mu)}. \tag{5.99}$$

Therefore, for the GBM ($\eta(t) = 1$) one finds

$$\langle x^2(t) \rangle = x_0^2 \, \mathcal{L}^{-1} \left[\frac{1}{s - (\sigma^2 + 2\mu)} \right] = x_0^2 \, e^{(\sigma^2 + 2\mu)t}. \tag{5.100}$$

5.2.4.2 *Log-moments of generalized GBM*

In finance, for geometric processes instead of the moments a more convenient measures are the log-moments defined as follows

$$\langle \log^n x(t) \rangle = \int_0^\infty \log^n x \, P(x, t) \, dx. \tag{5.101}$$

According to Eq. (5.88), these log-moments satisfy the following integral equation

$$\frac{\partial}{\partial t} \langle \log^n x(t) \rangle = \frac{\partial}{\partial t} \int_0^t \eta(t - t') \left[\left(\mu - \frac{\sigma^2}{2} \right) n \, \langle \log^{n-1} x(t') \rangle \right.$$
$$\left. + \frac{\sigma^2}{2} n(n - 1) \, \langle \log^{n-2} x(t') \rangle \right] dt'. \tag{5.102}$$

It is obvious that $\langle \log^0 x(t) \rangle = 1$. For the mean value ($n = 1$), we find

$$\frac{\partial}{\partial t} \langle \log x(t) \rangle = \left(\mu - \frac{\sigma^2}{2} \right) \frac{\partial}{\partial t} \int_0^t \eta(t - t') \underbrace{\langle \log^0 x(t') \rangle}_{=1} dt', \tag{5.103}$$

which yields

$$\langle \log x(t) \rangle = \log x_0 + \left(\mu - \frac{\sigma^2}{2} \right) \int_0^t \eta(t') \, dt'. \tag{5.104}$$

For the GBM ($\eta(t) = 1$), one finds that the expectation of the logarithm of the particle position has a linear dependence on time,

$$\langle \log x(t) \rangle = \log x_0 + \left(\mu - \frac{\sigma^2}{2} \right) t. \tag{5.105}$$

For geometric processes of high importance is the behavior of the expectation of the log return, given by

$$\frac{1}{\Delta t} \langle \log \left(x(t + \Delta t)/x(t) \right) \rangle,$$

which in asset pricing terms represents the continuously compounded return of the asset [32]. For the expectation of the periodic log return with period Δt, we find

$$\frac{1}{\Delta t}\langle \log\left(x(t+\Delta t)/x(t)\right)\rangle = \left(\mu - \frac{\sigma^2}{2}\right)\frac{1}{\Delta t}\int_t^{t+\Delta t} \eta(t')\,dt'$$

$$= \left(\mu - \frac{\sigma^2}{2}\right)\frac{I(t+\Delta t) - I(t)}{\Delta t}$$

$$\underset{\Delta t \to 0}{\sim} \left(\mu - \frac{\sigma^2}{2}\right)\eta(t), \tag{5.106}$$

where $I(t) = \int \eta(t)\,dt$, i.e., $I'(t) = \eta(t)$. Therefore, the expectation of the periodic log returns behaves as the rate of the first log-moment,

$$\frac{1}{\Delta t}\langle \log\left(x(t+\Delta t)/x(t)\right)\rangle \underset{\Delta t \to 0}{\sim} \frac{d}{dt}\langle \log x(t)\rangle, \tag{5.107}$$

which for the GBM reads

$$\frac{1}{\Delta t}\langle \log\left(x(t+\Delta t)/x(t)\right)\rangle \underset{\Delta t \to 0}{\sim} \mu - \frac{\sigma^2}{2}. \tag{5.108}$$

For $n = 2$ we obtain the second log-moment [32]

$$\frac{\partial}{\partial t}\langle \log^2 x(t)\rangle = \frac{\partial}{\partial t}\int_0^t \eta(t - t')\left[2\left(\mu - \frac{\sigma^2}{2}\right)\langle \log x(t')\rangle + \sigma^2\,\langle \log^0 x(t')\rangle\right]dt' \tag{5.109}$$

i.e.,

$$\langle \log^2 x(t)\rangle = \log^2 x_0 + \int_0^t \eta(t - t')\left\{2\left(\mu - \frac{\sigma^2}{2}\right)\right.$$

$$\times \left[\log x_0 + \left(\mu - \frac{\sigma^2}{2}\right)\int_0^{t'} \eta(t'')\,dt''\right] + \sigma^2\right\}dt', \tag{5.110}$$

and the log-variance is

$$\langle \log^2 x(t)\rangle - \langle \log x(t)\rangle^2 = \sigma^2\int_0^t \eta(t')\,dt' + \left(\mu - \frac{\sigma^2}{2}\right)^2$$

$$\times \left[2\int_0^t \eta(t - t')\left(\int_0^{t'} \eta(t'')\,dt''\right)dt' - \left(\int_0^t \eta(t')\,dt'\right)^2\right]. \tag{5.111}$$

For the GBM ($\eta(t) = 1$) the second log-moment and log-variance are reduced to

$$\langle \log^2 x(t)\rangle = \log^2 x_0 + \left[2\left(\mu - \frac{\sigma^2}{2}\right)\log x_0 + \sigma^2\right]t + \left(\mu - \frac{\sigma^2}{2}\right)^2 t^2, \tag{5.112}$$

$$\langle \log^2 x(t)\rangle - \langle \log x(t)\rangle^2 = \sigma^2 t, \tag{5.113}$$

respectively.

References

[1] W. Horsthemke and R. Lefever. *Noise-Induced Transitions. Theory and Applications in Physics, Chemistry, and Biology*. Berlin: Springer-Verlag, 1984 (cit. on p. 119).

[2] C. W. Gardiner. *Handbook of Stochastic Methods for Physics, Chemistry and the Natural Sciences*. 2nd. Berlin: Springer-Verlag, 1990 (cit. on p. 119).

[3] C. Loverdo et al. "Quantifying hopping and jumping in facilitated diffusion of DNA-binding proteins". In: *Physical Review Letters* 102.18 (2009), p. 188101 (cit. on p. 119).

[4] B. O'Shaughnessy and I. Procaccia. "Analytical solutions for diffusion on fractal objects". In: *Physical Review Letters* 54.5 (1985), p. 455 (cit. on p. 119).

[5] T. Srokowski. "Non-Markovian Lévy diffusion in nonhomogeneous media". In: *Physical Review E* 75.5 (2007), p. 051105 (cit. on p. 119).

[6] T. Srokowski. "Multiplicative lévy processes: Itô versus stratonovich interpretation". In: *Physical Review E* 80.5 (2009), p. 051113 (cit. on p. 119).

[7] A. G. Cherstvy, A. V. Chechkin, and R. Metzler. "Anomalous diffusion and ergodicity breaking in heterogeneous diffusion processes". In: *New Journal of Physics* 15.8 (2013), p. 083039 (cit. on pp. 119, 122).

[8] S. Regev, N. Grønbech-Jensen, and O. Farago. "Isothermal Langevin dynamics in systems with power-law spatially dependent friction". In: *Physical Review E* 94.1 (2016), p. 012116 (cit. on pp. 119, 123).

[9] K. S. Fa and E. K. Lenzi. "Power law diffusion coefficient and anomalous diffusion: analysis of solutions and first passage time". In: *Physical Review E* 67.6 (2003), p. 061105 (cit. on p. 119).

[10] M. F. De Andrade et al. "Anomalous diffusion and fractional diffusion equation: anisotropic media and external forces". In: *Physics Letters A* 347.4-6 (2005), pp. 160–169 (cit. on p. 119).

[11] A. T. Silva et al. "Exact propagator for a Fokker-Planck equation, first passage time distribution, and anomalous diffusion". In: *Journal of Mathematical Physics* 52.8 (2011), p. 083301 (cit. on p. 119).

[12] A. W. C. Lau and T. C. Lubensky. "State-dependent diffusion: Thermodynamic consistency and its path integral formulation". In: *Physical Review E* 76.1 (2007), p. 011123 (cit. on p. 119).

[13] A. Fuliński. "Anomalous diffusion and weak nonergodicity". In: *Physical Review E* 83.6 (2011), p. 061140 (cit. on p. 119).

[14] M. A. F. Dos Santos et al. "Critical patch size reduction by heterogeneous diffusion". In: *Physical Review E* 102.4 (2020), p. 042139 (cit. on p. 119).

[15] N. Leibovich and E. Barkai. "Infinite ergodic theory for heterogeneous diffusion processes". In: *Physical Review E* 99.4 (2019), p. 042138 (cit. on p. 119).

[16] B. P. English et al. "Single-molecule investigations of the stringent response machinery in living bacterial cells". In: *Proceedings of the National Academy of Sciences* 108.31 (2011), E365–E373 (cit. on p. 119).

[17]   M. Platani et al. "Cajal body dynamics and association with chromatin are ATP-dependent". In: *Nature Cell Biology* 4.7 (2002), pp. 502–508 (cit. on p. 119).

[18]   R. Haggerty and S. M. Gorelick. "Multiple-rate mass transfer for modeling diffusion and surface reactions in media with pore-scale heterogeneity". In: *Water Resources Research* 31.10 (1995), pp. 2383–2400 (cit. on p. 119).

[19]   M. Dentz et al. "Diffusion and trapping in heterogeneous media: An inhomogeneous continuous time random walk approach". In: *Advances in Water Resources* 49 (2012), pp. 13–22 (cit. on p. 119).

[20]   H. G. E. Hentschel and I. Procaccia. "Fractal nature of turbulence as manifested in turbulent diffusion". In: *Physical Review A* 27.2 (1983), p. 1266 (cit. on p. 124).

[21]   H. G. E. Hentschel and I. Procaccia. "Relative diffusion in turbulent media: the fractal dimension of clouds". In: *Physical Review A* 29.3 (1984), p. 1461 (cit. on p. 124).

[22]   A. M. Obukhov. "Description of turbulence in terms of Lagrangian variables". In: *Advances in Geophysics* 6 (1959), pp. 113–116 (cit. on p. 124).

[23]   F. Black and M. Scholes. "The Pricing of Options and Corporate Liabilities". In: *Journal of Political Economy* 81.3 (1973), pp. 637–654 (cit. on p. 124).

[24]   E. Baskin and A. Iomin. "Superdiffusion on a comb structure". In: *Physical Review Letters* 93.12 (2004), p. 120603 (cit. on p. 124).

[25]   S. M. Ross. *Introduction to probability models*. Academic press, 2014 (cit. on p. 124).

[26]   R. Metzler and J. Klafter. "The random walk's guide to anomalous diffusion: a fractional dynamics approach". In: *Physics Reports* 339.1 (2000), pp. 1–77 (cit. on pp. 126, 128).

[27]   M. Heidernätsch. *On the diffusion in inhomogeneous systems*. PhD Thesis, Fakultät für Naturwissenschaften der Technischen Universität Chemnitz, 2015 (cit. on p. 127).

[28]   G. Matheron and G. De Marsily. "Is transport in porous media always diffusive? A counterexample". In: *Water Resources Research* 16.5 (1980), pp. 901–917 (cit. on p. 129).

[29]   T. Sandev, A. Iomin, and L. Kocarev. "Hitting times in turbulent diffusion due to multiplicative noise". In: *Physical Review E* 102.4 (2020), p. 042109 (cit. on pp. 129, 133).

[30]   M. Magdziarz and J. Gajda. "Anomalous dynamics of Black-Scholes model time-changed by inverse subordinators". In: *Acta Physica Polonica B* 43.5 (2012) (cit. on p. 135).

[31]   C. Li. "Option pricing with generalized continuous time random walk models". PhD thesis. Queen Mary University of London, 2016 (cit. on p. 135).

[32]   V. Stojkoski et al. "Generalised geometric Brownian motion: Theory and applications to option pricing". In: *Entropy* 22.12 (2020), p. 1432 (cit. on pp. 135, 136, 138).

[33]   T. Sandev, R. Metzler, and A. Chechkin. "From continuous time random walks to the generalized diffusion equation". In: *Fractional Calculus and Applied Analysis* 21.1 (2018), pp. 10–28 (cit. on p. 136).

Chapter 6

Diffusion processes with stochastic resetting

In this chapter, we consider anomalous diffusion with reset processes. This issue of stochastic resetting is extensively explored in various search processes [1], such as foraging [2, 3], population dynamics [4], Michaelis–Menten enzymatic reactions [5], human behaviour of finding resources [6], different diffusion processes [7, 8, 9], geometric Brownian motion [10, 11, 12], one dimensional lattices [13, 14] and complex networks [15, 16], as well as in quantum systems [17, 18, 19], to mention but a few. Experimental realisations of the first-passage under stochastic resetting has been demonstrated using holographic optical tweezers [20] or laser traps [21].

Another important task is understanding of resetting in molecular reaction systems, where the resetting dynamics of a Brownian particle under external potentials is analyzed in detail [22, 23, 24, 25, 26, 27], and this issue is discussed in the chapter, as well. One dimensional Brownian motion affected by Poissonian resetting with a constant resetting rate r is introduced by Evans and Majumdar [28], see also the review paper [29] for more details. Describing anomalous diffusion with resetting, Brownian motion on two dimensional and three dimensional combs in the presence of stochastic resetting, stands for analytical considerations [30, 31, 32, 33, 34], as well. Following review [29], we first describe the main features of the one dimensional Brownian transport with stochastic resetting.

6.1 Introduction to diffusion with Poissonian resetting

To define one dimensional Brownian motion with Poissonian resetting, we consider a single particle dynamics affected by white noise $\xi(t)$ and resetting with the rate r to the fixed position X_r. Let us suppose that a particle, with the initial coordinate $x_0 = X(0)$ at time $t = 0$ is at the position $X(t)$

at time t. Then its dynamics for the next infinitesimal time interval dt is defined either by reset to the position X_r with the probability rdt, or by Brownian motion $X(t) + \xi(t)(dt)^{\frac{1}{2}}$ with the probability $(1 - rdt)$. Then the PDF $P(x, t + dt)$ of finding the particle at the position $x = X(t + dt)$ at time $t + dt$ consists of two parts and according to Eq. (3.131) reads

$$P(x, t + dt) = \langle \delta \left(x - X(t + dt) \right) \rangle$$

$$= rdt \langle \delta \left(x - X_r \right) \rangle + (1 - rdt) \left\langle \delta \left(x - X(t) - \xi(t)(dt)^{\frac{1}{2}} \right) \right\rangle$$

$$= rdt\delta \left(x - X_r \right) + (1 - rdt) \int_{-\infty}^{\infty} [d\rho(\xi)] P \left(x - \xi(dt)^{\frac{1}{2}} \right), \qquad (6.1)$$

where in the last line we use the definition of the PDF in Eq. (3.131) and the independence of the first delta function of the noise is accounted as well. Performing expansion with respect to $dt \to 0$ in Eq. (6.1), we obtain the Fokker-Planck equation for diffusion with Poissonian resetting

$$\frac{\partial}{\partial t} P(x, t) = \mathcal{D} \frac{\partial^2}{\partial x^2} P(x, t) - r\, P(x, t) + r\, \delta(x - X_r). \qquad (6.2)$$

The properties of the Gaussian white noise, defined in Eqs. (3.127) and (3.128) are also taken into account. In particular, we used that the Gaussian white noise is completely described by its mean $\langle \xi(t) \rangle = 0$ and correlation function $\langle \xi(t)\xi(t') \rangle = 2\mathcal{D}\, \delta(t - t')$. Note that X_r can be an arbitrary point on the x axis. In sequel, without restriction of generality, the resetting point is taken at the initial position: $X_r = x_0$.

6.1.1 *Renewal equation*

After every (or the last) resetting, Brownian motion starts from the very beginning, that is it renews. Therefore, diffusion with the reset processes can be described by renewal equations, which account two independent processes: normal diffusion with the PDF $P_0(x, t|x_0)$ and resets at random times described by, e,g., Poisson distribution $\phi_P(t) = re^{-rt}$. Therefore, the PDF of a renewal process consists of two terms: the PDF $e^{-rt} P_0(x, t|x_0)$ of Brownian motion with the probability of no resets until time t, which is $\int_0^t \phi_P(t')dt' = e^{-rt}$, and the PDF $r \int_0^t e^{r\tau} P_0(x, \tau|x_0)d\tau$, which describes no rests after time τ with the probability $re^{-r\tau}$ and all possible times $\tau \in [0, t]$ are accounted. Then the renewal equation for the transition PDF reads

$$P_r(x, t|x_0) = e^{-rt} P_0(x, t|x_0) + \int_0^t re^{-r\tau} P_0(x, \tau|x_0)\, d\tau, \qquad (6.3)$$

where $P_0(x,t|x_0)$ is the Gaussian kernel of Brownian motion in Eq. (3.93) and the initial condition $P_0(x,0|x_0) = \delta(x - x_0)$.

By the Laplace transform of renewal equation (6.3), one finds

$$\hat{P}_r(x,s|x_0) = \hat{P}_0(x,s+r|x_0) + \frac{r}{s}\hat{P}_0(x,s+r|x_0) = \frac{s+r}{s}\hat{P}_0(x,s+r|x_0),$$

(6.4)

where

$$\hat{P}_0(x,s|x_0) = \mathcal{L}\left[\frac{1}{\sqrt{4\pi\mathcal{D}t}}\exp\left(-\frac{(x-x_0)^2}{4\mathcal{D}t}\right)\right] = \frac{s^{-1/2}}{2\sqrt{\mathcal{D}}}e^{-\sqrt{\frac{s}{\mathcal{D}}}|x-x_0|}. \quad (6.5)$$

Thus, the PDF in Laplace space is

$$\hat{P}_r(x,s|x_0) = \frac{s+r}{s}\frac{(s+r)^{-1/2}}{2\sqrt{\mathcal{D}}}e^{-\sqrt{\frac{s+r}{\mathcal{D}}}|x-x_0|} = \frac{s^{-1}(s+r)^{1/2}}{2\sqrt{\mathcal{D}}}e^{-\sqrt{\frac{s+r}{\mathcal{D}}}|x-x_0|}.$$

(6.6)

It is interesting to admit that in the long time limit ($t \to \infty$, i.e., $s \to 0$) the PDF approaches the stationary Laplace distribution,

$$\lim_{t\to\infty} P(x,t|x_0) = \lim_{s\to 0} s\hat{P}(x,s|x_0) = \frac{1}{2}\sqrt{\frac{r}{\mathcal{D}}}\exp\left(-\sqrt{\frac{r}{\mathcal{D}}}|x-x_0|\right), \quad (6.7)$$

and we can write

$$P_r^{\text{st}}(x) = \frac{\alpha_0}{2}\exp\left(-\alpha_0|x-x_0|\right) \quad (6.8)$$

where $\alpha_0 = \sqrt{r/\mathcal{D}}$ is the inverse length scale, i.e., the typical distance diffused by the particle between resets. That is the resetting mechanism changes the distribution form Gaussian to Laplace in the long time limit. In Fig. 6.1, we compare the PDFs with and without resetting. The distribution $P_0(x,t)$ is Gaussian (left panel), while the distribution $P_r(x,t)$ is non-Gaussian (right panel), with a cusp at $x = x_0$. The reset processes create a probability source at x_0, which is absent for all $x \neq x_0$, where the created probability relaxes due to diffusion between resets. For the long time limit, this self-organization process leads eventually to the stationary Laplace distribution (6.8), which is a *nonequilibrium stationary state* (NESS).

6.2 Fokker-Planck equation for diffusion with resetting

The diffusion process with stochastic resetting can be also described by the Fokker-Planck equation

$$\frac{\partial}{\partial t}P_r(x,t|x_0) = \mathcal{D}\frac{\partial^2}{\partial x^2}P_r(x,t|x_0) - r\,P_r(x,t|x_0) + r\,\delta(x-x_0), \quad (6.9)$$

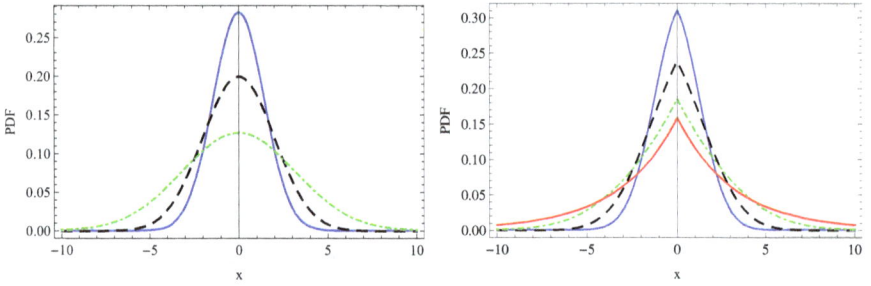

Fig. 6.1 Left panel: The Gaussian PDF; Right panel: The PDF in presence of resetting. Here we use $x_0 = 0$; $t = 1$ (blue solid line); $t = 2$ (black dashed line), $t = 5$ (green dot-dashed line), and the Laplace distribution (red solid line).

where the second term from the rhs of the equation is the loss of the probability from position x due to the reset to the initial position x_0, and the third term from the rhs of the equation is the gain of the probability at x_0 due to resetting from all other positions.

6.2.1 *Subordination approach*

One can solve this equation by means of the Laplace transformation, from where it follows

$$s\,\hat{P}_r(x, s|x_0) - \delta(x - x_0) = \mathcal{D}\frac{\partial^2}{\partial x^2}\hat{P}_r(x, s|x_0) - r\,\hat{P}_r(x, s|x_0) + \frac{r}{s}\,\delta(x - x_0),$$
(6.10)

which can be rewritten in the form

$$s\,\hat{P}_r(x, s|x_0) - \delta(x - x_0) = \frac{s}{s + r}\mathcal{D}\frac{\partial}{\partial x^2}\hat{P}_r(x, s|x_0).$$
(6.11)

The inverse Laplace transform yields the generalized diffusion equation

$$\frac{\partial}{\partial t}P_r(x, t|x_0) = \mathcal{D}\frac{\partial}{\partial t}\int_0^t \eta(t - \tau)\frac{\partial^2}{\partial x^2}P_r(x, \tau|x_0)\,d\tau,$$
(6.12)

with the exponential memory kernel $\eta(t) = e^{-rt}$, which is an equivalent representation of the diffusion process with stochastic resetting.

From the subordination integral $P_r(x, t|x_0) = \int_0^\infty P_0(x, u|x_0)h(u, t)du$, one finds the subordination function

$$h(u, t) = \mathcal{L}^{-1}\left[\frac{1}{s\hat{\eta}(s)}e^{-u/\hat{\eta}(s)}\right] = \mathcal{L}^{-1}\left[\frac{s + r}{s}e^{-u(s+r)}\right]$$

$$= e^{-ru}\delta(t - u) + re^{-ru}\theta(t - u).$$
(6.13)

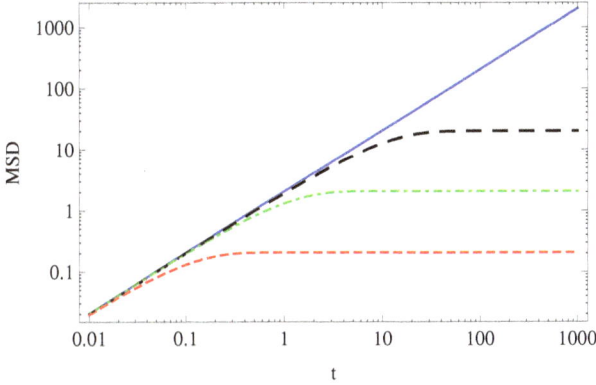

Fig. 6.2   MSD (6.15) for $\mathcal{D} = 1$ and $r = 0$ (blue solid line), $r = 0.1$ (black dashed line), $r = 1$ (green dot-dashed line), $r = 10$ (red dotted line).

Thus, the PDF becomes

$$P_r(x, t|x_0) = \int_0^\infty P_0(x, u|x_0) \left[e^{-ru} \delta(t - u) + re^{-ru} \theta(t - u) \right] du$$

$$= e^{-rt} P_0(x, t|x_0) + \int_0^t re^{-ru} P_0(x, u|x_0) \, du, \tag{6.14}$$

which is the renewal Eq. (6.3), discussed above. Note also that the both variants of diffusion with stochastic resetting are equivalent.

Multiplying the renewal equation by x^2 and integrating with respect to x, we obtain the MSD,

$$\langle x^2(t) \rangle_r = e^{-rt} \langle x^2(t) \rangle_0 + \int_0^t re^{-ru} \langle x^2(u) \rangle_0 \, du, \tag{6.15}$$

where $\langle x^2(t) \rangle_0 = x_0^2 + 2\mathcal{D}t$ is the MSD for Brownian motion. Substituting this in Eq. (6.15), we obtain

$$\langle x^2(t) \rangle_r = e^{-rt} \left(x_0^2 + 2\mathcal{D}t \right) + \int_0^t re^{-ru} \left(x_0^2 + 2\mathcal{D}u \right) du$$

$$= x_0^2 + \frac{2\mathcal{D}}{r} \left[1 - e^{-rt} \right]. \tag{6.16}$$

In the short time limit the MSD is $\langle x^2(t) \rangle_r \sim x_0^2 + 2\mathcal{D}t$, while in the long time limit the MSD is constant, $\langle x^2(t) \rangle_r = x_0^2 + \frac{2\mathcal{D}}{r}$. Graphical representation of the MSD (6.15) and (6.16) for different values of the resetting rate r is presented in Fig. 6.2.

6.2.2   *Numerical simulations: Langevin equation approach*

Numerical verification of the analytical approach for the renewal equation and corresponding FPE (6.9) can be performed in the framework of the Langevin equation. The discretized version of the Langevin equation is as follows

$$x(t + \Delta t) = \begin{cases} x(0), & \text{with prob. } r\Delta t, \\ x(t) + \sqrt{2\mathcal{D}\Delta t}\,\xi(t), & \text{with prob. } (1 - r\Delta t), \end{cases} \tag{6.17}$$

see discussion in the introductory Sec. 6.1. Here, the first line represents a reset process to the initial position x_0, while the second line describes the diffusion process. Recall that $\xi(t)$ is the Gaussian white noise with zero mean $\langle \xi(t) \rangle = 0$ and correlation $\langle \xi(t)\xi(t') \rangle = \delta(t - t')$. A graphical representation of trajectories of Brownian motion in absence and presence of stochastic resetting is given in Fig. 6.3.

6.2.3   *Transition to the steady state*

Let us now analyze the transition dynamics of the system to the NESS. To this end, we analyze the renewal equation for the large but finite time. Thus, we have

$$P_r(x, t | x_0) = e^{-rt} P_0(x, t | x_0) + \underbrace{\int_0^t r e^{-r\tau} P_0(x, \tau | x_0)\, d\tau}_{\text{dominant term for large } t}$$

$$= \frac{e^{-rt - \frac{(x - x_0)^2}{4\mathcal{D}t}}}{\sqrt{4\pi\mathcal{D}t}} + \int_0^t r \frac{e^{-r\tau - \frac{(x - x_0)^2}{4\mathcal{D}\tau}}}{\sqrt{4\pi\mathcal{D}\tau}}\, d\tau. \tag{6.18}$$

Note that in the limit $t \to \infty$, the NESS has been obtained, that is

$$\lim_{t \to \infty} P_r(x, t | x_0) = \int_0^\infty r e^{-r\tau} P_0(x, \tau | x_0)\, d\tau$$

$$= r\, \mathcal{L}[P_r(x, t | x_0)](r) = r\, \hat{P}(x, r | x_0)$$

$$= \frac{1}{2}\sqrt{\frac{r}{\mathcal{D}}}\, \exp\left(-\sqrt{\frac{r}{\mathcal{D}}} |x - x_0| \right) = P_r^{\text{st}}(x). \tag{6.19}$$

For evaluation of the integral in the renewal equation (6.18) we use the Laplace approximation. Performing the variable change $\tau = t\tau'$ ($d\tau = t\, d\tau'$), we obtain

$$P_r(x, t | x_0) = \frac{e^{-t\Phi(1, w)}}{\sqrt{4\pi\mathcal{D}t}} + \frac{r\sqrt{t}}{\sqrt{4\pi\mathcal{D}}} \int_0^1 e^{-t\Phi(\tau', w)} \frac{d\tau'}{\sqrt{\tau'}}, \tag{6.20}$$

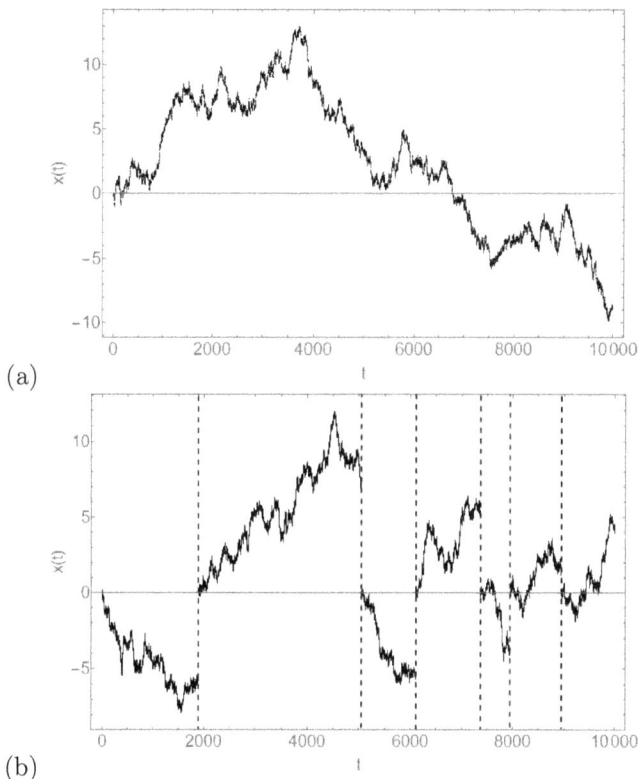

(a)

(b)

Fig. 6.3 Graphical representation of the trajectory: (a) no resetting, (b) with resetting. We set $x_0 = 0$, $\mathcal{D} = 0.005$, $\Delta t = 1$ and $r = 0.0005$.

where a new function $\Phi(\tau, w)$ is introduced as follows

$$\Phi(\tau', w) = r\,\tau' + \frac{w^2}{4\mathcal{D}\tau'}, \quad w = \frac{|x - x_0|}{t}. \qquad (6.21)$$

Then according to Eqs. (1.82) and (1.83), the extremum point z_* corresponds to equation $f'(z_*) = 0$. If $z_* < 1$, the Laplace approximation yields

$$\int_0^1 e^{-t\,f(z)} g(z)\,dz \approx e^{-tf(z_*)} g(z_*) \sqrt{\frac{2\pi}{t|f''(z_*)|}}. \qquad (6.22)$$

If the extremum point z_* is outside the integration limits ($z_* > 1$), then the approximation result is calculated at $z_* = 1$. Thus, it follows

$$\frac{\partial}{\partial \tau'}\Phi(\tau' = \tau_*, w) = r - \frac{w^2}{4\mathcal{D}\tau_*^2} = 0, \quad \text{where} \quad \tau_* = \frac{w}{\sqrt{4\mathcal{D}r}}. \qquad (6.23)$$

If $\tau_* < 1$, then

$$\Phi(\tau_*, w) = r\,\frac{w}{\sqrt{4\mathcal{D}r}} + \frac{w^2}{4\mathcal{D}\frac{w}{\sqrt{4\mathcal{D}r}}} = \sqrt{\frac{r}{\mathcal{D}}}\,w. \tag{6.24}$$

If $\tau_* > 1$, then

$$\Phi(1, w) = r + \frac{w^2}{4\mathcal{D}}. \tag{6.25}$$

Eventually, the PDF reads as follows

$$P_r(x, t|x_0) \sim \exp\left(-t\,I\left(\frac{|x - x_0|}{t}\right)\right), \tag{6.26}$$

where the large deviation function (LDF) obtained from Eqs. (6.25) and (6.26) is [35]

$$I\left(\frac{|x - X_r|}{t}\right) = \begin{cases} \sqrt{r/\mathcal{D}}\,\frac{|x - X_r|}{t}, & |x - X_r| \leq \sqrt{4\mathcal{D}r}\,t, \\ r + \frac{1}{4\mathcal{D}}\left(\frac{|x - x_0|}{t}\right)^2, & |x - x_0| \geq \sqrt{4\mathcal{D}r}\,t. \end{cases} \tag{6.27}$$

This result means that for the large but finite time t, the NESS has been achieved in the spatial region $|x - x_0| \leq \sqrt{4\mathcal{D}t}$, where the PDF is time independent, i.e., $P_r(x, t|x_0) \sim e^{-\sqrt{r/\mathcal{D}}|x - x_0|}$. Outside this region, the particles are not relaxed to the NESS, that is, they are in a transient regime. The boundary between these two regions moves at a constant speed $v = \sqrt{4\mathcal{D}r}$.

6.3 Diffusion-advection equation with resetting

Let us add a drift term with a constant velocity V to Eq. (6.9). Then the corresponding Fokker-Planck equation reads [33]

$$\frac{\partial}{\partial t}P_r(x, t|x_0) = \left[\mathcal{D}\frac{\partial^2}{\partial x^2} - V\frac{\partial}{\partial x}\right]P_r(x, t|x_0) - rP_r(x, t|x_0) + r\delta(x - x_0) \tag{6.28}$$

with the same initial condition $P_r(x, t = 0|x_0) = \delta(x - x_0)$.

The Laplace transform of Eq. (6.28) reads

$$s\hat{P}_r(x, s|x_0) - \delta(x - x_0) = \frac{s}{s + r}\left[\mathcal{D}\frac{\partial^2}{\partial x^2} - V\frac{\partial}{\partial x}\right]\hat{P}_r(x, s|x_0). \tag{6.29}$$

Then the inverse Laplace transform yields an equivalent form to Eq. (6.28)

$$\frac{\partial}{\partial t}P_r(x, t|x_0) = \frac{d}{dt}\int_0^t \eta(t - t')\left[\mathcal{D}\frac{\partial^2}{\partial x^2} - V\frac{\partial}{\partial x}\right]P_r(x, t'|x_0)\,dt', \tag{6.30}$$

where the memory kernel now is $\eta(t) = e^{-rt}$ with its Laplace image $\hat{\eta}(s) = \frac{1}{s+r}$.

This equation stands for the subordination approach, considered above. Then performing the Laplace transform of Eq. (6.30), we have

$$s\hat{P}_r(x, s|x_0) - \delta(x - x_0) = s\hat{\eta}(s)\left[\mathcal{D}\frac{\partial^2}{\partial x^2} - V\frac{\partial}{\partial x}\right]\hat{P}_r(x, s|x_0). \quad (6.31)$$

Following the standard procedure, we present the solution to Eq. (6.30) by means of the subordination integral

$$P_r(x, t|x_0) = \int_0^\infty P_0(x, u|x_0)h(u, t)\, du, \quad (6.32)$$

where $P_0(x, t|x_0)$ is the solution in Eq. (4.77). Here the subordination function $h(u, t)$ subordinates the process governed by Eq. (6.30) to the process governed by Eq. (4.75) with the solution (4.77). In Laplace space the subordination function reads

$$\hat{h}(u, s) = \frac{1}{s\hat{\eta}(s)}e^{-u/\hat{\eta}(s)}. \quad (6.33)$$

Therefore, the Laplace transform of Eq. (6.32) yields

$$\begin{aligned}
\hat{P}_r(x, s|x_0) &= \int_0^\infty P_0(x, u|x_0)\hat{h}(u, s)\, du \\
&= \frac{1}{s\hat{\eta}(s)}\int_0^\infty P_0(x, u|x_0)e^{-u/\hat{\eta}(s)}\, du \\
&= \frac{1}{s\hat{\eta}(s)}\hat{P}_0(x, 1/\hat{\eta}(s)|x_0).
\end{aligned} \quad (6.34)$$

Performing the replacement $s \to 1/\hat{\eta}(s)$ in Eq. (4.76), the latter has the form

$$\frac{1}{\hat{\eta}(s)}\hat{P}_0(x, 1/\hat{\eta}(s)|x_0) - \delta(x - x_0) = \left[\mathcal{D}\frac{\partial^2}{\partial x^2} - V\frac{\partial}{\partial x}\right]\hat{P}_0(x, 1/\hat{\eta}(s)|x_0). \quad (6.35)$$

Therefore, combining Eqs. (6.34) and (6.35), we arrive at Eq. (6.31). Eventually, from Eqs. (4.78) and (6.34), we obtain the PDF in the presence of resetting,

$$\hat{P}_r(x, s|x_0) = \frac{s+r}{s}\frac{1}{2\mathcal{D}}\frac{1}{\sqrt{\frac{s+r}{\mathcal{D}} + \frac{V^2}{4\mathcal{D}^2}}}e^{\frac{V}{2\mathcal{D}}(x-x_0) - \sqrt{\frac{s+r}{\mathcal{D}} + \frac{V^2}{4\mathcal{D}^2}}|x-x_0|}. \quad (6.36)$$

Then the Laplace inversion yields the solution in the form of the renewal equation

$$P_r(x, t|x_0) = e^{-rt}P_0(x, t|x_0) + r\int_0^t e^{-rt'}P_0(x, t'|x_0)\, dt', \quad (6.37)$$

where $P_0(x, t|x_0)$ is defined by Eq. (4.77).

From Eqs. (5.84) and (4.80) we find that the PDF $P_r(x,t)$ is normalised ($\langle x^0(t)\rangle_r = 1$) since

$$\langle \hat{x}^0(s)\rangle_r = \int_{-\infty}^{\infty} \hat{P}_r(x,s|x_0)\, dx = \frac{1}{s\hat{\eta}(s)} \int_{-\infty}^{\infty} \hat{P}_0(x,1/\hat{\eta}(s)|x_0)\, dx$$

$$= \frac{1}{s\hat{\eta}(s)} \langle \hat{x}^0(1/\hat{\eta}(s))\rangle_0 = \frac{\hat{\eta}(s)}{s\hat{\eta}(s)} = \frac{1}{s}. \tag{6.38}$$

From Eq. (5.84) for the MSD we find

$$\langle \hat{x}^2(s)\rangle_r = \int_{-\infty}^{\infty} x^2\, \hat{P}_r(x,s|x_0)\, dx$$

$$= \frac{1}{s\hat{\eta}(s)} \int_{-\infty}^{\infty} x^2\, \hat{P}_0(x,1/\hat{\eta}(s)|x_0)\, dx = \frac{1}{s\hat{\eta}(s)} \langle \hat{x}^2(1/\hat{\eta}(s))\rangle_0$$

$$= \frac{x_0^2\, \hat{\eta}(s) + 2(\mathcal{D} + x_0 V)\, \hat{\eta}^2(s) + 2V^2\, \hat{\eta}^3(s)}{s\hat{\eta}(s)}$$

$$= \frac{x_0^2}{s} + \frac{2(\mathcal{D} + x_0 V)\hat{\eta}(s)}{s} + \frac{2V^2\, \hat{\eta}^2(s)}{s}, \tag{6.39}$$

which results in

$$\langle x^2(t)\rangle_r = \mathcal{L}^{-1}\left[\frac{x_0^2}{s} + \frac{2(\mathcal{D} + x_0 V)}{s(s+r)} + \frac{2V^2}{s(s+r)^2} \right]$$

$$= x_0^2 + \frac{2(\mathcal{D} + x_0 V)(1 - e^{-rt})}{r} + \frac{2V^2(1 - e^{-rt} - rte^{-rt})}{r^2}. \tag{6.40}$$

Then the long time limit yields saturation of the MSD,

$$\langle x^2(t)\rangle_r \sim x_0^2 + \frac{2(\mathcal{D} + x_0 V)}{r} + \frac{2V^2}{r^2},$$

while the short time limit corresponds to the result without resetting, Eq. (4.82). In the absence of the drift, the MSD reads $\langle x^2(t)\rangle_r = x_0^2 + \frac{2\mathcal{D}}{r}(1 - e^{-rt})$, see Eq. (6.16).

Remark 6.1. We note that one can also consider diffusion-advection equation on a comb with resetting [33]

$$\frac{\partial}{\partial t} P_r(x,y,t|x_0,0) = \delta(y)\left[\mathcal{D}_x \frac{\partial^2}{\partial x^2} - v\frac{\partial}{\partial x} \right] P_r(x,y,t|x_0,0)$$

$$+ \mathcal{D}_y \frac{\partial^2}{\partial y^2} P_r(x,y,t|x_0,0)$$

$$- r P_r(x,y,t|x_0,0) + r\delta(x - x_0)\delta(y) \tag{6.41}$$

with the initial condition $P_r(x, y, t = 0|x_0, 0) = \delta(x - x_0)\delta(y)$. The corresponding equation for the marginal PDF along the backbone becomes

$$\frac{\partial}{\partial t} P_{1,r}(x, t|x_0) = \frac{1}{2\sqrt{\mathcal{D}_y}} \frac{d}{dt} \int_0^t \eta_r(t - t') \left[\mathcal{D}_x \frac{\partial^2}{\partial x^2} - v \frac{\partial}{\partial x} \right] P_{1,r}(x, t'|x_0) \, dt',$$

(6.42)

where $\eta_r(t) = \mathcal{L}^{-1} \left[(s + r)^{-1/2} \right] = e^{-rt} \frac{t^{-1/2}}{\Gamma(1/2)}$ (for more details on tempered operators, see Ref. [36] and Eq. (1.221) in Sec. 1.4.3). From the form of Eq. (6.42) it is clear that one can apply the subordination approach for analysis of the marginal PDF and MSD.

6.3.1 *Transition to the steady state*

To analyze the transition to the NESS, we consider the renewal Eq. (6.37). Using arguments of Sec. 6.2.3, we arrive at the equation for the LDF, which accounts the advection term with the velocity V and now reads

$$I = r \int_0^t e^{-rt'} \frac{1}{\sqrt{4\pi \mathcal{D} t'}} e^{-\frac{(x - V t')^2}{4 \mathcal{D} t'}} \, dt'.$$

(6.43)

Performing the variable change $t' = t\tau$ $(dt' = td\tau)$, we have

$$I = \frac{r}{\sqrt{4\pi \mathcal{D}}} \sqrt{t} \int_0^1 e^{-rt\tau} e^{-\frac{(x - V t\tau)^2}{4 \mathcal{D} t\tau}} \, d\tau = \frac{r}{\sqrt{4\pi \mathcal{D}}} \sqrt{t} \int_0^1 e^{-t\left[r\tau + \frac{(w - V\tau)^2}{4 \mathcal{D} \tau} \right]} \, d\tau,$$

(6.44)

where $w = x/t$ and [10]

$$\Phi(\tau, w) = r\tau + \frac{(w - V\tau)^2}{4 \mathcal{D} \tau},$$

(6.45)

and Eq. (6.44) takes the form

$$I = \frac{r}{\sqrt{4\pi \mathcal{D}}} \sqrt{t} \int_0^1 e^{-t\Phi(\tau, w)} \, d\tau.$$

(6.46)

Again the Laplace approximation is applied, and the extremum point τ_* results from the equation $\frac{\partial}{\partial \tau} \Phi(\tau, w)|_{\tau_*} = 0$, and it reads

$$\tau_* = \frac{|w|}{\sqrt{V^2 + 4 \mathcal{D} r}}.$$

(6.47)

Substituting τ^* into Eq. (6.45), we find

$$\Phi(\tau_*, w) = -\frac{V}{2\mathcal{D}} w + \frac{\sqrt{V^2 + 4 \mathcal{D} r}}{2\mathcal{D}} |w|, \quad \text{for} \quad \tau_* < 1.$$

(6.48)

Outside the region, we have

$$\Phi(1, w) = r + \frac{(w - V)^2}{4D}, \quad \text{for} \quad \tau_* > 1. \tag{6.49}$$

Therefore, the PDF is

$$P_r(x, t) \sim e^{-t\, I_r(x/t)}, \tag{6.50}$$

where $I_r(x/t)$ is the LDF

$$I_r(x/t) = \begin{cases} -\frac{V}{2D}\frac{x}{t} + \frac{\sqrt{V^2 + 4Dr}}{2D}\frac{|x|}{t}, & |x| < \sqrt{V^2 + 4Dr}\, t, \\ r + \frac{(x/t - V)^2}{4D}, & |x| > \sqrt{V^2 + 4Dr}\, t. \end{cases} \tag{6.51}$$

The result for the LDF in Eq. (6.51) describes two separated regions. In the first, defined by $\tau_* < 1$, relaxation takes place, while the second region is the transient one with $\tau_* = 1$, where relaxation is not achieved yet.

6.4   Diffusion in comb with confining branches with resetting

As already discussed in previous chapters, the comb geometry affects strongly the transport properties due to possible diffusion inside fingers. This situation influences the reset phenomena as well [31, 32, 33, 34]. In this section, we follow Ref. [34] and consider this phenomenon, introducing an additional control over the finger's transport by means of a confinement potential $V(y)$. Therefore, for a test particle performing Brownian motion in the two dimensional comb structure with the potential $V = V(y)$ along the branches, the Fokker–Planck equation describing the dynamics of the PDF $P_r(x, y, t|x_0, 0)$ under the resetting rate r reads

$$\frac{\partial}{\partial t} P_r(x, y, t|x_0, 0) = D_x\, \delta(y)\, \frac{\partial^2}{\partial x^2} P_r(x, y, t|x_0, 0)$$

$$+ \left(\frac{\partial}{\partial y} V'(y) + D_y \frac{\partial^2}{\partial y^2} \right) P_r(x, y, t|x_0, 0)$$

$$- r\, P_r(x, y, t|x_0, 0) + r\, \delta(x - x_0)\delta(y) \tag{6.52}$$

with $P_r(x, y, t = 0|x_0, 0) = \delta(x - x_0)\delta(y)$, and we choose the potential function to be a piecewise linear function

$$V(y) = \begin{cases} -U_0\, y, & y \leq 0 \\ U_0\, y, & y \geq 0 \end{cases} \tag{6.53}$$

Here we assume that the particle is reset to its initial position $(x_0, 0)$, with a constant rate r. By complete analogy with the one dimensional case (6.9), each resetting event to the initial position $(x_0, 0)$ renews the process at a rate r, i.e., between two consecutive renewal events, the particle undergoes diffusion on the comb in non-monotonic potential (6.53) along the fingers. The last two terms from the rhs of Eq. (6.52) represent the loss of the probability from the position (x, y) due to reset to the initial position $(x_0, 0)$, and the gain of the probability at $(x_0, 0)$ due to resetting from all other positions, respectively.

By the Laplace transform, we obtain

$$s \hat{P}_r(x, y, s|x_0, 0) - \delta(x - x_0)\,\delta(y) = D_x\,\delta(y)\,\frac{\partial^2}{\partial x^2}\hat{P}_r(x, y, s|x_0, 0)$$
$$+ \left(U_0\,\mathrm{sgn}(y)\frac{\partial}{\partial y} + 2U_0\,\delta(y) + D_y\frac{\partial^2}{\partial y^2}\right)\hat{P}_r(x, y, s|x_0, 0)$$
$$- r\,\hat{P}_r(x, y, s|x_0, 0) + \frac{r}{s}\delta(x - x_0)\delta(y). \qquad (6.54)$$

Note that Eq. (6.54) is symmetric with respect to y-inversion, $y \to -y$. Thus, performing substitution $z = |y|$ and separating terms with delta function, we arrive at the following system of equations

$$\begin{cases} (s + r)\,\hat{P}_r(x, z, s|x_0, 0) = U_0\frac{\partial}{\partial z}\hat{P}_r(x, z, s|x_0, 0) + D_y\frac{\partial^2}{\partial z^2}\hat{P}_r(x, z, s|x_0, 0), \\[2mm] -s^{-1}(s + r) = \left[D_x\frac{\partial^2}{\partial x^2} + 2U_0 + 2D_y\frac{\partial}{\partial z}\right]\hat{P}_r(x, z, s|x_0, 0)\Big|_{z=0}. \end{cases}$$
$$(6.55)$$

We look for the solution to the system (6.55) in the form

$$\hat{P}_r(x, y, s|x_0, 0) = \hat{P}(x, z = 0, s|x_0, 0) \times \exp\left(-\frac{U_0}{2D_y}\,[1 + \Delta_{s+r}]\,z\right),$$
$$(6.56)$$

where $\Delta_{s+r} = \sqrt{1 + \frac{4D_y(s+r)}{U_0^2}}$. Therefore, for the marginal PDF along the backbone we have

$$\hat{p}_r(x, s|x_0) = \int_{-\infty}^{\infty} \hat{P}_r(x, y, s|x_0, 0)\,dy$$
$$= 2\int_0^{\infty} \hat{P}_r(x, z, s|x_0, 0)\,dz = \frac{4D_y}{U_0}\frac{\hat{P}(x, z = 0, s|x_0, 0)}{1 + \Delta_{s+r}}. \qquad (6.57)$$

From Eqs. (6.55), (6.56) and (6.57), we find

$$s\,\hat{p}_r(x, s|x_0) - \delta(x - x_0) = \frac{D_x}{4D_y}U_0\,s \times \frac{1 + \Delta_{s+r}}{s + r}\frac{\partial^2}{\partial x^2}\hat{p}_r(x, s|x_0), \qquad (6.58)$$

and by the inverse Laplace transform we obtain a generalized diffusion equation

$$\frac{\partial}{\partial t} p_r(x, t|x_0) = \frac{D_x}{2\sqrt{D_y}} \frac{\partial}{\partial t} \int_0^t \eta(t - t') \frac{\partial^2}{\partial x^2} p_r(x, t'|x_0) \, dt', \qquad (6.59)$$

with the memory kernel $\eta(t)$ which is determined by the inverse Laplace transform from the Laplace image

$$\hat{\eta}(s) = \frac{1}{s + r} \left[\frac{U_0}{2\sqrt{D_y}} + \left(s + r + \frac{U_0^2}{4 D_y} \right)^{1/2} \right]. \qquad (6.60)$$

Therefore, this memory kernel reads

$$\eta(t) = \frac{U_0}{2\sqrt{D_y}} e^{-rt} + e^{-rt} \left[\frac{\exp\left(-\frac{U_0^2}{4 D_y} t\right)}{\sqrt{\pi t}} + \frac{U_0}{2\sqrt{D_y}} \operatorname{erf}\left(\frac{U_0}{2\sqrt{D_y}} \sqrt{t} \right) \right],$$

$$(6.61)$$

where $\operatorname{erf}(z)$ is the error function (1.125). Performing the Fourier-Laplace transform of Eq. (6.59), we find

$$\tilde{p}_r(k, s|x_0) = \frac{\frac{1}{s\,\hat{\eta}(s)}}{\frac{1}{\hat{\eta}(s)} + \frac{D_x}{2\sqrt{D_y}} k^2} e^{ik_x x_0}, \qquad (6.62)$$

which by the inverse Fourier transform yields

$$\hat{p}_r(x, s|x_0) = \frac{1}{2s} \sqrt{\frac{2\sqrt{D_y}}{D_x \hat{\eta}(s)}} \times \exp\left(-\sqrt{\frac{2\sqrt{D_y}}{D_x \hat{\eta}(s)}} |x - x_0| \right), \qquad (6.63)$$

and which is

$$\hat{p}_r(x, s|x_0) = \frac{1}{2} \sqrt{\frac{2\sqrt{D_y}}{D_x}} \frac{s^{-1}(s + r)^{1/2}}{\sqrt{\left(s + r + \frac{U_0^2}{4 D_y} \right)^{1/2} + \frac{U_0}{2\sqrt{D_y}}}}$$

$$\times \exp\left(-\sqrt{\frac{2\sqrt{D_y}}{D_x}} \frac{(s + r)^{1/2} |x - x_0|}{\sqrt{\left(s + r + \frac{U_0^2}{4 D_y} \right)^{1/2} + \frac{U_0}{2\sqrt{D_y}}}} \right). \qquad (6.64)$$

According to the final value theorem for the long time limit ($s \to 0$), the stationary distribution is defined by the Laplace image, namely

$$p_{r,\text{st}}(x) = \lim_{t \to \infty} p_r(x, t|x_0) = \lim_{s \to 0} s\hat{p}_r(x, s|x_0)$$

$$= \frac{1}{2}\sqrt{\frac{2\sqrt{D_y}}{D_x}} \frac{r^{1/2}}{\sqrt{\left(r + \frac{U_0^2}{4D_y}\right)^{1/2} + \frac{U_0}{2\sqrt{D_y}}}}$$

$$\times \exp\left(-\sqrt{\frac{2\sqrt{D_y}}{D_x}} \frac{r^{1/2}|x - x_0|}{\sqrt{\left(r + \frac{U_0^2}{4D_y}\right)^{1/2} + \frac{U_0}{2\sqrt{D_y}}}}\right). \tag{6.65}$$

In the absence of the potential, when $U_0 = 0$, the result (6.62) reduces to the Laplace image of the marginal PDF of the comb with stochastic resetting, namely it reads

$$\hat{p}_r(x, s|x_0) = \frac{s^{-1}(s + r)^{1/4}}{2}\sqrt{\frac{2\sqrt{D_y}}{D_x}} \times \exp\left(-\sqrt{\frac{2\sqrt{D_y}}{D_x}}(s + r)^{1/4}|x - x_0|\right), \tag{6.66}$$

and the corresponding stationary distribution is

$$p_{r,\text{st}}(x) = \frac{1}{2}\sqrt{\frac{2\sqrt{D_y}}{D_x}}r^{1/4} \times \exp\left(-\sqrt{\frac{2\sqrt{D_y}}{D_x}}r^{1/4}|x - x_0|\right). \tag{6.67}$$

Note that in absence of resetting, the system does not reach the stationary solution. It follows from Eq. (6.64) by setting $r = 0$. Thus, we have

$$\hat{p}_0(x, s|x_0) = \frac{1}{2}\sqrt{\frac{2\sqrt{D_y}}{D_x}} \frac{s^{-1/2}}{\sqrt{\left(s + \frac{U_0^2}{4D_y}\right)^{1/2} + \frac{U_0}{2\sqrt{D_y}}}}$$

$$\times \exp\left(-\sqrt{\frac{2\sqrt{D_y}}{D_x}} \frac{s^{1/2}|x - x_0|}{\sqrt{\left(s + \frac{U_0^2}{4D_y}\right)^{1/2} + \frac{U_0}{2\sqrt{D_y}}}}\right)$$

$$\underset{s \to 0}{\sim} \frac{1}{2}\sqrt{\frac{2D_y}{D_x U_0}}s^{-1/2} \times \exp\left(-\sqrt{\frac{2D_y}{D_x}U_0}s^{1/2}|x - x_0|\right). \tag{6.68}$$

Performing the inverse Laplace transform, we obtain the Gaussian distribution,

$$p_0(x, t|x_0) = \frac{1}{\sqrt{2\pi \frac{D_x U_0}{D_y}t}} \times \exp\left(-\frac{|x - x_0|}{2\frac{D_x U_0}{D_y}t}\right). \tag{6.69}$$

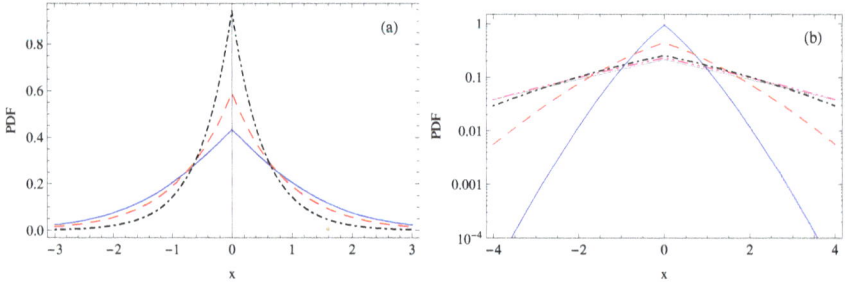

Fig. 6.4   PDF (6.64) for (a) $t = 1$ and $r = 0.1$ (blue solid line), $r = 1$ (red dashed line), $r = 5$ (black dot-dashed line); (b) for $r = 0.1$ and $t = 0.1$ (blue solid line), $t = 1$ (red dashed line), $t = 5$ (black dot-dashed line), $t = 10$ (violet dot-dot-dashed line), which approaches the stationary distribution (6.65) (solid thin grey line). We set $D_x = 1$, $D_y = 1$ and $U_0 = 1$. Republished with permission of IOP Publishing, LTD, from J. Phys. A: Math. Theor. R. K. Singh, T. Sandev, A. Iomin and R. Metzler, 54(40), 404006 (2021).

Graphical representation of the PDF and the transition to the steady state is shown in Fig. 6.4.

From Eq. (6.62) for the MSD $\langle x^2(t) \rangle = \mathcal{L}^{-1} \left\{ -\frac{\partial^2}{\partial k^2} \tilde{p}_r(k, s|x_0) \right\} \Big|_{k=0}$, we find

$$
\begin{aligned}
\langle x^2(t) \rangle &= 2 \left(\frac{D_x}{2\sqrt{D_y}} \right) \mathcal{L}^{-1} \left\{ s^{-1} \hat{\eta}(s) \right\} \\
&= \frac{D_x}{2\,D_y} U_0 \frac{1 - e^{-r\,t}}{r} - \frac{D_x}{2\,D_y} U_0 \frac{e^{-r\,t}}{r} \operatorname{erf} \left(\frac{U_0}{2\sqrt{D_y}} \sqrt{t} \right) \\
&\quad + \frac{D_x}{2\,D_y} U_0 \frac{\Delta_r}{r} \operatorname{erf} \left(\frac{U_0}{2\sqrt{D_y}} \Delta_r \sqrt{t} \right),
\end{aligned}
\tag{6.70}
$$

where $\Delta_r = \sqrt{1 + \frac{4 D_y r}{U_0^2}}$.

Let us consider several limiting cases. For $U_0 = 0$, the MSD is

$$
\langle x^2(t) \rangle = \frac{D_x}{\sqrt{D_y}} \frac{\operatorname{erf} \left(\sqrt{r\,t} \right)}{\sqrt{r}},
\tag{6.71}
$$

which corresponds to diffusion in the comb with stochastic resetting in absence of the potential. In absence of resetting, the MSD (6.70) turns to

$$\langle x^2(t)\rangle = \frac{D_x}{2\,D_y}U_0\left[t + \frac{2\sqrt{D_y}}{U_0}\frac{t^{1/2}}{\Gamma(1/2)}\exp\left(-\frac{U_0^2}{4\,D_y}t\right)\right.$$
$$\left. + \left(t + \frac{2\,D_y}{U_0^2}\right)\operatorname{erf}\left(\frac{U_0}{2\sqrt{D_y}}\sqrt{t}\right)\right], \tag{6.72}$$

which in the long time limit behaves as $\langle x^2(t)\rangle \sim t$. This means that due to the confining potential along the fingers, the particle returns back to the backbone more frequently, resulting in normal diffusion along the backbone. Therefore, the confining potential is a part of finger's transport mechanism, which affects the backbone diffusion strongly, namely it constantly returns the particle to the backbone.

From the final result for the MSD (6.70), for the long time limit we observe saturation

$$\langle x^2(t)\rangle \sim \frac{D_x}{2D_y}U_0\frac{1+\Delta_r}{r}, \tag{6.73}$$

which takes place due to the reset processes. In contrast, for the short time limit, both resetting and the potential do not affect the backbone transport, which is subdiffusion with the MSD

$$\langle x^2(t)\rangle \sim 2\frac{D_x}{\sqrt{D_y}}\frac{t^{1/2}}{\Gamma(1/2)}, \tag{6.74}$$

see Chapter 4. Graphical representation of the MSD is shown in Fig. 6.5. In Fig. 6.5 (a) one observes a transition from subdiffusion ($\sim t^{1/2}$) to saturation due to resetting for different values of the potential energy U_0, while Fig. 6.5 (b) presents the MSDs for the fixed potential energy ($U_0 = 1$) and different values of the resetting rate r. For $r = 0$ normal diffusion is observed in the long time limit (blue solid line), which occurs due to the confining potential in the fingers.

The Fokker-Planck equation for the marginal PDF along the fingers, $p_{r,2}(y,t|0) = \int_{-\infty}^{\infty} P_r(x,y,t)\,dx$, reads

$$\frac{\partial}{\partial t}p_{r,2}(y,t|0) = \left(\frac{\partial}{\partial y}V'(y) + D_y\frac{\partial^2}{\partial y^2}\right)p_{r,2}(y,t|0) - r\,p_{r,2}(y,t|0) + r\delta(y), \tag{6.75}$$

which corresponds to the diffusion-advection equation considered in Sec. 6.3.

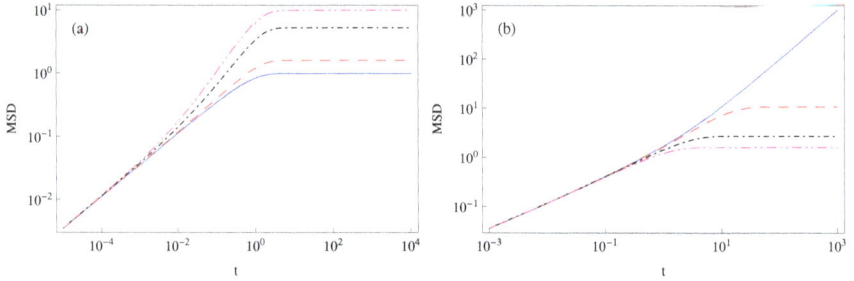

Fig. 6.5   MSD (6.70) for (a) $r = 1$ and $U_0 = 0$ (blue solid line), $U_0 = 1$ (red dashed line), $U_0 = 5$ (black dot-dashed line), $U_0 = 10$ (violet dot-dot-dashed line); (b) for $U_0 = 1$ and $r = 0$ (blue solid line), $r = 0.1$ (red dashed line), $r = 0.5$ (black dot-dashed line), $r = 1$ (violet dot-dot-dashed line). We set $D_x = 1$ and $D_y = 1$. Republished with permission of IOP Publishing, LTD, from J. Phys. A: Math. Theor. R. K. Singh, T. Sandev, A. Iomin and R. Metzler, 54(40), 404006 (2021).

6.5   Diffusion in 3D comb with resetting

As shown in Sec. 4.3.2, the transport in the 3D comb is described by the FPE (4.52) with the 3D Laplace, or Fokker-Planck operator of the form

$$L_{FP} = \mathcal{D}_x \delta(y)\delta(z) \frac{\partial^2}{\partial x^2} + \mathcal{D}_y \delta(z) \frac{\partial^2}{\partial y^2} + \mathcal{D}_z \frac{\partial^2}{\partial z^2}. \qquad (6.76)$$

Therefore, diffusion in a three dimensional comb with global resetting can be considered in the framework of the following three dimensional Fokker-Planck equation [31]

$$\frac{\partial}{\partial t} P(x, y, z, t|x_0, 0, 0) = L_{\mathrm{FP}} P(x, y, z, t|x_0, 0, 0)$$
$$- rP(x, y, z, t|x_0, 0, 0) + r\delta(x - x_0)\delta(y)\delta(z), \quad (6.77)$$

with the initial position $P(x, y, z, t = 0|x_0) = \delta(x - x_0)\delta(y)\delta(z)$. In complete analogy with the 2D comb, this equation for the transition PDF has the same interpretation in terms of the renewal process, when each reset event to the initial position $(x_0, y_0, z_0) = (x_0, 0, 0)$ renews the process at the rate r. Between any two consecutive renewal events, the particle undergoes diffusion on the xyz-comb structure.

To find the solution to Eq. (6.77), we apply the Fourier transformations with respect to x, y and z, and the Laplace transformation with respect to t. The Laplace transformation yields

$$s\hat{P}(x,y,z,s|x_0,0,0) - \delta(x-x_0)\delta(y)\delta(z)$$

$$= \mathcal{D}_x \delta(y)\delta(z)\frac{\partial^2}{\partial x^2}\hat{P}(x,y,z,s|x_0,0,0)$$

$$+ \mathcal{D}_y \delta(z)\frac{\partial^2}{\partial y^2}\hat{P}(x,y,z,s|x_0,0,0)$$

$$+ \mathcal{D}_z \frac{\partial^2}{\partial z^2}\hat{P}(x,y,z,s|x_0,0,0)$$

$$- r\hat{P}(x,y,z,s|x_0,0,0) + \frac{r}{s}\delta(x-x_0)\delta(y)\delta(z). \quad (6.78)$$

while, by the Fourier transform with respect to x, y and z, we obtain[1]

$$s\hat{\hat{P}}(k_x,k_y,k_z,s|x_0,0,0) - e^{\imath k_x x_0} = -\mathcal{D}_x k_x^2 \hat{\hat{P}}(k_x,y=0,z=0,s|x_0,0,0)$$

$$- \mathcal{D}_y k_y^2 \hat{\hat{P}}(k_x,k_y,z=0,s|x_0,0,0)$$

$$- \mathcal{D}_z k_z^2 \hat{\hat{P}}(k_x,k_y,k_z,s|x_0,0,0)$$

$$- r\hat{\hat{P}}(k_x,k_y,k_z,s|x_0,0,0) + \frac{r}{s}e^{\imath k_x x_0}. \quad (6.79)$$

This yields

$$\hat{\hat{P}}(k_x,k_y,k_z,s|x_0,0,0) = \frac{s^{-1}(s+r)e^{\imath k_x x_0}}{(s+r+\mathcal{D}_z k_z^2)}$$

$$- \frac{\mathcal{D}_x k_x^2 \hat{\hat{P}}(k_x,y=0,z=0,s|x_0,0,0)}{(s+r+\mathcal{D}_z k_z^2)}$$

$$- \frac{\mathcal{D}_y k_y^2 \hat{\hat{P}}(k_x,k_y,z=0,s|x_0,0,0)}{(s+r+\mathcal{D}_z k_z^2)}. \quad (6.80)$$

Let us now perform the inverse Fourier transform with respect to k_z that yields

$$\hat{\hat{P}}(k_x,k_y,z,s|x_0,0,0) = \frac{(s+r)^{-1/2}}{2\sqrt{\mathcal{D}_z}}e^{-\sqrt{\frac{s+r}{\mathcal{D}_z}}|z|}\left[s^{-1}(s+r)e^{\imath k_x x_0}\right.$$

$$-\mathcal{D}_x k_x^2 \hat{\hat{P}}(k_x,y=0,z=0,s|x_0,0,0)$$

$$\left.-\mathcal{D}_y k_y^2 \hat{\hat{P}}(k_x,k_y,z=0,s|x_0,0,0)\right], \quad (6.81)$$

---

[1]Note that the notation $\hat{\hat{F}}(k_x,k_y,k_z,s)$ is used for the Fourier-Laplace images with respect to x, y and z. However, in order to avoid complication with different notations, the same notation $\hat{\hat{F}}$ is used also for $\hat{\hat{F}}(k_x,y=0,z=0,s)$ and $\hat{\hat{F}}(k_x,k_y,z=0,s)$ to define the Laplace transform with respect to t and the Fourier transform with respect to x and y, only.

where we use $\mathcal{F}\left[e^{-a|\xi|}\right] - \frac{2a}{a^2+k_\xi^2}$, i.e., $\mathcal{F}^{-1}\left[\frac{1}{a^2+k_\xi^2}\right] - \frac{1}{2a}e^{-a|\xi|}$, i.e.,

$\frac{1}{s+r+\mathcal{D}_z k_z^2} = \frac{1}{\mathcal{D}_z}\frac{1}{\frac{s+r}{\mathcal{D}_z}+k_z^2}$, and thus $a^2 = \frac{s+r}{\mathcal{D}_z}$, $a = \sqrt{\frac{s+r}{\mathcal{D}_z}}$. Then, we find

$$\hat{\tilde{P}}(k_x, k_y, z = 0, s|x_0, 0, 0)$$

$$= \frac{s^{-1}(s+r)e^{ik_x x_0} - \mathcal{D}_x k_x^2 \hat{\tilde{P}}(k_x, y = 0, z = 0, s|x_0, 0, 0)}{\mathcal{D}_y\left[\frac{2\sqrt{\mathcal{D}_z}}{\mathcal{D}_y}(s+r)^{1/2} + k_y^2\right]}.$$

$$(6.82)$$

The inverse Fourier transform with respect to k_y, yields

$$\hat{\tilde{P}}(k_x, y, z = 0, s|x_0, 0, 0) = \frac{(s+r)^{-1/4}}{2\sqrt{2\mathcal{D}_y\sqrt{D_z}}}\left[s^{-1}(s+r)e^{ik_x x_0}\right.$$

$$\left. - \mathcal{D}_x k_x^2 \hat{\tilde{P}}(k_x, y = 0, z = 0, s|x_0, 0, 0)\right]e^{-\sqrt{\frac{2\sqrt{\mathcal{D}_z}}{\mathcal{D}_y}}}, \quad (6.83)$$

that yields

$$\hat{P}(k_x, y = 0, z = 0, s|x_0, 0, 0) = \frac{s^{-1}(s+r)e^{ik_x x_0}}{\mathcal{D}_x\left[\frac{2\sqrt{2\mathcal{D}_y\sqrt{D_z}}}{\mathcal{D}_x}(s+r)^{1/4} + k_x^2\right]}. \quad (6.84)$$

Performing the inverse Fourier transform with respect to k_x, we find

$$\hat{P}(x, y = 0, z = 0, s|x_0, 0, 0) = \frac{s^{-1}(s+r)^{7/8}}{2\sqrt{\mathcal{D}_x}\sqrt{2\sqrt{\mathcal{D}_y}\sqrt{2\sqrt{\mathcal{D}_z}}}}$$

$$\times \exp\left(-\frac{(s+r)^{1/8}}{2\sqrt{\mathcal{D}_x}\sqrt{2\sqrt{\mathcal{D}_y}\sqrt{2\sqrt{\mathcal{D}_z}}}}|x - x_0|\right), \quad (6.85)$$

and thus

$$\hat{P}(x = 0, y = 0, z = 0, s|x_0, 0, 0) = \frac{s^{-1}(s+r)^{7/8}}{2\sqrt{\mathcal{D}_x}\sqrt{2\sqrt{\mathcal{D}_y}\sqrt{2\sqrt{\mathcal{D}_z}}}}$$

$$\times \exp\left(-\frac{(s+r)^{1/8}}{2\sqrt{\mathcal{D}_x}\sqrt{2\sqrt{\mathcal{D}_y}\sqrt{2\sqrt{\mathcal{D}_z}}}}|x_0|\right). \quad (6.86)$$

Eventually, the PDF reads

$$\hat{\tilde{P}}(k_x, k_y, k_z, s|x_0, 0, 0) = \frac{1}{s} \times \frac{(s+r)^{1/4}}{(s+r)^{1/4} + \frac{\mathcal{D}_x}{2\sqrt{2\mathcal{D}_y\sqrt{\mathcal{D}_z}}}k_x^2}$$

$$\times \frac{(s+r)^{1/2}}{(s+r)^{1/2} + \frac{\mathcal{D}_y}{2\sqrt{\mathcal{D}_z}}k_y^2} \times \frac{(s+r)}{(s+r) + \mathcal{D}_z k_z^2} \times e^{ik_x x_0}. \quad (6.87)$$

In order to analyze the motion along all three directions, we analyze the marginal PDFs (see Eqs. (4.54), (4.55) and (4.56))

$$p_1(x, t|x_0) = \int_{-\infty}^{\infty} \int_{-\infty}^{\infty} P(x, y, z, t|x_0, 0, 0) \, dy \, dz \qquad (6.88)$$

$$p_2(y, t|0) = \int_{-\infty}^{\infty} \int_{-\infty}^{\infty} P(x, y, z, t|x_0, 0, 0) \, dx \, dz, \qquad (6.89)$$

$$p_3(z, t|0) = \int_{-\infty}^{\infty} \int_{-\infty}^{\infty} P(x, y, z, t|x_0, 0, 0) \, dx \, dy. \qquad (6.90)$$

The marginal PDF $p_1(x, t|x_0)$ describes the transport along the backbone. The marginal PDF $p_2(y, t|0)$ describes the transport along the y-fingers, which are "backbones" for the z-fingers. We call this direction by the main fingers. The marginal PDF $p_3(z, t|0)$ describes the transport along the z-fingers (secondary fingers).

In Fourier-Laplace space, the marginal PDFs are

$$\hat{\bar{p}}_1(k_x, s|x_0) = \hat{\bar{P}}(k_x, k_y = 0, k_z = 0, s|x_0, 0, 0), \qquad (6.91)$$

$$\hat{\bar{p}}_2(k_y, s|0) = \hat{\bar{P}}(k_x = 0, k_y, k_z = 0, s|x_0, 0, 0), \qquad (6.92)$$

$$\hat{\bar{p}}_3(k_z, s|0) = \hat{\bar{P}}(k_x = 0, k_y = 0, k_z, s|x_0, 0, 0). \qquad (6.93)$$

Therefore, from Eqs. (6.87) and (6.91), for the marginal PDF along the backbone we have

$$\hat{\bar{p}}_1(k_x, s|x_0) = \frac{s^{-1}(s + r)^{1/4}}{(s + r)^{1/4} + \mathcal{D}_1 \, k_x^2} e^{\imath k_x x_0}, \qquad (6.94)$$

where $\mathcal{D}_1 = \frac{\mathcal{D}_x}{2\sqrt{2\mathcal{D}_y}\sqrt{\mathcal{D}_z}}$. Applying the inverse Fourier transform, we obtain

$$\hat{\bar{p}}_1(x, s|x_0) = \frac{1}{2\sqrt{\mathcal{D}_1}} s^{-1}(s + r)^{1/8} e^{-\frac{(s+r)^{1/8}}{\sqrt{\mathcal{D}_1}}|x - x_0|}. \qquad (6.95)$$

From Eq. (6.94), by means of the inverse Fourier and Laplace transforms we arrive at the generalized (non-Markovian) diffusion equation along the backbone

$$\int_0^t \gamma_1(t - t') \frac{\partial}{\partial t'} p_1(x, t'|x_0) \, dt' = \mathcal{D}_1 \frac{\partial^2}{\partial x^2} p_1(x, t|x_0), \qquad (6.96)$$

with the memory kernel (see Example 1.15)

$$\gamma_1(t) = \mathcal{L}^{-1}\left[s^{-1}(s+r)^{1/4}\right] = \mathcal{L}^{-1}\left[\frac{s+r}{s}(s+r)^{-3/4}\right]$$

$$= e^{-rt}\frac{t^{-1/4}}{\Gamma(3/4)} + \int_0^t re^{-rt'}\frac{t'^{-1/4}}{\Gamma(3/4)}dt'$$

$$= e^{-rt}\frac{t^{-1/4}}{\Gamma(3/4)} + \sqrt[4]{r}\frac{\Gamma(3/4) - \Gamma(3/4, rt)}{\Gamma(3/4)}$$

$$= e^{-rt}\frac{t^{-1/4}}{\Gamma(3/4)} + \sqrt[4]{r}\frac{\gamma(3/4, rt)}{\Gamma(3/4)}$$

$$= t^{-1/4}E_{1,3/4}^{-1/4}(-rt). \tag{6.97}$$

Here $\Gamma(a,z)$ and $\gamma(a,z)$ are the upper (see Eq. (1.161)) and lower (see Eq. (1.160)) incomplete gamma functions, and $E_{\alpha,\beta}^{\delta}(z)$ is the three parameter M-L function (1.165). The initial condition is $p_1(x, t = 0|x_0) = \delta(x - x_0)$. By substitution of the memory kernel (6.97) with three parameter M-L function in Eq. (6.96), we obtain the equation with the regularized Prabhakar derivative $_C\mathcal{D}_{\rho,\omega,a+}^{\gamma,\mu}$ (1.226) with $n = 1$,

$$_C\mathcal{D}_{1,-r,0+}^{1/4,1/4}p_1(x, t|x_0) = \mathcal{D}_1\frac{\partial^2}{\partial x^2}p_1(x, t|x_0). \tag{6.98}$$

Using the connection between the kernels $\gamma(t)$ and $\eta(t)$ in Laplace space (4.16), Eq. (6.96) can be written in the form of Eq. (4.18),

$$\frac{\partial}{\partial t}p_1(x, t|x_0) = \mathcal{D}_1\frac{d}{dt}\int_0^t \eta_1(t-t')\frac{\partial^2}{\partial x^2}p_1(x, t'|x_0)dt', \tag{6.99}$$

where the memory kernel reads

$$\eta_1(t) = \mathcal{L}^{-1}\left[\frac{1}{s\hat{\gamma}_1(s)}\right] = \mathcal{L}^{-1}\left[\frac{1}{(s+r)^{1/4}}\right] = e^{-rt}\frac{t^{-3/4}}{\Gamma(1/4)}.$$

Substituting this memory kernel in Eq. (6.99), and using the definition of the tempered R-L fractional derivative $_{TRL}D_{a+}^{\mu}$ (1.221), we arrive at the following generalized Fokker-Planck equation

$$\frac{\partial}{\partial t}p_1(x, t|x_0) = \mathcal{D}_1\,_{TRL}D_{0+}^{1/4}\frac{\partial^2}{\partial x^2}p_1(x, t'|x_0). \tag{6.100}$$

The stationary PDF along the backbone can be also obtained in the long time limit, which is

$$\lim_{t\to\infty} p_1(x, t|x_0) = \lim_{s\to0} s\hat{p}_1(x, s|x_0)$$

$$= \frac{1}{\sqrt{4\mathcal{D}_1/\sqrt[4]{r}}}e^{-\frac{|x-x_0|}{\sqrt{\mathcal{D}_1/\sqrt[4]{r}}}} = p_{1,\text{st}}(x). \tag{6.101}$$

The marginal PDF along the y fingers is obtained from Eqs. (6.87) and (6.92) and reads

$$\hat{\bar{p}}_2(k_y, s|0) = \frac{s^{-1}(s+r)^{1/2}}{(s+r)^{1/2} + \mathcal{D}_2\, k_y^2}, \tag{6.102}$$

where $\mathcal{D}_2 = \frac{\mathcal{D}_y}{2\sqrt{\mathcal{D}_x}}$. Then by means of the inverse Fourier transform we obtain

$$\hat{p}_2(y, s|0) = \frac{(s+r)^{1/4}}{2s\sqrt{\mathcal{D}_2}} e^{-\frac{(s+r)^{1/4}}{\sqrt{\mathcal{D}_2}}|y|}. \tag{6.103}$$

The marginal PDF $p_2(y, t|0)$ is governed by the equation

$$\int_0^t \gamma_2(t - t')\frac{\partial}{\partial t'}p_2(y, t'|0)\, dt' = \mathcal{D}_2\frac{\partial^2}{\partial y^2}p_2(y, t|0), \tag{6.104}$$

with the initial condition $p_2(y, t = 0|0) = \delta(y)$ and the memory kernel, which is

$$\gamma_2(t) = \mathcal{L}^{-1}\left[s^{-1}(s+r)^{1/2}\right] = \mathcal{L}^{-1}\left[\frac{s+r}{s}(s+r)^{-1/2}\right]$$

$$= \mathcal{L}^{-1}\left[(s+r)^{-1/2} + rs^{-1}(s+r)^{-1/2}\right]$$

$$= e^{-rt}\frac{t^{-1/2}}{\Gamma(1/2)} + \int_0^t re^{-rt'}\frac{t'^{-1/2}}{\Gamma(1/2)}dt'$$

$$= \frac{1}{\sqrt{\pi t}}e^{-rt} + \sqrt{r}\operatorname{erf}\left(\sqrt{rt}\right)$$

$$= t^{-1/2}E_{1,1/2}^{-1/2}(-rt), \tag{6.105}$$

where $\operatorname{erf}(z)$ is the error function (1.125). Let us substitute the memory kernel (6.105) with the three parameter M-L function in Eq. (6.104). Thus, we obtain the equation with the regularized Prabhakar derivative $_C\mathcal{D}_{\rho,\omega,a+}^{\gamma,\mu}$ (1.226) with $n = 1$, as follows

$$_C\mathcal{D}_{1,-r,0+}^{1/2,1/2}p_2(y, t|0) = \mathcal{D}_2\frac{\partial^2}{\partial y^2}p_2(y, t|0). \tag{6.106}$$

Again, from the relation (4.16), Eq. (6.105) can be written in the form

$$\frac{\partial}{\partial t}p_2(y, t|0) = \mathcal{D}_2\frac{d}{dt}\int_0^t \eta_2(t - t')\frac{\partial^2}{\partial y^2}p_2(y, t'|0)dt', \tag{6.107}$$

where the memory kernel is

$$\eta_2(t) = \mathcal{L}^{-1}\left[\frac{1}{s\hat{\gamma}_2(s)}\right] = \mathcal{L}^{-1}\left[\frac{1}{(s+r)^{1/2}}\right] = e^{-rt}\frac{t^{-1/2}}{\Gamma(1/2)}.$$

Substituting this memory kernel in Eq. (6.107), from the definition of the tempered Riemann-Liouville fractional derivative $_{\mathrm{TRL}}D_{a+}^{\mu}$ (1.221), we obtain the following generalized Fokker-Planck equation

$$\frac{\partial}{\partial t}p_2(y,t|0) = \mathcal{D}_2 \, _{\mathrm{TRL}}D_{0+}^{1/2}\frac{\partial^2}{\partial y^2}p_2(y,t'|0).\tag{6.108}$$

An equivalent representation can be written if we use the memory kernel (6.105) in the form $\gamma_2(t) = \frac{1}{\sqrt{\pi t}}e^{-rt} + \sqrt{r}\,\mathrm{erf}\left(\sqrt{rt}\right)$ and substitute it in Eq. (6.104). Thus, we obtain

$$_{\mathrm{TC}}D_r^{1/2}p_2(y,t|0) = \mathcal{D}_2\frac{\partial^2}{\partial y^2}p_2(y,t|0)$$

$$- \sqrt{r}\int_0^t \mathrm{erf}\left(\sqrt{r(t-t')}\right)\frac{\partial}{\partial t'}p_2(y,t'|0)\,dt',\tag{6.109}$$

where $_{\mathrm{TC}}D_b^{\alpha}f(t)$ is the tempered Caputo derivative (1.222), where $b > 0$ is the truncation parameter.

For the stationary PDF along the y direction, we find

$$\lim_{t\to\infty}p_2(y,t|0) - \lim_{s\to 0}s\hat{p}_2(y,s|0)$$

$$= \frac{1}{\sqrt{4\mathcal{D}_2/\sqrt{r}}}e^{-\frac{|y|}{\sqrt{\mathcal{D}_2/\sqrt{r}}}} = p_{2,\mathrm{st}}(y).\tag{6.110}$$

For the z direction, we have

$$\hat{p}_3(k_z,s|0) = \frac{s^{-1}(s+r)}{(s+r)+\mathcal{D}_z k_z^2},\tag{6.111}$$

that yields

$$\hat{p}_3(z,s|0) = \frac{1}{2\sqrt{\mathcal{D}_z}}s^{-1}(s+r)^{1/2}e^{-\frac{(s+r)^{1/2}}{\sqrt{\mathcal{D}_3}}|z|},\tag{6.112}$$

where $\mathcal{D}_3 = \mathcal{D}_z$. The corresponding equation for the transport along secondary fingers reads

$$\int_0^t \gamma_3(t-t')\frac{\partial}{\partial t'}p_3(z,t'|0)\,dt' = \mathcal{D}_3\frac{\partial^2}{\partial z^2}p_3(z,t|0),\tag{6.113}$$

where

$$\gamma_3(t) = \delta(t) + r,\tag{6.114}$$

and it can be rewritten in the equivalent form

$$\frac{\partial}{\partial t}p_3(z,t|0) = \mathcal{D}_3\frac{\partial^2}{\partial z^2}p_3(z,t|0) - rp_3(z,t|0) + r\delta(z).\tag{6.115}$$

From the relation (4.16), Eq. (6.113) can be written in the form

$$\frac{\partial}{\partial t} p_3(z,t|0) = \mathcal{D}_3 \frac{d}{dt} \int_0^t \eta_3(t-t') \frac{\partial^2}{\partial z^2} p_3(z,t'|0) dt', \qquad (6.116)$$

where the memory kernel is

$$\eta_3(t) = \mathcal{L}^{-1} \left[\frac{1}{s\hat{\gamma}_3(s)} \right] = \mathcal{L}^{-1} \left[\frac{1}{s+r} \right] = e^{-rt}.$$

The stationary PDF along the z direction is

$$\lim_{t\to\infty} p_3(z,t|0) = \lim_{s\to 0} s\hat{p}_3(z,s|0)$$

$$= \frac{1}{\sqrt{4\mathcal{D}_3/r}} e^{-\frac{|z|}{\sqrt{\mathcal{D}_3/r}}} = p_{3,\text{st}}(z). \qquad (6.117)$$

It should be admitted that the underlying diffusion processes for the 1D marginal PDFs are different. In particular, $p_{1,\text{st}}(x|x_0)$ in Eq. (6.100) and $p_{2,\text{st}}(y|0)$ in Eq. (6.108) correspond to subdiffusion while only $p_{3,\text{st}}(z|0)$ in Eq. (6.115) describes normal diffusion. Therefore, according to the finite value theorem, the stationary marginal PDFs reflect just the underlying transport properties in Laplace space. These different behaviors are reflected in Fig. 6.6, where the marginal PDFs are presented for the finite time $t = 1$ and fix values of $\mathcal{D}$ and r for all three distributions. In particular, the cusps reflect subdiffusion along the backbone and the main fingers, while normal diffusion along the z direction is reflected by the smooth Gaussian curve.

Let us now analyze the MSDs along all three directions,

$$\langle \zeta_i^2(t) \rangle = \int_{-\infty}^{\infty} \zeta_i^2 \, p_i(\zeta_i, t|x_{i,0}) \, d\zeta_i, \quad \zeta_i = \{x,y,z\}, \quad i = 1,2,3.$$

Taking into account the corresponding solutions for the marginal PDFs, we have

$$\langle x^2(t) \rangle = x_0^2 + 2\mathcal{D}_1 \mathcal{L}^{-1} \left[\frac{s^{-1}}{(s+r)^{1/4}} \right]$$

$$= x_0^2 + 2\mathcal{D}_1 t^{1/4} E_{1,5/4}^{1/4}(-rt)$$

$$= x_0^2 + r^{-1/4} \frac{\gamma(1/4, rt)}{\Gamma(1/4)},$$

$$= x_0^2 + 2\mathcal{D}_1 t^{1/4} \left[\frac{1}{\sqrt[4]{rt}} - \frac{\bar{E}_{3/4}(rt)}{\Gamma(1/4)} \right], \qquad (6.118)$$

see relation (1.180). Here

$$\bar{E}_n(z) = \int_1^\infty \frac{e^{-z\tau}}{\tau^n} d\tau, \qquad (6.119)$$

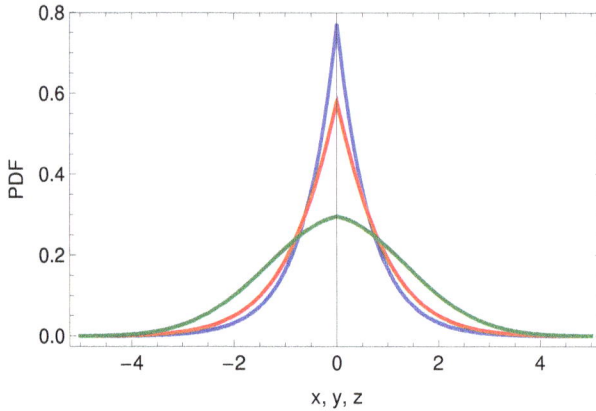

Fig. 6.6   Marginal PDFs for global resetting with $r = 0.05$ at $t = 1$, and reset position $x_0 = 0$. More specifically, we show $p_1(x, t|0)$ (blue line), $p_2(y, t|0)$ (red line) and $p_3(z, t|0)$ (green line) for $\mathcal{D}_x = \mathcal{D}_y = \mathcal{D}_z = 1$.

is the generalized exponential integral function (it is implemented in Wolfram Language as $\mathtt{ExpIntegralE[n, z]}$), which relates to the upper and lower incomplete gamma function as follows

$$\bar{E}_n(z) = z^{n-1}\Gamma(1 - n, z) = z^{n-1}\left[\Gamma(1 - n) - \gamma(1 - n, z)\right]. \qquad (6.120)$$

This corresponds to the transition from subdiffusion to localization

$$\langle x^2(t)\rangle \sim x_0^2 + 2\,\mathcal{D}_1 \begin{cases} \frac{t^{1/4}}{\Gamma(5/4)}, & rt \ll 1, \\ \frac{1}{\sqrt[4]{r}}, & rt \gg 1, \end{cases}$$

where for $rt \ll 1$ we use asymptotics of the associated three parameter M-L function (1.177), while for $rt \gg 1$ we use asymptotic formula (1.175). For the y fingers we have

$$\langle y^2(t)\rangle = 2\,\mathcal{D}_2\,\mathcal{L}^{-1}\left[\frac{s^{-1}}{(s + r)^{1/2}}\right]$$

$$= 2\,\mathcal{D}_2\,t^{1/2}E_{1,3/2}^{1/2}(-rt)$$

$$= 2\,\mathcal{D}_2\,\frac{\mathrm{erf}(\sqrt{rt})}{\sqrt{r}}, \qquad (6.121)$$

where we use relation (1.182). This MSD corresponds to the transition from subdiffusion to localization as well,

$$\langle y^2(t)\rangle \sim 2\,\mathcal{D}_2 \begin{cases} \frac{t^{1/2}}{\Gamma(3/2)}, & rt \ll 1, \\ \frac{1}{\sqrt{r}}, & rt \gg 1, \end{cases}$$

where we again use (1.177) for $rt \ll 1$ and (1.175) for $rt \gg 1$. Eventually, the MSD for the z fingers reads

$$\langle z^2(t) \rangle = 2\mathcal{D}_3 \mathcal{L}^{-1} \left[\frac{s^{-1}}{s+r} \right] = 2\mathcal{D}_3 \frac{1 - e^{-rt}}{r}, \tag{6.122}$$

that corresponds to saturation in the long time limit,

$$\langle z^2(t) \rangle \sim 2\mathcal{D}_3 \begin{cases} t, & rt \ll 1 \\ \frac{1}{r}, & rt \gg 1. \end{cases}$$

Therefore, unlike the initial transient behavior, all the MSDs saturate towards constant values (exhibiting stochastic localization) as in the scenario of one dimensional diffusion with resets. This confirms the existence of the NESS.

References

[1]   D. Campos and V. Méndez. "Phase transitions in optimal search times: How random walkers should combine resetting and flight scales". In: *Physical Review E* 92.6 (2015), p. 062115 (cit. on p. 141).

[2]   F. Bartumeus and J. Catalan. "Optimal search behavior and classic foraging theory". In: *Journal of Physics A: Mathematical and Theoretical* 42.43 (2009), p. 434002 (cit. on p. 141).

[3]   A. Pal, Ł. Kuśmierz, and S. Reuveni. "Search with home returns provides advantage under high uncertainty". In: *Physical Review Research* 2.4 (2020), p. 043174 (cit. on p. 141).

[4]   P. Visco et al. "Switching and growth for microbial populations in catastrophic responsive environments". In: *Biophysical Journal* 98.7 (2010), pp. 1099–1108 (cit. on p. 141).

[5]   S. Reuveni, M. Urbakh, and J. Klafter. "Role of substrate unbinding in Michaelis–Menten enzymatic reactions". In: *Proceedings of the National Academy of Sciences* 111.12 (2014), pp. 4391–4396 (cit. on p. 141).

[6]   W. J. Bell. *The behavioural ecology of finding resources*. 1991 (cit. on p. 141).

[7]   A. Pal, A. Kundu, and M. R. Evans. "Diffusion under time-dependent resetting". In: *Journal of Physics A: Mathematical and Theoretical* 49.22 (2016), p. 225001 (cit. on p. 141).

[8]   Ł. Kuśmierz and E. Gudowska-Nowak. "Subdiffusive continuous-time random walks with stochastic resetting". In: *Physical Review E* 99.5 (2019), p. 052116 (cit. on p. 141).

[9]   G. Tucci et al. "Controlling particle currents with evaporation and resetting from an interval". In: *Physical Review Research* 2.4 (2020), p. 043138 (cit. on p. 141).

[10] V. Stojkoski et al. "Autocorrelation functions and ergodicity in diffusion with stochastic resetting". In: *Journal of Physics A: MAthematical and Theoretical* 55 (2021), p. 104003 (cit. on pp. 141, 151).

[11] Deepak Vinod et al. "Nonergodicity of reset geometric Brownian motion". In: *Physical Review E* 105.1 (2022), p. L012106 (cit. on p. 141).

[12] V. Stojkoski et al. "Income inequality and mobility in geometric Brownian motion with stochastic resetting: theoretical results and empirical evidence of non-ergodicity". In: *Philosophical Transactions of the Royal Society A* 380.2224 (2022), p. 20210157 (cit. on p. 141).

[13] L. N. Christophorov. "Resetting random walks in one-dimensional lattices with sinks". In: *Journal of Physics A: Mathematical and Theoretical* 55.15 (2022), p. 155006 (cit. on p. 141).

[14] O. L. Bonomo and A. Pal. "First passage under restart for discrete space and time: Application to one-dimensional confined lattice random walks". In: *Physical Review E* 103.5 (2021), p. 052129 (cit. on p. 141).

[15] A. P Riascos et al. "Random walks on networks with stochastic resetting". In: *Physical Review E* 101.6 (2020), p. 062147 (cit. on p. 141).

[16] F. Huang and H. Chen. "Random walks on complex networks with first-passage resetting". In: *Physical Review E* 103.6 (2021), p. 062132 (cit. on p. 141).

[17] B. Mukherjee, K. Sengupta, and S. N. Majumdar. "Quantum dynamics with stochastic reset". In: *Physical Review B* 98.10 (2018), p. 104309 (cit. on p. 141).

[18] D. C. Rose et al. "Spectral properties of simple classical and quantum reset processes". In: *Physical Review E* 98.2 (2018), p. 022129 (cit. on p. 141).

[19] G. Perfetto et al. "Designing nonequilibrium states of quantum matter through stochastic resetting". In: *Physical Review B* 104.18 (2021), p. L180302 (cit. on p. 141).

[20] O. Tal-Friedman et al. "Experimental realization of diffusion with stochastic resetting". In: *Journal of Physical Chemistry Letters* 11.17 (2020), pp. 7350–7355 (cit. on p. 141).

[21] B. Besga et al. "Optimal mean first-passage time for a Brownian searcher subjected to resetting: experimental and theoretical results". In: *Physical Review Research* 2.3 (2020), p. 032029 (cit. on p. 141).

[22] S. Ray, D. Mondal, and S. Reuveni. "Péclet number governs transition to acceleratory restart in drift-diffusion". In: *Journal of Physics A: Mathematical and Theoretical* 52.25 (2019), p. 255002 (cit. on p. 141).

[23] S. Ray and S. Reuveni. "Diffusion with resetting in a logarithmic potential". In: *Journal of Chemical Physics* 152.23 (2020), p. 234110 (cit. on p. 141).

[24] A. Masó-Puigdellosas, D. Campos, and V. Méndez. "Transport properties and first-arrival statistics of random motion with stochastic reset times". In: *Physical Review E* 99.1 (2019), p. 012141 (cit. on p. 141).

[25] S. Ahmad et al. "First passage of a particle in a potential under stochastic resetting: A vanishing transition of optimal resetting rate". In: *Physical Review E* 99.2 (2019), p. 022130 (cit. on p. 141).

[26] A. Pal. "Diffusion in a potential landscape with stochastic resetting". In: *Physical Review E* 91.1 (2015), p. 012113 (cit. on p. 141).

[27] R. K. Singh, R. Metzler, and T. Sandev. "Resetting dynamics in a confining potential". In: *Journal of Physics A: Mathematical and Theoretical* 53.50 (2020), p. 505003 (cit. on p. 141).

[28] M. R. Evans and S. N. Majumdar. "Diffusion with stochastic resetting". In: *Physical Review Letters* 106.16 (2011), p. 160601 (cit. on p. 141).

[29] M. R. Evans, S. N. Majumdar, and G. Schehr. "Stochastic resetting and applications". In: *Journal of Physics A: Mathematical and Theoretical* 53.19 (2020), p. 193001 (cit. on p. 141).

[30] A. A. Tateishi et al. "Quenched and annealed disorder mechanisms in comb models with fractional operators". In: *Physical Review E* 101.2 (2020), p. 022135 (cit. on p. 141).

[31] V. Domazetoski et al. "Stochastic resetting on comblike structures". In: *Physical Review Research* 2.3 (2020), p. 033027 (cit. on pp. 141, 152, 158).

[32] M. A. F. Dos Santos. "Comb model with non-static stochastic resetting and anomalous diffusion". In: *Fractal and Fractional* 4.2 (2020), p. 28 (cit. on pp. 141, 152).

[33]   T. Sandev et al. "Diffusion–Advection Equations on a Comb: Resetting and Random Search". In: *Mathematics* 9.3 (2021), p. 221 (cit. on pp. 141, 148, 150, 152).

[34]   R. K. Singh et al. "Backbone diffusion and first-passage dynamics in a comb structure with confining branches under stochastic resetting". In: *Journal of Physics A: Mathematical and Theoretical* 54.40 (2021), p. 404006 (cit. on pp. 141, 152).

[35]   S. N. Majumdar, S. Sabhapandit, and G. Schehr. "Dynamical transition in the temporal relaxation of stochastic processes under resetting". In: *Physical Review E* 91.5 (2015), p. 052131 (cit. on p. 148).

[36]   T. Sandev. "Generalized Langevin equation and the Prabhakar derivative". In: *Mathematics* 5.4 (2017), p. 66 (cit. on p. 151).

Chapter 7

Random search

Another example, where the Fox H-functions are extensively employed is search problems in topologically constrained areas, like combs, where a random search is due to a non-Markovian diffusion process. The first studies on random search have employed Brownian motion of a searcher as a default strategy, which is also considered as a part of a foraging theory with incomplete information [1]. Shlesinger and Klafter have also proposed Lévy flights [2] as an efficient strategy in searching for sufficiently sparse targets. In recent years, optimization of the search processes has been extensively discussed for random blind searches of sparse targets [3]. In this respect, various combinations of the Brownian search and Lévy flights have been introduced [4, 5, 6], to mention a few. We shall confine ourselves to considering random search processes on comb geometry, following our results [7, 8, 9, 10] on the study of impact of geometry on the search optimization, as the first passage problem [11, 12, 13, 14, 15, 16].

7.1 One dimensional search

We start our description of random search processes from the main characteristics of the one dimensional random search process [3, 17], which can be described by the one dimensional reaction-diffusion equation for the non-normalized density function $f(x,t)$ with a δ-sink of a strength $\mathcal{P}_{\mathrm{fa}}$

$$\frac{\partial}{\partial t} f(x,t) = \mathcal{D} \frac{\partial^2}{\partial x^2} f(x,t) - \mathcal{P}_{\mathrm{fa}}(t)\delta(x - X), \qquad (7.1)$$

where $\mathcal{D}$ is a diffusion coefficient. The δ-sink means that the random searcher positioned at the beginning at $x = x_0$ ($f(x, t = 0) = \delta(x - x_0)$), will be removed at the first arrival at $x = X$. That is $f(x = X, t) = 0$.

From Eq. (7.1) one concludes that $\mathcal{P}_{\mathrm{fa}}(t)$ represents the *first arrival time distribution* (FATD) [3, 17], i.e., the negative time derivative of the *survival probability* $\mathcal{S}(t)$, $\mathcal{S}(t) = \int_{-\infty}^{\infty} f(x, t)\, dx$,

$$\mathcal{P}_{\text{fa}}(t) = -\frac{d}{dt} \int_{-\infty}^{\infty} f(x, t)\, dx = -\frac{d}{dt} \mathcal{S}(t). \tag{7.2}$$

Following the definitions of the *search reliability* (the cumulative arrival probability) [3], we have

$$\mathcal{P} = \int_0^{\infty} \mathcal{P}_{\text{fa}}(t)\, dt = \hat{\mathcal{P}}_{\text{fa}}(s = 0), \tag{7.3}$$

where $\hat{\mathcal{P}}_{\text{fa}}(s) = \mathcal{L}\left[\mathcal{P}_{\text{fa}}(t)\right](s)$ is the Laplace image of $\mathcal{P}_{\text{fa}}(t)$, and the search *efficiency* is defined as the average over inverse search times,

$$\mathcal{E} = \left\langle \frac{1}{t} \right\rangle = \int_0^{\infty} \frac{\mathcal{P}_{\text{fa}}(t)}{t}\, dt. \tag{7.4}$$

Using the properties of the Laplace transform

$$\mathcal{L}\left[\frac{f(t)}{t}\right](s) = \int_0^{\infty} e^{-st} \frac{f(t)}{t}\, dt = \int_s^{\infty} \hat{f}(u)\, du, \tag{7.5}$$

one also finds the following expression for the search efficiency

$$\mathcal{E} = \int_0^{\infty} \hat{\mathcal{P}}_{\text{fa}}(s)\, ds. \tag{7.6}$$

By the Laplace-Fourier transformation of Eq. (7.1) one finds

$$s\tilde{f}(k, s) - e^{ikx_0} = -\mathcal{D}k^2 \tilde{f}(k, s) - \hat{\mathcal{P}}_{\text{fa}}(s)e^{ikX}, \tag{7.7}$$

from where it follows

$$\tilde{f}(k, s) = \frac{e^{ikx_0} - \hat{\mathcal{P}}_{\text{fa}}(s)e^{ikX}}{s + \mathcal{D}k^2}. \tag{7.8}$$

By the inverse Fourier transform we arrive at

$$\hat{f}(x, s) = \frac{s^{-1/2}}{2\sqrt{\mathcal{D}}} \left[e^{-s^{1/2} \frac{|x - x_0|}{\sqrt{\mathcal{D}}}} - \hat{\mathcal{P}}_{\text{fa}}(s)\, e^{-s^{1/2} \frac{|x - X|}{\sqrt{\mathcal{D}}}} \right]. \tag{7.9}$$

From the condition that the particle will be removed at the first arrival at $x = X$, which is $\hat{f}(x = X, s) = 0$, for the FATD one obtains

$$\hat{\mathcal{P}}_{\text{fa}}(s) = e^{-\frac{s^{1/2}|X - x_0|}{\sqrt{\mathcal{D}}}}, \tag{7.10}$$

while $f(x, t)$ in Laplace space becomes

$$\hat{f}(x, s) = \frac{s^{-1/2}}{2\sqrt{\mathcal{D}}} \left[e^{-\frac{s^{1/2}|x - x_0|}{\sqrt{\mathcal{D}}}} - e^{-\frac{s^{1/2}|x - X| + |X - x_0|}{\sqrt{\mathcal{D}}}} \right]. \tag{7.11}$$

The inverse Laplace transform of $\hat{\mathcal{P}}_{\text{fa}}(s)$ yields, see Example 1.2,

$$\mathcal{P}_{\text{fa}}(t) = \frac{|X - x_0|}{\sqrt{4\pi \mathcal{D}t^3}} e^{-\frac{(X - x_0)^2}{4\mathcal{D}t}}, \tag{7.12}$$

which is the Lévy-Smirnov distribution with the power-law decay $t^{-3/2}$ in the long time limit. The reliability equals unity, since $\mathcal{P} = \hat{\mathcal{P}}_{\text{fa}}(s = 0) = 1$, while the search efficiency is

$$\mathcal{E} = \int_0^{\infty} e^{-\frac{s^{1/2}}{\sqrt{\mathcal{D}}}|X - x_0|}\, ds = \frac{2\mathcal{D}}{(X - x_0)^2}. \tag{7.13}$$

7.1.1 *One dimensional Brownian search with drift*

The random Brownian search with the drift in the one dimension is described by the Fokker-Planck equation[1]

$$\frac{\partial}{\partial t} f(x,t) = \left[\mathcal{D} \frac{\partial^2}{\partial x^2} - V \frac{\partial}{\partial x} \right] f(x,t) - \mathcal{P}_{\text{fa}}(t) \delta(x - X), \qquad (7.14)$$

with the same initial condition as in Eq. (7.1). By means of the Fourier-Laplace transform and from the condition $f(x = X, s) = 0$, one finds the FATD in Laplace space

$$\hat{\mathcal{P}}_{\text{fa}}(s) = e^{\frac{V}{2\mathcal{D}}(X-x_0) - \sqrt{\frac{s}{\mathcal{D}} + \frac{V^2}{4\mathcal{D}^2}}|X-x_0|}. \qquad (7.15)$$

Then, the inverse Laplace transform yields the final form of the FATD

$$\mathcal{P}_{\text{fa}}(t) = \frac{|X - x_0|}{\sqrt{4\pi \mathcal{D} t^3}} e^{-\frac{(X-x_0-Vt)^2}{4\mathcal{D} t}}. \qquad (7.16)$$

The search reliability according to Eq. (7.3) is

$$P = \hat{\mathcal{P}}_{\text{fa}}(s = 0) = e^{\frac{V(X-x_0)}{2\mathcal{D}} - \frac{|V(X-x_0)|}{2\mathcal{D}}}$$

$$= \begin{cases} 1, & \text{for } V(X - x_0) > 0, \\ e^{-\frac{V(x_0-X)}{\mathcal{D}}}, & \text{for } V(X - x_0) < 0, \end{cases} \qquad (7.17)$$

while the search efficiency according to Eq. (7.4) has the form

$$\mathcal{E} = \frac{2\mathcal{D} + |V(X - x_0)|}{(X - x_0)^2} \times \begin{cases} 1, & \text{for } V(X - x_0) > 0, \\ e^{-\frac{V(x_0-X)}{\mathcal{D}}}, & \text{for } V(X - x_0) < 0. \end{cases} \qquad (7.18)$$

For $V = 0$, one recovers the known result for the random Brownian search (7.13). The FATD and the search efficiency are depicted in Fig. 7.1.

7.1.2 *One dimensional turbulent diffusion search*

Next, we consider the one dimensional turbulent diffusion search [9]

$$\frac{\partial}{\partial t} f(x,t) = \mathcal{D} \left(\frac{\partial}{\partial x} x \right)^2 f(x,t) - \mathcal{P}_{\text{fa}}(t) \delta(x - X), \qquad (7.19)$$

with the initial position at $x_0 > 0$, $f(x, t = 0) = \delta(x - x_0)$. From the δ-sink at $x = X > 0$, one has $f(x = X, t) = 0$. Performing the Laplace-Mellin transform of Eq. (7.19), we obtain

$$s \bar{\hat{f}}(q, s) - x_0^{q-1} = \mathcal{D} q^2 \bar{\hat{f}}(q, s) - \mathcal{P}_{\text{fa}}(s) X^{q-1}, \qquad (7.20)$$

---

[1] In this example, we follow Refs. [3, 4, 11].

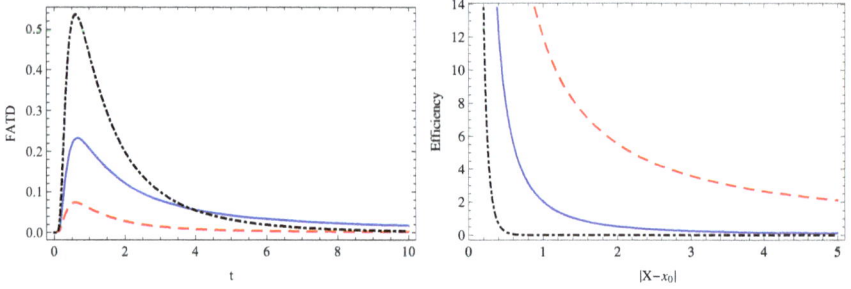

Fig. 7.1   Left panel: FATD (7.16) for $\mathcal{D} = 1$, $V = 0$ and $X - x_0 = 2$ (blue solid line), $V = 1$, $X - x_0 = -2$ (red dashed line) and $V = 1$, $X - x_0 = 2$ (black dot-dashed line). Right panel: efficiency (7.18), for $\mathcal{D} = 1$, $V = 0$ (blue solid line), $v = 10$, $X > x_0$ (red dashed line), and $v = 10$, $X < x_0$ (black dot-dashed line), see Ref. [10].

from where it follows

$$\bar{\tilde{f}}(q, s) = \frac{x_0^{q-1} - \hat{\mathcal{P}}_{\text{fa}}(s)\, X^{q-1}}{s - \mathcal{D}\, q^2}. \tag{7.21}$$

The inverse Laplace transform reads (see Eq. (1.35) and the Laplace transform formula (1.22) for convolution)

$$\bar{f}(q, t) = x_0^{q-1}\, e^{\mathcal{D}\, t\, q^2} - X^{q-1} \int_0^t \mathcal{P}_{\text{fa}}(t - t')\, e^{\mathcal{D}\, t'\, q^2}\, dt', \tag{7.22}$$

while the inverse Mellin transform (see Eq. (1.67) and the Mellin transform formula (1.63) for convolution) yields

$$f(x, t) = \frac{e^{-\frac{\log^2 \frac{x}{x_0}}{4\,\mathcal{D}\, t}}}{x\sqrt{4\pi\mathcal{D}t}} - \int_0^t \mathcal{P}_{\text{fa}}(t') \frac{e^{-\frac{\log^2 \frac{x}{X}}{4\,\mathcal{D}\,(t - t')}}}{x\sqrt{4\pi\mathcal{D}(t - t')}}\, dt'. \tag{7.23}$$

After the Laplace transform of Eq. (7.23), and using that $\hat{f}(x = X, s) = 0$, we obtain

$$\hat{\mathcal{P}}_{\text{fa}}(s) = e^{-\sqrt{\frac{s}{\mathcal{D}}}\left|\log \frac{X}{x_0}\right|}, \tag{7.24}$$

which yields the FATD in the form of the Lévy-Smirnov distribution

$$\mathcal{P}_{\text{fa}}(t) = \frac{\left|\log \frac{X}{x_0}\right|}{\sqrt{4\pi\mathcal{D}t^3}} \times e^{-\frac{\log^2 \frac{X}{x_0}}{4\,\mathcal{D}\, t}}. \tag{7.25}$$

In the long time limit, it scales as follows $\mathcal{P}_{\text{fa}}(t) \sim \frac{\log \frac{X}{x_0}}{\sqrt{4\pi\mathcal{D}}} \times t^{-3/2}$. Graphical representation of the FATD is given in Fig. 7.2.

The reliability equals unity, $\mathcal{P} = \hat{\mathcal{P}}_{\text{fa}}(s = 0) = 1$, while the efficiency becomes

$$\mathcal{E} = \int_0^\infty \hat{\mathcal{P}}_{\text{fa}}(s)\, ds = \frac{2\mathcal{D}}{\log^2 \frac{X}{x_0}}. \tag{7.26}$$

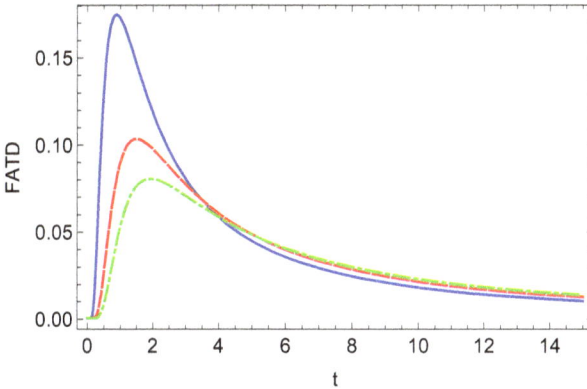

Fig. 7.2 Graphical representation of the FATD, Eq. (7.25), for $x_0 = 1$, $\mathcal{D} = 1$, and $X = 10$ (blue solid line), $X = 20$ (red dashed line), and $X = 30$ (green dot-dashed line). Reprinted figure with permission from T. Sandev, A. Iomin and L. Kocarev, Phys. Rev. E, 102, 042109 (2020). Copyright (2020) by the American Physical Society.

7.2 Random search on comb-like structures

7.2.1 *Brownian search with drift on comb*

In the section, we consider a Brownian random search process on the comb-like structure in the presence of an external drift along the backbone. The Fokker-Planck equation then reads

$$\frac{\partial}{\partial t} f(x,y,t) = \delta(y) \left[\mathcal{D}_x \frac{\partial^2}{\partial x^2} - v \frac{\partial}{\partial x} \right] f(x,y,t) + \mathcal{D}_y \frac{\partial^2}{\partial y^2} f(x,y,t)$$
$$- \wp_{\text{fa}}(t)\delta(x-X)\delta(y) \tag{7.27}$$

with the initial condition $f(x,y,t=0) = \delta(x-x_0)\delta(y)$. The δ-sink with strength $\wp_{\text{fa}}(t)$ means that the searcher is annihilated by reaching the point $(x,y) = (X,0)$. According to the definition in Eq. (7.2), the FATD results from the double integration[2]

$$\wp_{\text{fa}}(t) = -\frac{d}{dt} \int_{-\infty}^{\infty} \int_{-\infty}^{\infty} f(x,y,t)\,dx\,dy = -\frac{d}{dt}\mathcal{S}(t), \tag{7.28}$$

where $\mathcal{S}(t) = \int_{-\infty}^{\infty} \int_{-\infty}^{\infty} f(x,y,t)\,dx\,dy$ is the survival probability.

---

[2]We suggest another notation of the FATD for the 2D and 3D Brownian random search on the combs.

By the Laplace transform of Eq. (7.28), we find

$$s\hat{f}(x,y,s) - \delta(x-x_0)\delta(y) = \delta(y)\left[\mathcal{D}_x\frac{\partial^2}{\partial x^2} - v\frac{\partial}{\partial x}\right]\hat{f}(x,y,s)$$
$$+ \mathcal{D}_y\frac{\partial^2}{\partial y^2}\hat{f}(x,y,s) - \hat{\wp}_{\mathrm{fa}}(s)\delta(x-X)\delta(y).$$

$$(7.29)$$

Again, using the substitution

$$\hat{f}(x,y,s) = g(x,s)\,e^{-\sqrt{\frac{s}{\mathcal{D}_y}}|y|}$$

we introduce the marginal distribution

$$\hat{f}_1(x,s) = \int_{-\infty}^{\infty}\hat{f}(x,y,s)\,dy = 2\sqrt{\frac{\mathcal{D}_y}{s}}g(x,s).$$

Then, Eq. (7.29) reduces to

$$s\hat{f}_1(x,s) - \delta(x-x_0) = \frac{1}{2\sqrt{\mathcal{D}_y}}\,s\times s^{-1/2}\left[\mathcal{D}_x\frac{\partial^2}{\partial x^2} - v\frac{\partial}{\partial x}\right]\hat{f}_1(x,s)$$
$$- \hat{\wp}_{\mathrm{fa}}(s)\delta(x-X).$$

$$(7.30)$$

By the Fourier transform with respect to x, we find

$$s\tilde{\hat{f}}_1(k,s) - e^{\imath k x_0} = \frac{1}{2\sqrt{\mathcal{D}_y}}\,s\times s^{-1/2}\left[-\mathcal{D}_x k^2 - \imath v k\right]\tilde{\hat{f}}_1(k,s)$$
$$- \hat{\wp}_{\mathrm{fa}}(s)e^{\imath k X},$$

$$(7.31)$$

that yields

$$\tilde{\hat{f}}_1(k,s) = \frac{s^{-1/2}}{s^{1/2} + \frac{\mathcal{D}_x}{2\sqrt{\mathcal{D}_y}}k^2 + \imath\frac{v}{2\sqrt{\mathcal{D}_y}}k}\left[e^{\imath k x_0} - \hat{\wp}_{\mathrm{fa}}(s)\,e^{\imath k X}\right],\qquad(7.32)$$

which is the solution of the equation

$$\frac{\partial}{\partial t}f_1(x,t) = \frac{1}{2\sqrt{\mathcal{D}_y}}{}_{\mathrm{RL}}D_{0+}^{1/2}\left[\mathcal{D}_x\frac{\partial^2}{\partial x^2} - v\frac{\partial}{\partial x}\right]f_1(x,t) - \wp_{\mathrm{fa}}(t)\delta(x-X),$$

$$(7.33)$$

where ${}_{\mathrm{RL}}D_{0+}^{\mu}$ is the Riemann-Liouville fractional derivative (1.194). The inverse Fourier transform of Eq. (7.32) with respect to k gives

$$\hat{f}_1(x,s) = \frac{1}{2\sqrt{\mathcal{D}}}\frac{s^{-1/2}}{\left(s^{1/2}+\frac{V^2}{4\mathcal{D}}\right)^{1/2}}\left[e^{\frac{V}{2\mathcal{D}}(x-x_0)-\sqrt{\frac{s^{1/2}}{\mathcal{D}}+\frac{V^2}{4\mathcal{D}^2}}|x-x_0|}\right.$$
$$\left. - \hat{\wp}_{\mathrm{fa}}(s)\,e^{\frac{V}{2\mathcal{D}}(x-X)-\sqrt{\frac{s^{1/2}}{\mathcal{D}}+\frac{V^2}{4\mathcal{D}^2}}|x-X|}\right],\qquad(7.34)$$

where $\mathcal{D} = \frac{\mathcal{D}_x}{2\sqrt{\mathcal{D}_y}}$ and $V = \frac{v}{2\sqrt{\mathcal{D}_y}}$. From the condition

$$\hat{f}(x = X, y = 0, s) = g(x = X, s) = 0,$$

we obtain

$$\hat{f}_1(x = X, s) = \frac{2\sqrt{\mathcal{D}_y}}{s^{1/2}} g(x = X, s) = 0,$$

which means that $f_1(x = X, t) = 0$. Thus, in Laplace space, the FATD reads

$$\hat{\wp}_{\mathrm{fa}}(s) = e^{\frac{v}{2\mathcal{D}_x}(X - x_0) - \sqrt{\frac{2\sqrt{\mathcal{D}_y}}{\mathcal{D}_x} s^{1/2} + \frac{v^2}{4\mathcal{D}_x^2}} |X - x_0|}. \tag{7.35}$$

We also note that the FATD (7.35) can be directly obtained from the FATD for the one dimensional search with the drift, see Eq. (7.15). Then, we have

$$\hat{\wp}_{\mathrm{fa}}(s) = \hat{P}_{\mathrm{fa}}(s^{1/2}), \tag{7.36}$$

where we use $\mathcal{D} = \frac{\mathcal{D}_x}{2\sqrt{\mathcal{D}_y}}$ and $V = \frac{v}{2\sqrt{\mathcal{D}_y}}$. The search reliability becomes

$$P = \hat{\wp}_{\mathrm{fa}}(s = 0) = e^{\frac{v(X - x_0)}{2\mathcal{D}_x} - \frac{|v(X - x_0)|}{2\mathcal{D}_x}}$$

$$= \begin{cases} 1, & \text{for } v(X - x_0) > 0, \\ e^{-\frac{v(x_0 - X)}{\mathcal{D}_x}}, & \text{for } v(X - x_0) < 0, \end{cases} \tag{7.37}$$

while the efficiency is given by

$$\mathcal{E} = \frac{4\left(\frac{\mathcal{D}_x}{2\sqrt{\mathcal{D}_y}}\right)}{(X - x_0)^4} e^{\frac{v(X - x_0)}{2\mathcal{D}_x} - \frac{|v(X - x_0)|}{2\mathcal{D}_x}}$$

$$\times \left[6\frac{\mathcal{D}_x}{2\sqrt{\mathcal{D}_y}}\left(1 + \frac{|v(X - x_0)|}{2\mathcal{D}_x}\right) + \frac{v^2(X - x_0)^2}{2\sqrt{\mathcal{D}_y}} \right]$$

$$= \frac{4\left(\frac{\mathcal{D}_x}{2\sqrt{\mathcal{D}_y}}\right)}{(X - x_0)^4}$$

$$\times \begin{cases} 6\frac{\mathcal{D}_x}{2\sqrt{\mathcal{D}_y}}\left(1 + \frac{v(X - x_0)}{2\mathcal{D}_x}\right) + \frac{v^2(X - x_0)^2}{2\sqrt{\mathcal{D}_y}}, & \text{for } v(X - x_0) > 0, \\ e^{\frac{v(X - x_0)}{\mathcal{D}_x}}\left[6\frac{\mathcal{D}_x}{2\sqrt{\mathcal{D}_y}}\left(1 - \frac{v(X - x_0)}{2\mathcal{D}_x}\right) + \frac{v^2(X - x_0)^2}{2\sqrt{\mathcal{D}_y}}\right], & \text{for } v(X - x_0) < 0. \end{cases} \tag{7.38}$$

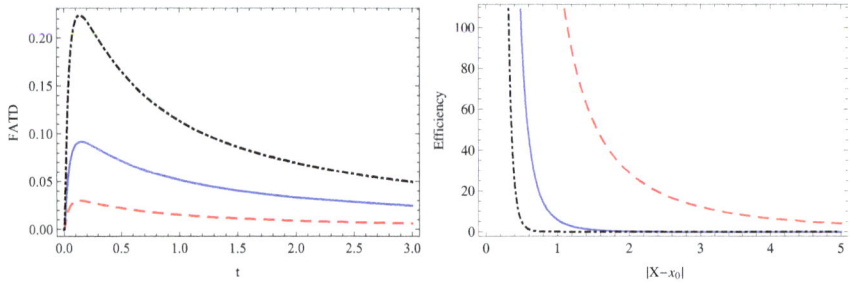

Fig. 7.3 Left panel: FATD (7.35) for $\mathcal{D}_x = 1$, $\mathcal{D}_y = 1$, $v = 0$ and $X - x_0 = 2$ (blue solid line), $v = 1$, $X - x_0 = -2$ (red dashed line) and $v = 1$, $X - x_0 = 2$ (black dot-dashed line). Right panel: efficiency (7.38), for $\mathcal{D}_x = 1$, $\mathcal{D}_y = 1$, $v = 0$ (blue solid line), $v = 10$, $X > x_0$ (red dashed line), and $v = 10$, $X < x_0$ (black dot-dashed line), see Ref. [10].

For $v = 0$ we recover the known result for the random search on comb with the efficiency $\mathcal{P} = 1$. The FATD image in Eq. (7.35) can be expressed by means of the Fox H-function

$$\wp_{\text{fa}}(s) = e^{-\sqrt{\frac{2\sqrt{\mathcal{D}_y}}{\mathcal{D}_x}} s^{1/4}|X-x_0|} = H_{0,1}^{1,0}\left[\sqrt{\frac{2\sqrt{\mathcal{D}_y}}{\mathcal{D}_x}} s^{1/4}|X - x_0| \,\middle|\, \begin{matrix} - \\ (0,1) \end{matrix}\right],$$
(7.39)

which by the inverse Laplace transform becomes

$$\wp_{\text{fa}}(t) = \frac{1}{t} H_{1,1}^{1,0}\left[\frac{|X - x_0|}{\sqrt{\frac{2\sqrt{\mathcal{D}_y}}{\mathcal{D}_x}} t^{1/4}} \,\middle|\, \begin{matrix} (0,1/4) \\ (0,1) \end{matrix}\right].$$
(7.40)

The long time behavior of the FATD is of the power-law form, $\wp_{\text{fa}}(t) \sim t^{-5/4}$, in contrast to the one dimensional case for which the FATD scales as $t^{-3/2}$. The reliability equals to one, while the efficiency reads

$$\mathcal{E} = \int_0^\infty e^{-\sqrt{\frac{2\sqrt{\mathcal{D}_y}}{\mathcal{D}_x}} s^{1/4}|X-x_0|}ds = \frac{24\left(\frac{2\sqrt{\mathcal{D}_y}}{\mathcal{D}_x}\right)^2}{|X - x_0|^4}.$$
(7.41)

The FATD and the efficiency for the random search with the drift on the comb are depicted in Fig. 7.3.

7.2.2   *Random search on three dimensional comb*

A random search process in the three dimensional comb is described by the equation

$$\frac{\partial}{\partial t}F(x,y,z,t) = \mathcal{D}_x\delta(y)\delta(z)\frac{\partial^2}{\partial x^2}F(x,y,z,t) + \mathcal{D}_y\delta(z)\frac{\partial^2}{\partial y^2}F(x,y,z,t)$$

$$+ \frac{\partial^2}{\partial z^2}F(x,y,z,t) - \wp_{\mathrm{fa}}(t)\delta(x)\delta(y)\delta(z). \qquad (7.42)$$

We use the initial condition $F(x,y,z,t=0) = \delta(x-x_0)\delta(y-y_0)\delta(z-z_0)$, and the δ-sink means that the searcher is annihilated by reaching the point $(x,y,z)=(0,0,0)$, i.e., $F(x=0,y=0,z=0,t)=0$. By the triple space integration of Eq. (7.42), we define the FATD,

$$\wp_{\mathrm{fa}}(t) = -\frac{d}{dt}\int_{-\infty}^{\infty}\int_{-\infty}^{\infty}\int_{-\infty}^{\infty} F(x,y,z,t)\,dx\,dy\,dz = -\frac{d}{dt}S(t). \qquad (7.43)$$

We can also calculate the *search reliability* and efficiency, as defined in Eqs. (7.3) and (7.6), respectively.

By means of the Fourier transforms with respect to x, y and z, and the Laplace transform with respect to time of Eq. (7.42), we find[3]

$$s\tilde{\hat{F}}(k_x,k_y,k_z,s) - e^{-\imath(k_x x_0 + k_y y_0 + k_z z_0)} = -\mathcal{D}_x k_x^2\,\tilde{\hat{F}}(k_x,y=0,z=0,s)$$

$$- \mathcal{D}_y k_y^2\,\tilde{\hat{F}}(k_x,k_y,z=0,s) - \mathcal{D}_z k_z^2\,\tilde{\hat{F}}(k_x,k_y,k_z,s) - \hat{\wp}_{\mathrm{fa}}(s). \qquad (7.44)$$

This yields

$$\tilde{\hat{F}}(k_x,k_y,k_z,s) = \frac{e^{-\imath(k_x x_0 + k_y y_0 + k_z z_0)} - \hat{\wp}_{\mathrm{fa}}(s)}{s + \mathcal{D}_z k_z^2} - \frac{\mathcal{D}_x k_x^2\,\tilde{\hat{F}}(k_x,y=0,z=0,s)}{s + \mathcal{D}_z k_z^2}$$

$$- \frac{\mathcal{D}_y k_y^2\,\tilde{\hat{F}}(k_x,k_y,z=0,s)}{s + \mathcal{D}_z k_z^2}. \qquad (7.45)$$

Then, performing the Fourier inversions and solving the corresponding equation at $x=y=z=0$, we obtain the FATD in Laplace space

$$\hat{\wp}_{\mathrm{fa}}(s) = \exp\left(-\sqrt{\frac{2\sqrt{2\mathcal{D}_y\sqrt{\mathcal{D}_z}}}{\mathcal{D}_x}}\,s^{1/8}|x_0| - \sqrt{\frac{2\sqrt{\mathcal{D}_z}}{\mathcal{D}_y}}\,s^{1/4}|y_0| - \frac{1}{\sqrt{\mathcal{D}_z}}\,s^{1/2}|z_0|\right). \qquad (7.46)$$

---

[3]Note that the notation $\tilde{\hat{F}}(k_x,k_y,k_z,s)$ is used for the Fourier-Laplace images with respect to x, y and z. However, in order to avoid complication with different notations, the same notation $\tilde{\hat{F}}$ is used also for $\tilde{\hat{F}}(k_x,y=0,z=0,s)$ and $\tilde{\hat{F}}(k_x,k_y,z=0,s)$ to define the Laplace transform with respect to t and the Fourier transform with respect to x and y, only.

Correspondingly, the inverse Laplace transformation yields the FATD in the form of the convolution integral

$$\wp_{\text{fa}}(t) = \int_0^t \wp_{\text{fa},x}(t - t')\wp_{\text{fa},yz}(t')\,dt', \tag{7.47}$$

where

$$\wp_{\text{fa},x}(t) = \mathcal{L}^{-1}\left[\exp\left(-\sqrt{\frac{2\sqrt{2\mathcal{D}_y\sqrt{\mathcal{D}_z}}}{\mathcal{D}_x}}\, s^{1/8}|x_0|\right)\right]$$

$$= \frac{1}{t}H_{1,1}^{1,0}\left[\sqrt{\frac{2\sqrt{2\mathcal{D}_y\sqrt{\mathcal{D}_z}}}{\mathcal{D}_x}}\,\frac{|x_0|}{t^{1/8}}\,\middle|\,\begin{matrix}(0,\frac{1}{8})\\(0,1)\end{matrix}\right], \tag{7.48}$$

and

$$\wp_{\text{fa},yz}(t) = \mathcal{L}^{-1}\left[\exp\left(-\sqrt{\frac{2\sqrt{\mathcal{D}_z}}{\mathcal{D}_y}}\,s^{1/4}|y_0| - \frac{1}{\sqrt{\mathcal{D}_z}}s^{1/2}|z_0|\right)\right]$$

$$= \int_0^t \frac{e^{-\frac{z_0^2}{4\mathcal{D}_z(t-t')}}}{\sqrt{4\pi\mathcal{D}_z(t-t')^3}}\,\frac{1}{t'}H_{1,1}^{1,0}\left[\sqrt{\frac{2\sqrt{\mathcal{D}_z}}{\mathcal{D}_y}}\,\frac{|y_0|}{t'^{1/4}}\,\middle|\,\begin{matrix}(0,\frac{1}{4})\\(0,1)\end{matrix}\right]\,dt'. \tag{7.49}$$

We also analyze the long time asymptotic behavior of the FATD $\wp_{\text{fa}}(t)$, by using the small s expansion of $\hat{\wp}_{\text{fa}}(s)$ in Eq. (7.46) and applying the Tauberian theorem, see Eqs. (1.68) and (1.69). Thus, we have

$$\hat{\wp}_{\text{fa}}(s) \sim \sum_{n=0}^4 \frac{(-1)^n}{n!}\left(as^{1/8}|x_0| + bs^{1/4}|y_0| + cs^{1/2}|z_0|\right)^n \tag{7.50}$$

Then, we find the small s behavior of $\frac{1}{s}\hat{\wp}_{\text{fa}}(s)$, which corresponds to the long time behavior of $\int^t \wp_{\text{fa}}(t)\,dt$, from where, by differentiation, we find

$$\wp_{\text{fa}}(t) \sim \frac{a|x_0|}{8}\frac{t^{-9/8}}{\Gamma(7/8)} + \left(b|y_0| - \frac{a^2 x_0^2}{2}\right)\frac{t^{-5/4}}{4\Gamma(3/4)} + \frac{3a^3|x_0|^3}{48}\frac{t^{-11/8}}{\Gamma(5/8)}$$

$$+ \left(c|z_0| - \frac{b^2 y_0^2}{2} - \frac{a^4 x_0^4}{24}\right)\frac{t^{-3/2}}{2\Gamma(1/2)} \sim \frac{a|x_0|}{8}\frac{t^{-9/8}}{\Gamma(7/8)}, \tag{7.51}$$

where $a = \sqrt{\frac{2\sqrt{2\mathcal{D}_y\sqrt{\mathcal{D}_z}}}{\mathcal{D}_x}}$, $b = \sqrt{\frac{2\sqrt{\mathcal{D}_z}}{\mathcal{D}_y}}$ and $c = \frac{1}{\sqrt{\mathcal{D}_z}}$. This result differs from the one dimensional case where the FATD has been found in the form $\mathcal{P}_{\text{fa}}(t) \sim t^{-3/2}$, as obtained from the Lévy-Smirnov distribution (7.12). It also differs from the two dimensional comb, where the FATD reads $\wp_{\text{fa}}(t) \sim t^{-5/4}$.

For the search reliability, from Eqs. (7.3) and (7.46), we conclude that

$$\mathcal{P} = \mathcal{P}_{\text{fa}}(s = 0) = 1.$$

It means that the searcher will find the target with probability 1. By integration of Eq. (7.46) we find the efficiency,

$$\mathcal{E} = \int_0^\infty e^{-\left(a|x_0|s^{1/8} + b|y_0|s^{1/4} + c|z_0|s^{1/2}\right)} ds.$$

By its definition, the main contribution to the efficiency is due to short and intermediate arrival times, which are dominated in Eq. (7.4). Here we note that for $y_0 = 0$ and $z_0 = 0$ the efficiency behaves as

$$\mathcal{E} \sim \int_0^\infty e^{-a|x_0|s^{1/8}} ds = \frac{\Gamma(9)}{a^8 |x_0|^8},$$

which differs from $\mathcal{E} \sim \frac{1}{|x_0|^4}$ for the two dimensional comb, and $\mathcal{E} \sim \frac{1}{|x_0|^2}$ for the one dimensional Brownian search.

7.2.3 Turbulent diffusion search on comb

For completeness of the analysis, we consider the turbulent diffusion search on the two dimensional comb governed by the Fokker-Planck equation [9]

$$\frac{\partial}{\partial t} F(x, y, t) = - \mathcal{D}_x \, \delta(y) \left(\frac{\partial}{\partial x} x\right)^2 F(x, y, t)$$

$$+ \mathcal{D}_y \frac{\partial^2}{\partial y^2} F(x, y, t) - \wp_{\text{fa}}(t) \, \delta(x - X) \, \delta(y), \qquad (7.52)$$

with the initial position $F(x, y, t = 0) = \delta(x - x_0)\delta(y)$, $x_0 > 0$, and δ-sink at $(X > 0, y = 0)$, $F(x = X, y = 0, t) = 0$. Performing the Laplace transform, and looking for the solution in the form $\hat{F}(x, y, s) = g(x, s) \, e^{-\sqrt{\frac{s}{\mathcal{D}}}|y|}$, we arrive at the corresponding Fokker-Planck equation along the backbone for the marginal distribution $f_1(x, t) = \int_{-\infty}^\infty F(x, y, t) \, dy$. The FFPE reads

$$\frac{\partial}{\partial t} f_1(x, t) = \frac{\mathcal{D}_x}{2\sqrt{\mathcal{D}_y}} \, {}_{\text{RL}}D_{0+}^{1/2} \left(\frac{\partial}{\partial x} x\right)^2 f_1(x, t) - \wp(t) \, \delta(x - X), \qquad (7.53)$$

where ${}_{\text{RL}}D_{a+}^\mu f(t)$ is the R-L fractional derivative (1.194). In the same way, as for the Brownian search on comb, we can show that the relation (7.36) is valid for the turbulent diffusion search on comb, as well. Therefore,

$$\hat{\wp}_{\text{fa}}(s) = \hat{\mathcal{P}}_{\text{fa}}(s^{1/2}) = e^{-\frac{s^{1/4}}{\sqrt{\mathcal{D}}} \left|\log \frac{X}{x_0}\right|}, \qquad (7.54)$$

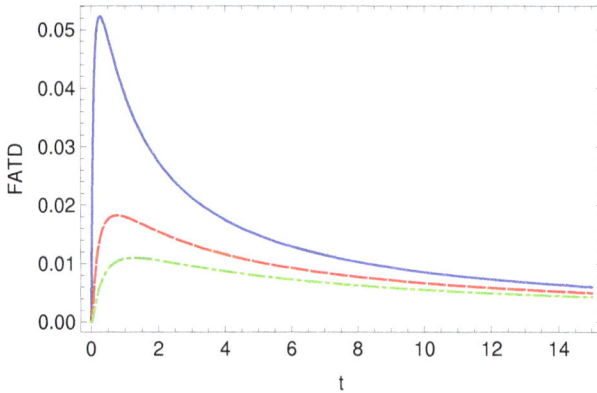

Fig. 7.4   Graphical representation of the FATD, Eq. (7.55), for $x_0 = 1$, $\mathcal{D}_x = 1$, $\mathcal{D}_y = 1$, and $X = 10$ (blue solid line), $X = 20$ (red dashed line), and $X = 30$ (green dot-dashed line). Reprinted figure with permission from T. Sandev, A. Iomin and L. Kocarev, Phys. Rev. E, 102, 042109 (2020). Copyright (2020) by the American Physical Society.

where $\mathcal{D} = \dfrac{\mathcal{D}_x}{2\sqrt{\mathcal{D}_y}}$ and $\hat{\mathcal{P}}_{\text{fa}}$ is the FATD (7.24) for the one dimensional turbulent diffusion search. Performing the Laplace inversion, we follow the details of Chapter 2 for the inverse Laplace transform of the Fox H-functions, we obtain the FATD again in terms of the Fox H-function,

$$\wp_{\text{fa}}(t) = \frac{2}{t} H_{1,1}^{1,0} \left[\frac{\log^2 \frac{X}{x_0}}{\mathcal{D}_1 \, t^{1/2}} \,\middle|\, \begin{array}{c} (0, 1/2) \\ (0, 2) \end{array} \right]. \tag{7.55}$$

The long time limit scales as follows

$$\wp_{\text{fa}}(t) \sim \frac{\left| \log \frac{X}{x_0} \right|}{4\,\Gamma(3/4)\sqrt{\mathcal{D}_1}} \times t^{-5/4}.$$

Graphical representation of the FATD (7.55) is given in Fig. 7.4.

The search reliability $\mathcal{P} = \wp_{\text{fa}}(s = 0) = 1$, indicates that the target will be found with the probability 1 with the search efficiency

$$\mathcal{E} = \frac{24\,\mathcal{D}_1^2}{\log^4 \frac{X}{x_0}}. \tag{7.56}$$

References

[1] D. W. Stephens and J. R. Krebs. *Foraging theory*. Princeton University Press, 2019 (cit. on p. 171).

[2] M. F. Shlesinger and J. Klafter. "Lévy walks versus Lévy flights". In: *On Growth and Form*. Springer, 1986, pp. 279–283 (cit. on p. 171).

[3] V. V. Palyulin, A. V. Chechkin, and R. Metzler. "Lévy flights do not always optimize random blind search for sparse targets". In: *Proceedings of the National Academy of Sciences* 111.8 (2014), pp. 2931–2936 (cit. on pp. 171–173).

[4] V. V. Palyulin, A. V. Chechkin, and R. Metzler. "Space-fractional Fokker–Planck equation and optimization of random search processes in the presence of an external bias". In: *Journal of Statistical Mechanics: Theory and Experiment* 2014.11 (2014), P11031 (cit. on pp. 171, 173).

[5] V. V. Palyulin et al. "Search reliability and search efficiency of combined Lévy–Brownian motion: long relocations mingled with thorough local exploration". In: *Journal of Physics A: Mathematical and Theoretical* 49.39 (2016), p. 394002 (cit. on p. 171).

[6] V. V. Palyulin et al. "Comparison of pure and combined search strategies for single and multiple targets". In: *European Physical Journal B* 90.9 (2017), pp. 1–16 (cit. on p. 171).

[7] T. Sandev, A. Iomin, and L. Kocarev. "Random search on comb". In: *Journal of Physics A: Mathematical and Theoretical* 52.46 (2019), p. 465001 (cit. on p. 171).

[8] E. K. Lenzi et al. "Anomalous diffusion and random search in xyz-comb: exact results". In: *Journal of Statistical Mechanics: Theory and Experiment* 2020.5 (2020), p. 053203 (cit. on p. 171).

[9] T. Sandev, A. Iomin, and L. Kocarev. "Hitting times in turbulent diffusion due to multiplicative noise". In: *Physical Review E* 102.4 (2020), p. 042109 (cit. on pp. 171, 173, 181).

[10] T. Sandev et al. "Diffusion–Advection Equations on a Comb: Resetting and Random Search". In: *Mathematics* 9.3 (2021), p. 221 (cit. on pp. 171, 174, 178).

[11] S. Redner. *A guide to first-passage processes*. Cambridge University Press, 2001 (cit. on pp. 171, 173).

[12] A. J. Bray, S. N. Majumdar, and G. Schehr. "Persistence and first-passage properties in nonequilibrium systems". In: *Advances in Physics* 62.3 (2013), pp. 225–361 (cit. on p. 171).

[13] M. R. Evans, S. N. Majumdar, and K. Mallick. "Optimal diffusive search: nonequilibrium resetting versus equilibrium dynamics". In: *Journal of Physics A: Mathematical and Theoretical* 46.18 (2013), p. 185001 (cit. on p. 171).

[14] M. R. Evans and S. N. Majumdar. "Diffusion with stochastic resetting". In: *Physical Review Letters* 106.16 (2011), p. 160601 (cit. on p. 171).

[15] M. C. Santos et al. "Optimization of random searches on defective lattice networks". In: *Physical Review E* 77.4 (2008), p. 041101 (cit. on p. 171).

[16]   R. Metzler, S. Redner, and G. Oshanin. *First-passage phenomena and their applications*. Vol. 35. World Scientific, 2014 (cit. on p. 171).

[17]   A. V. Chechkin et al. "First passage and arrival time densities for Lévy flights and the failure of the method of images". In: *Journal of Physics A: Mathematical and General* 36.41 (2003), p. L537 (cit. on p. 171).

Chapter 8

Diffusion on fractal tartan

The comb model, shown in Fig. 4.1, is one of the most simple structures, where anomalous diffusion can be realized due to the geometry. However, the comb geometry can be generalized. In particular, finite, or infinitely-uncountable number of backbones can be inside a finite-width strip. Such a comb structure is analogous to a grid and was named a grid comb [1]. Analogous fractal structure can be considered for the density of fingers-branches, as well. These fractional geometries lead to changing the transport exponent of subdiffusion [1, 2, 3]. More sophisticated (of higher fractal level) fractal structure in the form of orthogonally crossed Cantor sets of both backbones and fingers was named fractal mesh [3]. The latter structure, also known as a fractal tartan, is fabricated and employed as an ultrasonic array transducer [4], as well as used for description of anomalous transport through porous medium with various porous geometries [5, 6].

In this chapter we discuss transport properties of these various geometric structures (including orthogonally crossed), paying attention to the dimensionality of the medium, which is expressed in the form of a fractal dimension, and which plays an important role in the transport characteristics of random processes. This presentation is mainly based on our results [1, 2, 3, 7, 8] and it was also discussed in Ref. [9]. Here we mostly pay attention on the role of the Fox H-function and the related functions. We start our analysis with a grid comb, which is a simple generalization of the comb structure [1].

8.1   Grid comb model

In this section, we consider a generalization of Eq. (4.21) where anomalous diffusion along the x-direction may occur on many backbones, which can form infinite and fractal structures. We start our consideration with the simplest case of the finite number of backbones.

8.1.1   *Finite number of backbones*

Let us consider a comb with N backbones located at $y = l_j$, $j = 1, 2, \ldots, N$, $0 \le l_1 < l_2 < \cdots < l_N$, see Fig. 8.1. This means that we have a comb grid where N can be arbitrarily large, even infinity. The governing equation for such a structure is given by

$$\frac{\partial}{\partial t} P(x, y, t) = \mathcal{D}_x \sum_{j=1}^{N} w_j \delta(y - l_j) \frac{\partial^2}{\partial x^2} P(x, y, t) + \mathcal{D}_y \frac{\partial^2}{\partial y^2} P(x, y, t), \quad (8.1)$$

where w_j are structural constants such that $\sum_{j=1}^{N} w_j = 1$. The initial condition is given by

$$P(x, y, t = 0) = \delta(x)\delta(y), \quad (8.2)$$

and the boundary conditions for $P(x, y, t)$ and $\frac{\partial}{\partial q} P(x, y, t)$, $q = \{x, y\}$ are set to zero at infinity, $q = \pm\infty$. One can easily verify that for $l_1 = 0$, $w_1 = 1$, and $w_2 = w_3 = \cdots = w_N = 0$ Eq. (8.1) becomes (4.21). The physical dimensions of $\mathcal{D}_x$ and $\mathcal{D}_y$ for the finite number of backbones are the same as those in Eq. (4.21). The case of a fractal structure of backbones will be described by an appropriate generalization of Eq. (8.1). The motivation to introduce such model is to describe diffusion of solvents in thin porous films [10].

We apply the Laplace transform ($\mathcal{L}[f(t)] = \hat{f}(s)$) to Eq. (8.1), and then the Fourier transform with respect to x ($\mathcal{F}_x[f(x)] = \tilde{f}(\kappa_x)$) and y ($\mathcal{F}_y[f(y)] = \bar{f}(\kappa_y)$) variables. Thus, we obtain

$$\bar{\hat{\tilde{P}}}(\kappa_x, \kappa_y, s) = \frac{\bar{\tilde{P}}(\kappa_x, \kappa_y, t = 0)}{s + \mathcal{D}_y \kappa_y^2}$$
$$- \frac{\sum_{j=1}^{N} w_j \hat{\tilde{P}}(\kappa_x, y = l_j, s) \exp(i\kappa_y l_j)}{s + \mathcal{D}_y \kappa_y^2} \mathcal{D}_x \kappa_x^2, \quad (8.3)$$

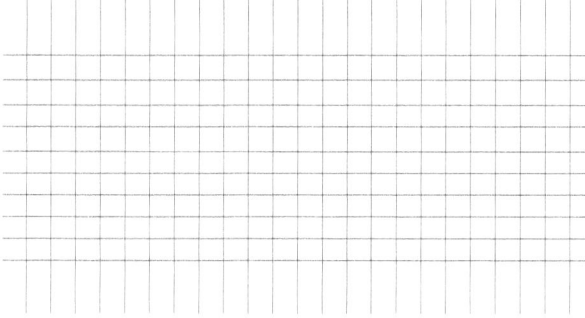

Fig. 8.1 Grid comb. It consists of a finite number of backbones and continuously distributed fingers along the x direction.

where $\tilde{\tilde{P}}(\kappa_x, \kappa_y, t = 0) = 1$. From relation (8.3), the inverse Fourier transform with respect to κ_y yields

$$\hat{\tilde{P}}(\kappa_x, y, s) = \frac{\exp\left(-\sqrt{\frac{s}{\mathcal{D}_y}}|y|\right)}{2\sqrt{\mathcal{D}_y}s^{1/2}}$$
$$-\frac{\mathcal{D}_x\kappa_x^2\sum_{j=1}^{N}w_j\hat{\tilde{P}}(\kappa_x, y = l_j, s)\exp\left(-\sqrt{\frac{s}{\mathcal{D}_y}}|y - l_j|\right)}{2\sqrt{\mathcal{D}_y}s^{1/2}}.$$

$$(8.4)$$

Let us estimate the MSD according to Eq. (8.4). To this end following the standard procedure to study the marginal PDF, we integrate Eq. (8.1) with respect to y. Then performing the Laplace transform with respect to time t, and the Fourier transform with respect to x, one obtains

$$\hat{\tilde{p}}_1(\kappa_x, s) = \frac{1}{s}\left[1 - \mathcal{D}_x\kappa_x^2\sum_{j=1}^{N}w_j\hat{\tilde{P}}(\kappa_x, y = l_j, s)\right]. \qquad (8.5)$$

Then taking into account Eq. (8.4), the MSD according to Eq. (3.7) reads

$$\langle x^2(t) \rangle = \mathcal{L}^{-1} \left[-\frac{\partial^2}{\partial \kappa_x^2} \hat{\bar{p}}_1 (\kappa_x, s) \right] \bigg|_{\kappa_x = 0}$$

$$= \frac{\mathcal{D}_x}{\sqrt{\mathcal{D}_y}} \mathcal{L}^{-1} \left[s^{-3/2} \sum_{j=1}^{N} w_j e^{-\sqrt{\frac{s}{\mathcal{D}_y}} |l_j|} \right]$$

$$= \frac{\mathcal{D}_x}{\sqrt{\mathcal{D}_y}} \sum_{j=1}^{N} w_j \left[\frac{2}{\sqrt{\pi}} t^{1/2} e^{-\frac{|l_j|^2}{4\mathcal{D}_y t}} - \frac{|l_j|}{\sqrt{\mathcal{D}_y}} \text{erfc} \left(\frac{|l_j|}{\sqrt{4\mathcal{D}_y t}} \right) \right], \quad (8.6)$$

where $\text{erfc}(x)$ is the complementary error function (1.127).

For $l_1 = 0$, it follows

$$\langle x^2(t) \rangle = \frac{2 w_1 \mathcal{D}_x}{\sqrt{\mathcal{D}_y}} \frac{t^{\frac{1}{2}}}{\Gamma \left(\frac{1}{2} \right)} + \frac{\mathcal{D}_x}{\sqrt{\mathcal{D}_y}}$$

$$\times \sum_{j=2}^{N} w_j \left[\frac{2}{\sqrt{\pi}} t^{1/2} e^{-\frac{|l_j|^2}{4\mathcal{D}_y t}} - \frac{|l_j|}{\sqrt{\mathcal{D}_y}} \text{erfc} \left(\frac{|l_j|}{\sqrt{4\mathcal{D}_y t}} \right) \right]. \quad (8.7)$$

For the long time scale when $\frac{|l_j|}{\sqrt{\mathcal{D}_y t}} \ll 1$, $j = 2, 3, \ldots, N$, the MSD reads

$$\langle x^2(t) \rangle = 2 \sum_{j=1}^{N} w_j \mathcal{D}_x \frac{1}{\sqrt{\mathcal{D}_y \pi}} t^{\frac{1}{2}}, \quad (8.8)$$

which means that all backbones contribute to the MSD. In contrast to this, on a short time scale, when $\frac{|l_j|}{\sqrt{\mathcal{D}_y t}} \gg 1$, $j = 2, 3, \ldots, N$, one finds that the main contribution to the MSD is due to the first backbone at $y = 0$, i.e.,

$$\langle x^2(t) \rangle \simeq 2 w_1 \mathcal{D}_x \frac{1}{\sqrt{\mathcal{D}_y \pi}} t^{\frac{1}{2}}. \quad (8.9)$$

This result is expected since for short times the particles move mainly in the first backbone because they had not enough time to reach the other ones by diffusion in the y direction. This can be easily verified by considering diffusion along the y-direction. We analyze the marginal PDF

$$p_2(y, t) = \int_{-\infty}^{\infty} P(x, y, t) \, dx, \quad (8.10)$$

for which we find that

$$\hat{\bar{p}}_2(\kappa_y, s) = \frac{1}{s + \mathcal{D}_y \kappa_y^2}, \quad (8.11)$$

i.e.,

$$p_2(y, t) = \frac{1}{\sqrt{4\pi \mathcal{D}_y t}} \exp\left(-\frac{y^2}{4\mathcal{D}_y t}\right). \tag{8.12}$$

For the MSD along the y-direction one finds a linear dependence on time $\langle y^2(t) \rangle = 2\mathcal{D}_y t$, i.e., normal diffusion along the y-direction. Therefore, the probability to find the particle at the first backbone ($l_1 = 0$) is

$$p_{2,1}(y, t) = \frac{1}{\sqrt{4\pi \mathcal{D}_y t}}, \tag{8.13}$$

while at the second backbone it is

$$p_{2,2}(y, t) = \frac{1}{\sqrt{4\pi \mathcal{D}_y t}} \exp\left(-\frac{l_2^2}{4\mathcal{D}_y t}\right) \tag{8.14}$$

and so on. Since for the short time scales, $p_{2,1}(y, t) \gg p_{2,2}(y, t) \gg ...$, we conclude that the main contribution to the MSD for short times is due to the displacements in the first backbone.

From relation (8.7) for $w_1 = 1$, $w_2 = w_3 = \cdots = w_N = 0$, and $l_1 = 0$, we obtain the MSD for the comb-like model (4.21)

$$\langle x^2(t) \rangle = \frac{2\mathcal{D}_x}{\sqrt{\mathcal{D}_y}} \frac{t^{1/2}}{\Gamma(1/2)}. \tag{8.15}$$

Graphical representation of the obtained results is depicted in Fig. 8.2, where the MSD is estimated for both two backbones and five backbones. It is assumed that the first backbone is at $y = 0$ and all the other backbones are at distances equal to L, $2L$, $3L$, $4L$.

From relations (8.8) and (8.9) we conclude that any finite number of backbones does not change the transport exponent in the short and long time limits. In the intermediate time scale there is more complicated behavior of the MSD given by relation (8.7). The cross-over time scales separating the behavior at short, intermediate, and long times are given by $t_{short} = \min\{l_j^2, j > 1\}/2\mathcal{D}_y = l_2^2/2\mathcal{D}_y$ and $t_{long} = \max\{l_j^2\}/2\mathcal{D}_y = l_N^2/2\mathcal{D}_y$.

In the presence of a constant external force F along the backbones we arrive at the following Fokker-Planck equation

$$\frac{\partial}{\partial t} P(x, y, t) = \sum_{j=1}^{N} w_j \delta(y - l_j) \left[-\eta_m F \frac{\partial}{\partial x} + \mathcal{D}_x \frac{\partial^2}{\partial x^2} \right] P(x, y, t)$$

$$+ \mathcal{D}_y \frac{\partial^2}{\partial y^2} P(x, y, t), \tag{8.16}$$

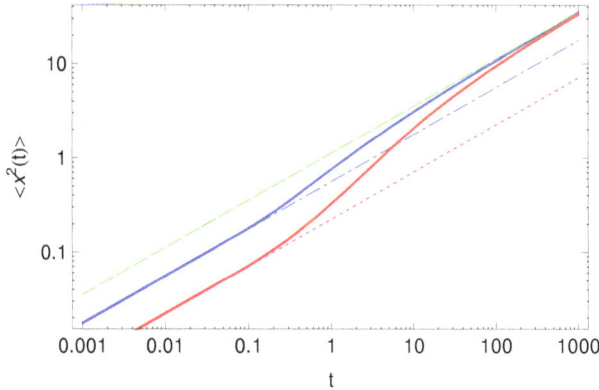

Fig. 8.2 MSD (8.7) on log-log scale. The blue solid line (upper solid line) corresponds to the MSD in the case of two backbones, $l_1 = 0$, $l_2 = L = 1$, and $w_1 = w_2 = 1/2$. The blue dot-dashed line describes the asymptotic behavior of the MSD for short times, given by (8.9). The red solid line (lower solid line) corresponds to the MSD in case of five backbones, $l_j = (j-1)L$, $j = 1, 2, \ldots, 5$, $L = 1$, $w_j = 1/5$. The red dotted line corresponds to its asymptotic behavior for short times, given by (8.9). The MSDs in both cases have the same asymptotic in the long time limit given by (8.8) (green dashed line). The diffusion coefficients are $\mathcal{D}_x = \mathcal{D}_y = 1$. Reprinted figure with permission from T. Sandev, A. Iomin and H. Kantz, Phys. Rev. E, 91, 032108 (2015). Copyright (2015) by the American Physical Society.

where η_m is the mobility. One can compute the first moment as a function of time

$$\langle x(t) \rangle_F = \frac{\eta_m F}{2\sqrt{\mathcal{D}_y}} \mathcal{L}^{-1} \left[s^{-3/2} \sum_{j=1}^{N} w_j e^{-\sqrt{\frac{s}{\mathcal{D}_y}}|l_j|} \right]. \tag{8.17}$$

Comparing Eq. (8.17) with relation (8.6) we conclude that the generalized Einstein relation is fulfilled [11]

$$\langle x(t) \rangle_F = \frac{F}{2k_B T} \langle x^2(t) \rangle_{F=0}, \tag{8.18}$$

where $\eta_m k_B T = \mathcal{D}_x$.

8.1.2 *Fractal grid: infinite number of backbones*

To introduce a fractal structure of the backbones we go back to Eq. (8.1) and replace the summation over the discrete set of the backbones with

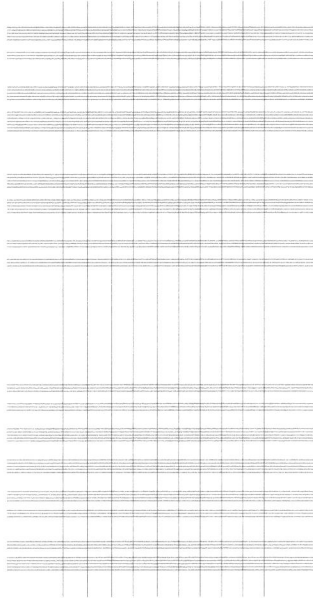

Fig. 8.3   Fractal grid comb. It consists of backbones with positions belonging to a fractal set along y direction, and continuously distributed fingers along x direction.

summation over a fractal set $\mathcal{S}_\nu$, i.e.,

$$\sum_{j=1}^{N} w_j \delta(y - l_j) \quad \rightarrow \quad \sum_{l_j \in \mathcal{S}_\nu} \delta(y - l_j). \tag{8.19}$$

That is, the backbones are at positions y belonging to the fractal set $\mathcal{S}_\nu$ with the fractal dimension $0 < \nu < 1$, see Fig. 8.3.

A simple example, which corresponds to an infinite fractal set, can be treated as follows. In relation (8.6), for the MSD, we have

$$\sum_{j=1}^{N} w_j e^{-\sqrt{\frac{s}{\mathcal{D}_y}}|l_j|} \rightarrow \sum_{l_j \in \mathcal{S}_\nu} e^{-\sqrt{\frac{s}{\mathcal{D}_y}}|l_j|}. \tag{8.20}$$

One should recognize that fractal sets (like a Cantor set) are uncountable. Therefore, the summation over the fractal set is formal and its mathematical realization corresponds to integration to fractal measure $\mu_\nu \sim l^\nu$ such that

$$\sum_{l_j \in \mathcal{S}_\nu} \delta(l - l_j) = \frac{1}{\Gamma(\nu)} l^{\nu - 1} \tag{8.21}$$

is the fractal density [12, 13], and $d\mu_\nu = \frac{1}{\Gamma(\nu)} l^{\nu-1} dl$. It is worth noting that $\mathcal{D}_x$ is a generalized diffusion coefficient with physical dimension $[\mathcal{D}_x] = \text{m}^{3-\nu} \text{s}^{-1}$ that absorbs dimension of the fractal volume/measure μ_ν. Eventually, one obtains

$$\sum_{l_j \in \mathcal{S}_\nu} e^{-\sqrt{\frac{s}{\mathcal{D}_y}}|l_j|} = \frac{1}{\Gamma(\nu)} \int_0^\infty l^{\nu-1} e^{-\sqrt{\frac{s}{\mathcal{D}_y}}l} \, dl = \left(\frac{\mathcal{D}_y}{s}\right)^{\nu/2}. \tag{8.22}$$

For the MSD, we obtain from Eq. (8.6)

$$\langle x^2(t) \rangle = \frac{\mathcal{D}_x}{\mathcal{D}_y^{\frac{1-\nu}{2}}} \frac{t^{\frac{1+\nu}{2}}}{\Gamma\left(1 + \frac{1+\nu}{2}\right)}, \tag{8.23}$$

That is, the transport exponent of anomalous diffusion now reads $\frac{1}{2} < \frac{1+\nu}{2} < 1$. Thus, the fractal set $\mathcal{S}_\nu$ of the backbones changes the transport exponent, from $1/2$ to $\frac{1+\nu}{2}$. For $\nu = 1$ the MSD becomes $\langle x^2(t) \rangle \simeq t$, which is consistent with normal diffusion, as expected, and for $\nu = 0$, the result reduces to the case of the finite number of backbones, where the fractal dimension of any finite number of discrete points is $\nu = 0$.

Let us consider a random fractal set $\mathcal{S}_\nu \in [a, b]$ embedded inside finite limits. From relation (8.6), the integration in Eq. (8.22), performed in the finite interval $[0, L]$, results in

$$\frac{1}{\Gamma(\nu)} \int_0^L l^{\nu-1} e^{-\sqrt{\frac{s}{\mathcal{D}_y}}l} \, dl = \left(\frac{\mathcal{D}_y}{s}\right)^{\nu/2} \frac{\gamma(\nu, L)}{\Gamma(\nu)}, \tag{8.24}$$

where $\gamma(a, x)$ is the lower incomplete gamma function (1.160). Thus, the MSD becomes

$$\langle x^2(t) \rangle = \frac{\mathcal{D}_x}{\mathcal{D}_y^{\frac{1-\nu}{2}}} \frac{\gamma(\nu, L)}{\Gamma(\nu)} \frac{t^{\frac{1+\nu}{2}}}{\Gamma\left(1 + \frac{1+\nu}{2}\right)}. \tag{8.25}$$

This result for the MSD (8.23) can be obtained in the framework of the fractional integration as well. Integrating Eq. (8.1) with respect to y and using the summation over the fractal set $\mathcal{S}_\nu$, one obtains the equation for the marginal PDF $p_1(x, t)$ as follows

$$\frac{\partial}{\partial t} p_1(x, t) = \mathcal{D}_x \sum_{l_j \in \mathcal{S}_\nu} \frac{\partial^2}{\partial x^2} P(x, y = l_j, t). \tag{8.26}$$

The Laplace transform of Eq. (8.26) yields

$$s\hat{p}_1(x, s) - p_1(x, t = 0) = \mathcal{D}_x \sum_{l_j \in \mathcal{S}_\nu} \frac{\partial^2}{\partial x^2} \hat{P}(x, y = l_j, s). \tag{8.27}$$

By representing the Laplace image of the solution $\hat{P}(x, y, s)$ in the form of the ansatz

$$\hat{P}(x, y, s) = \hat{g}(x, s)e^{-\sqrt{\frac{s}{\mathcal{D}_y}}|y|}, \tag{8.28}$$

i.e., $\hat{P}(x, y = l_j, s) = \hat{g}(x, s)e^{-\sqrt{\frac{s}{\mathcal{D}_y}}|l_j|}$, then the Laplace image of the marginal PDF $\hat{p}_1(x, s)$ reads

$$\hat{p}_1(x, s) = \int_{-\infty}^{\infty} \hat{P}(x, y, s)\, dy = 2\hat{g}(x, s)\sqrt{\frac{\mathcal{D}_y}{s}}. \tag{8.29}$$

Using summation (8.19) over the fractal set, we have

$$\sum_{l_j \in \mathcal{S}_\nu} \hat{P}(x, y = l_j, s) = \hat{g}(x, s)\frac{1}{\Gamma(\nu)}\int_0^{\infty} l^{\nu-1}e^{\sqrt{\frac{s}{\mathcal{D}_y}}l}\, dl$$

$$= \hat{g}(x, s)\left(\frac{\mathcal{D}_y}{s}\right)^{\nu/2} = \frac{1}{2\mathcal{D}_y^{\frac{1-\nu}{2}}}s^{\frac{1-\nu}{2}}\hat{p}_1(x, s). \tag{8.30}$$

By substituting relation (8.30) in Eq. (8.27), we obtain

$$s^{\frac{1+\nu}{2}}\hat{p}_1(x, s) - s^{\frac{1+\nu}{2}-1}p_1(x, t = 0) = \frac{\mathcal{D}_x}{2\mathcal{D}_y^{\frac{1-\nu}{2}}}\frac{\partial^2}{\partial x^2}\hat{p}_1(x, s). \tag{8.31}$$

Then performing the inverse Laplace transform, we obtain the following time fractional diffusion equation

$$_cD_{0+}^{\frac{1+\nu}{2}}p_1(x, t) = \frac{\mathcal{D}_x}{2\mathcal{D}_y^{\frac{1-\nu}{2}}}\frac{\partial^2}{\partial x^2}p_1(x, t), \tag{8.32}$$

where $_cD_{0+}^{\mu}$ is the Caputo time fractional derivative (1.196) of order $\frac{1}{2} < \mu = \frac{1+\nu}{2} < 1$. Then the MSD can be easily obtained from Eq. (8.23). The solution for the marginal PDF $p_1(x, t)$ can be represented in terms of the Fox H-function (2.17)

$$p_1(x, t) = \frac{1}{2|x|}H_{1,1}^{1,0}\left[\frac{|x|}{\sqrt{\mathcal{D}_\nu t^{(1+\nu)/2}}}\middle| \begin{array}{c}(1, (1+\nu)/4)\\(1, 1)\end{array}\right], \tag{8.33}$$

where $\mathcal{D}_\nu = \mathcal{D}_x/2\mathcal{D}_y^{\frac{1-\nu}{2}}$ is the generalized diffusion coefficient with physical dimension $[\mathcal{D}_\nu] = \mathrm{m}^2/\mathrm{s}^{(1+\nu)/2}$. Therefore, the fractal structure of the backbones changes the transport exponent, see Eq. (4.33) with the solution (4.35) for the comb structure.

The asymptotic behavior of the marginal PDF in Eq. (8.33) for $\frac{|x|}{\sqrt{\mathcal{D}_\nu t^{(1+\nu)/2}}} \gg 1$ is as follows

$$
p_1(x,t) \sim \frac{1}{\sqrt{2(3-\nu)\pi}} \left(\frac{1+\nu}{4}\right)^{\frac{\nu-1}{3-\nu}} |x|^{\frac{\nu-1}{3-\nu}} \left(\mathcal{D}_\nu t^{\frac{1+\nu}{2}}\right)^{-\frac{1}{3-\nu}}
$$

$$
\times \exp\left[-\frac{3-\nu}{4}\left(\frac{1+\nu}{4}\right)^{\frac{1+\nu}{3-\nu}} |x|^{\frac{4}{3-\nu}} \left(\mathcal{D}_\nu t^{\frac{1+\nu}{2}}\right)^{-\frac{2}{3-\nu}}\right], \quad (8.34)
$$

i.e., it has a non-Gaussian stretched exponential behavior, where we apply Eq. (2.39). For $\nu = 1$ it turns to the Gaussian behavior as expected.

Additionally to the MSD we can calculate the q-th moment, for which one finds

$$
\langle |x|^q \rangle = 2 \int_0^\infty x^q p_1(x,t)\,dx = \left(\mathcal{D}_\nu t^{\frac{1+\nu}{2}}\right)^{q/2} \frac{\Gamma(1+q)}{\Gamma\left(1+\frac{1+\nu}{2}\frac{q}{2}\right)}. \quad (8.35)
$$

For example, for the fourth moment it follows

$$
\langle |x|^4 \rangle = 24 \mathcal{D}_\nu^2 \frac{t^{1+\nu}}{\Gamma(2+\nu)} = 6 \frac{\mathcal{D}_x^2}{\mathcal{D}_y^{1-\nu}} \frac{t^{1+\nu}}{\Gamma(2+\nu)}. \quad (8.36)
$$

The calculation of the fourth moment is useful to discriminate subdiffusive processes with identical MSDs, e.g., subdiffusion due to different fractal structures or different mechanisms.[1] For the even moments we obtain

$$
\langle |x|^{2n} \rangle = (2n)! \frac{\mathcal{D}_\nu^n t^{\frac{(1+\nu)n}{2}}}{\Gamma\left(1+\frac{(1+\nu)n}{2}\right)}, \quad (8.37)
$$

from where we can find an interesting relation in the form of the one parameter M-L function

$$
\sum_{n=0}^\infty \frac{\langle |x|^{2n} \rangle}{(2n)!} = \sum_{n=0}^\infty \frac{\mathcal{D}_\nu^n t^{\frac{(1+\nu)n}{2}}}{\Gamma\left(1+\frac{(1+\nu)n}{2}\right)} = E_{(1+\nu)/2}\left(\mathcal{D}_\nu t^{\frac{1+\nu}{2}}\right). \quad (8.38)
$$

8.2   Fractal mesh

Generalizing the fractal grid model (8.1), we introduce a two dimensional current $\mathbf{j} = (j_x, j_y)$ along the fractal structures of both fingers and backbones, which reads

$$
j_x = -\mathcal{D}_x \sum_{l_j \in \mathcal{S}_\nu} \delta(y - l_j) \frac{\partial}{\partial x} P(x,y,t), \quad (8.39)
$$

$$
j_y = -\mathcal{D}_y \sum_{r_k \in \mathcal{S}_{\bar\nu}} \delta(x - r_k) \frac{\partial}{\partial y} P(x,y,t). \quad (8.40)
$$

---

[1]More details see in Refs. [14, 15].

The summations in Eqs. (8.39) and (8.40) are over the fractal sets $\mathcal{S}_\nu$ and $\mathcal{S}_{\bar\nu}$ with the fractal dimensions ν and $\bar\nu$, respectively.

Substituting this current in the Liouville equation,

$$\frac{\partial}{\partial t} P + \operatorname{div} \mathbf{j} = 0, \tag{8.41}$$

one obtains

$$\frac{\partial}{\partial t} P(x, y, t) = \mathcal{D}_x \sum_{l_j \in \mathcal{S}_\nu} \delta(y - l_j) \frac{\partial^2}{\partial x^2} P(x, y, t)$$

$$+ \mathcal{D}_y \sum_{r_k \in \mathcal{S}_{\bar\nu}} \delta(x - r_k) \frac{\partial^2}{\partial y^2} P(x, y, t). \tag{8.42}$$

This model represents a two dimensional grid structure, or *a fractal mesh* with infinite number of backbones and fingers at positions which belong to the fractal sets $\mathcal{S}_\nu$ and $\mathcal{S}_{\bar\nu}$, respectively. Here we note that $\mathcal{D}_x$ and $\mathcal{D}_y$ are generalized diffusion coefficients with physical dimensions $[\mathcal{D}_x] = \mathrm{m}^{2-\nu}/\mathrm{s}$ and $[\mathcal{D}_y] = \mathrm{m}^{2-\bar\nu}/\mathrm{s}$ that absorb the dimension of fractal volumes l^ν and $l^{\bar\nu}$, respectively [1].

Illustrations of fractal mesh construction is plotted in Fig. 8.4, where the fractal structure of fingers/backbones is a random form of a middle third Cantor set [16]. The algorithm of the construction is as follows. A given segment is randomly divided in three parts and the middle part is removed. Therefore, the first generation consists of two subsets. This middle third procedure is repeated for each subset to obtain the second generation with four random subsets of continuously distributed fingers, or backbones. Then, one obtains the third generation with eight random subsets, etc. In Fig. 8.4, we present a fractal structure of both backbones and fingers, which corresponds to the fractal mesh structure.[2]

While construction of the fractal set on the finite segment $[-L, L]$ is straightforward, construction of the fractal set on the infinite axis needs some care. Therefore, an algorithm of construction of a random third middle Cantor set on the infinite x axis has been suggested [3]. This algorithm of mapping of a random third middle Cantor set is only illustrative, however it should work for any fractal set. To this end we take into account that the power of the finite segment $[0, 1]$ is the same as the power of an infinite line. Figure 8.5 illustrates this algorithm, where we present the unit segment in the form of a circle by closing the end points (end points O on

---

[2]This structure is also known as a Cantor tartan. Interesting mathematical relations for functions constructed on this fractal structure can be found in Ref. [17].

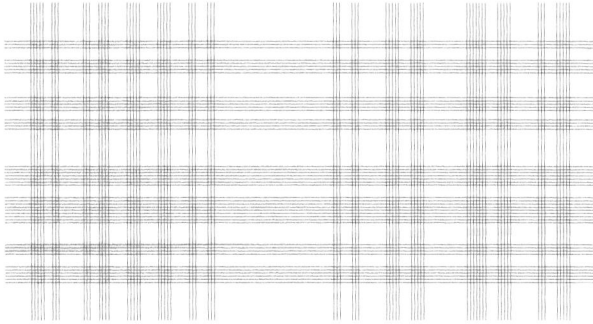

Fig. 8.4 A fractal mesh, where a random one third Cantor set of backbones (third generation of construction) can be distributed on either a finite strip, or entire y axis, while the fractal set of fingers is placed on the entire x axis with fifth generation of construction. Reprinted figure with permission from T. Sandev, A. Iomin and H. Kantz, Phys. Rev. E, 95, 052107 (2017). Copyright (2017) by the American Physical Society.

the top of the circle). Then we follow the previous procedure of a random one third Cantor set construction. We randomly divide the circle in three parts by points O, A_1 and B_1, and remove central segment A_1B_1. From point O we draw rays passing through A_1 and B_1, and intersect the horizontal line. Next, we divide each segment OA_1 and OB_1 randomly each in three parts by points A_2 and A_3, and B_2 and B_3, respectively, and remove central segments A_2A_3 and B_2B_3. In the same way, we draw rays from O passing through A_2, B_2, A_3 and B_3, which intersect the horizontal line. We follow this procedure of random division of the segments A_1A_2, B_1B_2, A_3O, B_3O in three parts, and we remove the middle segments. Then, we map the random middle third Cantor set of points on the unit circle onto the infinite axis by directing rays from O through the each point of the circle Cantor set to the x axis, producing a fractal set on the infinite real axis with the same fractal dimension [18]. Considering fingers on the horizontal line we construct a fractal set of fingers on the infinite x axis. It is worth noting that the ends O of the unit segment, belonging to the circle fractal set, also belong to the fractal set of the x axis since the intersections of the rays from O with the horizontal line are at infinities. Therefore, this projection also ensures the existence of the boundary conditions at infinities.

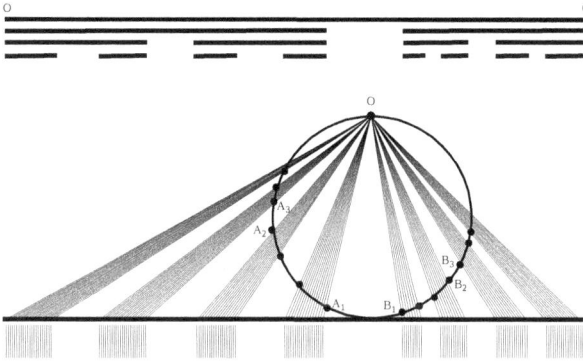

Fig. 8.5 Construction of a fractal grid of fingers on the infinite axis. The fractal set on the infinite axis is constructed by projecting the fractal set on a circle from point O, which belongs to the fractal set as well. Its projection on the x axis corresponds to $\pm\infty$, which ensures the boundary condition on infinities. Reprinted figure with permission from T. Sandev, A. Iomin and H. Kantz, Phys. Rev. E, 95, 052107 (2017). Copyright (2017) by the American Physical Society.

8.2.1 *From Weierstrass function to the power law*

Fractals can be described by means of the Weierstrass function [19, 20]. We rewrite the last term of Eq. (8.42), in the form

$$
\mathcal{D}_y \frac{\partial^2}{\partial y^2} \sum_{k=1}^{\infty} w_k \, \delta(x - r_k) P(x, y, t). \tag{8.43}
$$

It describes diffusion along the fingers located at $x = r_k$, $k = 1, 2, \ldots$, which belong to the fractal set $S_{\bar{\nu}}$ with the fractal dimension $0 < \bar{\nu} < 1$. Taking the structural constant in the form

$$
w_k = \frac{l - b}{b} \left(\frac{b}{l} \right)^k, \tag{8.44}
$$

where $l, b > 1$, $l - b \ll b$ (l and b are dimensionless scale parameters), we obtain that

$$
\sum_{k=1}^{\infty} w_k = \frac{l - b}{l} \sum_{k=0}^{\infty} \left(\frac{b}{l} \right)^k = 1. \tag{8.45}
$$

In the Fourier (κ_x) space, the last term of Eq. (8.42) (given by Eq. (8.43)) reads

$$\mathcal{D}_y \frac{\partial^2}{\partial y^2} \sum_{k=1}^{\infty} w_k e^{i\kappa_x r_k} P(x = r_k, y, t)$$

$$= \mathcal{D}_y \frac{\partial^2}{\partial y^2} \sum_{k=1}^{\infty} w_k e^{i\kappa_x r_k} \frac{1}{2\pi} \int_{-\infty}^{\infty} \tilde{P}(\kappa'_x, y, t) e^{-i\kappa'_x r_k} d\kappa'_x$$

$$= \mathcal{D}_y \frac{\partial^2}{\partial y^2} \frac{1}{2\pi} \int_{-\infty}^{\infty} \Psi(\kappa_x - \kappa'_x) \tilde{P}(\kappa'_x, y, t) d\kappa'_x. \tag{8.46}$$

Here we introduce a so called Weierstrass function (see *e.g.*, [20])

$$\Psi(z) = \frac{l - b}{b} \sum_{k=1}^{\infty} \left(\frac{b}{l}\right)^k \exp\left(i\frac{z}{l^k}\right), \tag{8.47}$$

where $r_k = L/l^k$, $z = (\kappa_x - \kappa_x') L$, and $L = 1$. One obtains it in the form of the scaling function

$$\Psi(z/l) = \frac{l}{b}\Psi(z) - \frac{l - b}{b} \exp\left(i\frac{z}{l}\right), \tag{8.48}$$

and by neglecting the last term ($l - b \ll b$), we arrive at the following scaling

$$\Psi(z/l) \simeq \frac{l}{b}\Psi(z). \tag{8.49}$$

This scaling corresponds to the approximation of the Weierstrass function by the power law form $\Psi(z) \sim \frac{1}{z^{1-\bar{\nu}}}$, where $\bar{\nu} = \log b / \log l$ and $0 < \bar{\nu} < 1$, is the fractal dimension [19, 20]. Eventually, relation (8.46) can be obtained in the convolution form

$$\mathcal{D}_y \frac{\partial^2}{\partial y^2} \frac{1}{2\pi} \int_{-\infty}^{\infty} \Psi(\kappa_x - \kappa'_x) \tilde{P}(\kappa'_x, y, t) d\kappa'_x$$

$$= \frac{\mathcal{D}_y}{2\pi} \frac{\partial^2}{\partial y^2} \int_{-\infty}^{\infty} \frac{\tilde{P}(\kappa'_x, y, t)}{|\kappa_x - \kappa'_x|^{1-\bar{\nu}}} d\kappa'_x, \tag{8.50}$$

which is the Riesz fractional integral (see Eq. (1.210) in the reciprocal Fourier space. It is worth noting the specific property of this construction of the fractal set of fingers. When $\bar{\nu} = 0$ the fractal dimension of fingers is one. Therefore, the fractal dimension of fingers in real space is $1 - \bar{\nu}$.

Performing the Fourier inversion, we arrive at the fractal mesh equation

$$\frac{\partial}{\partial t} P(x, y, t) = \mathcal{D}_x \sum_{l_j \in \mathcal{S}_\nu} \delta(y - l_j) \frac{\partial^2}{\partial x^2} P(x, y, t) + \mathcal{D}_y C_{\bar{\nu}} |x|^{-\bar{\nu}} \frac{\partial^2}{\partial y^2} P(x, y, t),$$

$$\tag{8.51}$$

where $C_{\bar{\nu}} = \Gamma(\bar{\nu})\cos\frac{\bar{\nu}\pi}{2}$, which appears from the definition of the Riesz fractional integral. By the Laplace transform, one finds

$$s\hat{P}(x,y,s) - P(x,y,t=0) = \mathcal{D}_x \sum_{l_j \in \mathcal{S}_\nu} \delta(y-l_j)\frac{\partial^2}{\partial x^2}P(x,y,s)$$

$$+ \mathcal{D}_y C_{\bar{\nu}}|x|^{-\bar{\nu}}\frac{\partial^2}{\partial y^2}\hat{P}(x,y,s). \qquad (8.52)$$

We look for the solution of Eq. (8.52) in the form of the ansatz

$$\hat{P}(x,y,s) = \hat{g}(x,s) \times \exp\left(-\sqrt{\frac{s}{\mathcal{D}_y C_{\bar{\nu}}}}|x|^{\bar{\nu}/2}|y|\right). \qquad (8.53)$$

Then, we calculate the marginal PDF along the backbones

$$p_1(x,t) = \int_{-\infty}^{\infty} P(x,y,t)\,dy. \qquad (8.54a)$$

Therefore, in Laplace space, we find

$$\hat{p}_1(x,s) = 2g(x,s)\sqrt{\frac{\mathcal{D}_y C_{\bar{\nu}}}{s}}|x|^{-\bar{\nu}/2}. \qquad (8.54b)$$

Integrating Eq. (8.52) over y, one obtains

$$s\hat{p}_1(x,s) - p_1(x,t=0) = \mathcal{D}_x\frac{\partial^2}{\partial x^2}\sum_{l_j \in \mathcal{S}_\nu}\hat{P}(x,y=l_j,s), \qquad (8.55)$$

where the initial condition $p_1(x,t=0) = \delta(x)$ results from Eq. (8.2). From Eq. (8.53) it follows

$$\hat{P}(x,y=l_j,s) = g(x,s)\exp\left(-\sqrt{\frac{s}{\mathcal{D}_y C_{\bar{\nu}}}}|x|^{\bar{\nu}/2}|l_j|\right). \qquad (8.56)$$

The summation in Eq. (8.55) is over the fractal set $\mathcal{S}_\nu$, which corresponds to integration over the fractal measure $\mu_\nu \sim l^\nu$. Therefore, according to Eq. (8.22), one finds

$$\sum_{l_j \in \mathcal{S}_\nu}\hat{P}(x,y=l_j,s) = g(x,s)\frac{1}{\Gamma(\nu)}\int_0^{\infty} l^{\nu-1}\exp\left(-\sqrt{\frac{s}{\mathcal{D}_y C_{\bar{\nu}}}}|x|^{\bar{\nu}/2}l\right)dl$$

$$= \hat{g}(x,s)\left(\frac{\mathcal{D}_y C_{\bar{\nu}}}{s|x|^{\bar{\nu}}}\right)^{\nu/2} = \frac{s^{(1-\nu)/2}|x|^{\bar{\nu}(1-\nu)/2}}{2\left(\mathcal{D}_y C_{\bar{\nu}}\right)^{(1-\nu)/2}}\hat{p}_1(x,s), \qquad (8.57)$$

where the finite result is obtained by means of Eq. (8.53). This result can be also obtained for the finite fractal set of backbones embedded inside a finite segment $[-L, L]$, such that integration in Eq. (8.57) is performed from 0 to L, which leads to the incomplete gamma function, see Eq. (8.24). Eventually taking the limit $L \to \infty$, we obtain the result in Eq. (8.57). Substituting result (8.57) in Eq. (8.55), one obtains

$$s^{-(1-\nu)/2} \left[s\hat{p}_1(x,s) - \delta(x) \right] = \frac{\mathcal{D}_x}{2 \left(\mathcal{D}_y C_{\bar{\nu}} \right)^{(1-\nu)/2}} \frac{\partial^2}{\partial x^2} \left(|x|^{\bar{\nu}(1-\nu)/2} \hat{p}_1(x,s) \right).$$

(8.58)

After the substitution $\hat{f}(x,s) = |x|^{\bar{\nu}(1-\nu)/2} \hat{p}_1(x,s)$, Eq. (8.58) becomes

$$s|x|^{-\bar{\nu}(1-\nu)/2} \hat{f}(x,s) - \frac{\mathcal{D}_x s^{(1-\nu)/2}}{2 \left(\mathcal{D}_y C_{\bar{\nu}} \right)^{(1-\nu)/2}} \frac{\partial^2}{\partial x^2} \hat{f}(x,s) = \delta(x), \qquad (8.59)$$

which is the Green's function equation [8]. First we consider the homogeneous part of the equation, which reads

$$\frac{2 \left(\mathcal{D}_y C_{\bar{\nu}} \right)^{(1-\nu)/2}}{\mathcal{D}_x} s^{(1+\nu)/2} |x|^{-\bar{\nu}(1-\nu)/2} \hat{G}(x,s) = \frac{\partial^2}{\partial x^2} \hat{G}(x,s). \qquad (8.60)$$

Equation (8.60) is the Lommel-type equation, and can be solved exactly, with the solution

$$\hat{G}(x,s) = \sqrt{x} K_{\frac{1}{\alpha}} \left(\frac{2 s^{\mu/2} x^{\alpha/2}}{\alpha} \sqrt{\frac{2 \left(\mathcal{D}_y C_{\bar{\nu}} \right)^{1-\mu}}{\mathcal{D}_x}} \right)$$

$$= \frac{\sqrt{x}}{2} H_{0,2}^{2,0} \left[\frac{s^\mu x^\alpha}{\alpha^2} \frac{2 \left(\mathcal{D}_y C_{\bar{\nu}} \right)^{1-\mu}}{\mathcal{D}_x} \middle| \begin{array}{c} - \\ (\frac{1}{2\alpha}, 1), (-\frac{1}{2\alpha}, 1) \end{array} \right], \qquad (8.61)$$

where

$$\alpha = \frac{4 - \bar{\nu}(1-\nu)}{2}, \qquad \mu = (1+\nu)/2,$$

$K_{1/\alpha}(z)$ is the modified Bessel function (of the third kind) and $H_{p,q}^{m,n}(z)$ is the Fox H-function (2.17).

Considering the inhomogeneous Lommel Eq. (8.59), we use the function $\hat{f}(|x|,s) = \mathcal{C}(s) \hat{G}(|x|,s) = \mathcal{C}(s) \hat{G}(y,s)$, where $\hat{G}(y,s)$ is obtained in Eq. (8.61), $y = |x|$, and $\mathcal{C}(s)$ is a function which depends on s. Thus, we find

$$-2 \left[\frac{\mathcal{D}_x}{2 \left(\mathcal{D}_y C_{\bar{\nu}} \right)^{1-\mu}} s^{1-\mu} \right] \frac{\partial}{\partial y} f(y=0,s) = 1. \qquad (8.62)$$

From this equation, by using series representation of the modified Bessel function (2.53), we find $\mathcal{C}(s)$, and by the inverse Laplace transform we

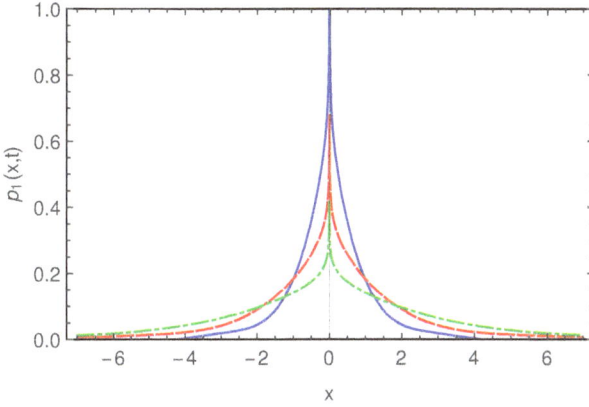

Fig. 8.6 PDF (8.63) for $\nu = \bar{\nu} = 1/2$, $\mathcal{D}_x = 1$, $\mathcal{D}_y = 1$, and $t = 1$ (solid blue line), $t = 5$ (red dashed line), $t = 20$ (green dot-dashed line). Reprinted figure with permission from T. Sandev, A. Iomin and H. Kantz, Phys. Rev. E, 95, 052107 (2017). Copyright (2017) by the American Physical Society.

finally obtain

$$p_1(x,t) = \frac{\left(\frac{2(\mathcal{D}_y C_{\bar{\nu}})^{1-\mu}}{\mathcal{D}_x}\right)^{1-\frac{1}{2\alpha}}}{2\alpha^{1-1/\alpha}\Gamma\left(1-1/\alpha\right)} \frac{|x|^{\alpha-3/2}}{t^{(1-\frac{1}{2\alpha})\mu}}$$

$$\times H_{1,2}^{2,0}\left[\frac{1}{\alpha^2}\frac{2\left(\mathcal{D}_y C_{\bar{\nu}}\right)^{1-\mu}}{\mathcal{D}_x}\frac{x^\alpha}{t^\mu} \middle| \begin{array}{c}\left(1-\mu+\frac{\mu}{2\alpha},\mu\right)\\ \left(\frac{1}{2\alpha},1\right),\left(-\frac{1}{2\alpha},1\right)\end{array}\right]. \quad (8.63)$$

Graphical representation of solution (8.63) is plotted in Fig. 8.6, where a heavy tailed behavior is explicitly shown, when the slop of the tail increases with time, together with the cusp at $x = 0$ it is a strong indication of the anomalous transport.

From Eq. (8.63), by using the Mellin transform of the Fox H-function, we calculate the MSD, which reads

$$\langle x^2(t)\rangle \simeq t^{2\mu/\alpha} = t^{\frac{2(1+\nu)}{4-\bar{\nu}(1-\nu)}}. \quad (8.64)$$

This corresponds to subdiffusion with the transport exponent $\frac{1}{2} < \frac{2(1+\nu)}{4-\bar{\nu}(1-\nu)} < 1$. It is larger than $1/2$, and one observes enhanced subdiffusion in comparison to the classical comb where the transport exponent is equal to $1/2$. This results from the fractal structure of the fingers and backbones. Graphical representation of the transport exponent $2\mu/\alpha$ is given in Fig. 8.7. It follows that for $\nu = \bar{\nu} = 0$, the transport exponent is equal to $1/2$. For $\nu = 0$, which corresponds to diffusion along the one

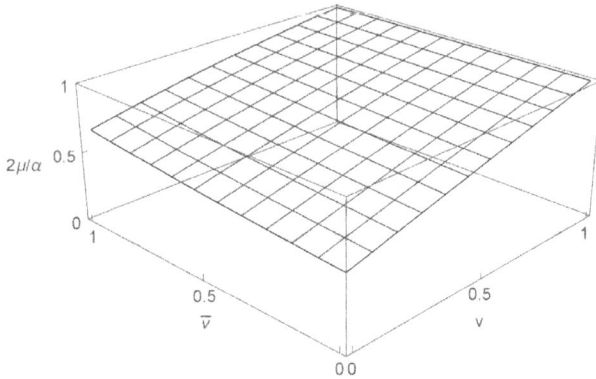

Fig. 8.7   The transport exponent $2\mu/\alpha$ vs ν and $\bar{\nu}$. Reprinted figure with permission from T. Sandev, A. Iomin and H. Kantz, Phys. Rev. E, 95, 052107 (2017). Copyright (2017) by the American Physical Society.

backbone only, the transport exponent is $\frac{2}{4-\bar{\nu}}$ and increases from $1/2$ to $2/3$ by increasing $\bar{\nu}$. For $\bar{\nu} = 0$, which means a continuous distribution of fingers, the transport exponent is $\frac{1+\nu}{2}$, and it increases from $1/2$ to 1 by increasing ν. For $\nu = 1$ (particle moves in a finite strip along the x axis) the transport exponent is equal to one for any value of $\bar{\nu}$, which corresponds to the two dimensional normal diffusion.

8.3   Fractal structure of fingers

A mathematical presentation of the fractal distribution of fingers in the fractal mesh model (8.42) can have a various realizations. In particular, it can be a realization of a convolution integral in real space [2]. It is convenient to present it by means of the inverse Fourier transform.[3] The correspondingly modified fractal mesh model reads

$$\frac{\partial}{\partial t}P(x,y,t) = \mathcal{D}_x \sum_{l_j \in \mathcal{S}_\nu} \delta(y - l_j)\frac{\partial^2}{\partial x^2}P(x,y,t)$$

$$+ \mathcal{D}_y \mathcal{F}_{\kappa_x}^{-1}\left[|\kappa_x|^{1-\bar{\nu}}\frac{\partial^2}{\partial y^2}\tilde{P}(\kappa_x,y,t)\right]. \qquad (8.65)$$

---

[3]The density of fingers is $\int dx\,\rho(x)$, where $\rho(x)$ is given by $\mathcal{F}[\rho(x)] = |\kappa_x|^{1-\bar{\nu}}$.

Applying the Laplace and Fourier transforms, one obtains Eq. (8.65) in the (κ_x, y, s) space

$$s\tilde{\hat{P}}(\kappa_x, y, s) - \delta(y) = -\mathcal{D}_x \kappa_x^2 \sum_{l_j \in S_\nu} \delta(y - l_j)\tilde{\hat{P}}(\kappa_x, y, s)$$

$$+ \mathcal{D}_y |\kappa_x|^{1-\bar{\nu}} \frac{\partial^2}{\partial y^2} \tilde{\hat{P}}(\kappa_x, y, s), \qquad (8.66)$$

where $\tilde{\hat{P}}(\kappa_x, y, t = 0) = \delta(y)$. By integrating over the y coordinate, we analyze the marginal PDF (8.54a) for the backbone dynamics. From Eq. (8.66) we find

$$\tilde{\hat{p}}_1(\kappa_x, s) = \frac{1}{s}\left[1 - \mathcal{D}_x \kappa_x^2 \sum_{l_j \in S_\nu} \tilde{\hat{P}}(\kappa_x, y = l_j, s)\right]. \qquad (8.67)$$

Presenting the image $\tilde{\hat{P}}(\kappa_x, y, s)$ in the form of the ansatz

$$\tilde{\hat{P}}(\kappa_x, y, s) = \tilde{\hat{g}}(\kappa_x, s)\exp\left(-\sqrt{\frac{s}{\mathcal{D}_y |\kappa_x|^{1-\bar{\nu}}}}|y|\right), \qquad (8.68)$$

and then integrating with respect to y, one obtains the marginal PDF as follows

$$\tilde{\hat{p}}_1(\kappa_x, s) = \int_{-\infty}^{\infty} \tilde{\hat{P}}(\kappa_x, y, s)\,dy = 2\tilde{\hat{g}}(\kappa_x, s)\sqrt{\frac{\mathcal{D}_y |\kappa_x|^{1-\bar{\nu}}}{s}}. \qquad (8.69)$$

The summation in Eq. (8.67) is over the fractal set, which corresponds to integration over the fractal measure $\mu_\nu \sim l^\nu$ $(d\mu_\nu = \frac{1}{\Gamma(\nu)}l^{\nu-1}dl)$. This yields

$$\sum_{l_j \in S_\nu} \tilde{\hat{P}}(\kappa_x, y = l_j, s) = \tilde{\hat{g}}(\kappa_x, s)\frac{1}{\Gamma(\nu)}\int_0^\infty l^{\nu-1}\exp\left(-\sqrt{\frac{s}{\mathcal{D}_y |\kappa_x|^{1-\bar{\nu}}}}l\right)dl$$

$$= \tilde{\hat{g}}(\kappa_x, s)\left(\frac{\mathcal{D}_y |\kappa_x|^{1-\bar{\nu}}}{s}\right)^{\nu/2}$$

$$= \frac{1}{2\mathcal{D}_y^{\frac{1-\nu}{2}}}\frac{s^{\frac{1-\nu}{2}}}{|\kappa_x|^{(1-\nu)(1-\bar{\nu})/2}}\tilde{\hat{p}}_1(\kappa_x, s), \qquad (8.70)$$

where solution (8.69) is taken into account. Substituting relation (8.70) in Eq. (8.67), we obtain

$$s^{\frac{-1+\nu}{2}}\left[s\tilde{\hat{p}}_1(\kappa_x, s) - \tilde{p}_1(\kappa_x, t = 0)\right] = -\frac{\mathcal{D}_x}{2\mathcal{D}_y^{\frac{1-\nu}{2}}}|\kappa_x|^\alpha \tilde{\hat{p}}_1(\kappa_x, s), \qquad (8.71)$$

where

$$\alpha = \frac{4 - (1 - \nu)(1 - \bar{\nu})}{2}.$$

The inverse Fourier and Laplace transforms yield the generalized diffusion equation

$$\int_0^t \gamma(t - t') \frac{\partial}{\partial t'} p_1(x, t') \, dt' = \frac{\mathcal{D}_x}{2\mathcal{D}_y^{\frac{1-\nu}{2}}} \frac{\partial^\alpha}{\partial |x|^\alpha} p_1(x, t),$$

$$(8.72)$$

with the space fractional Riesz derivative $\frac{\partial^\alpha}{\partial |x|^\alpha}$ (1.208), and the convolution kernel

$$\gamma(t) = \mathcal{L}^{-1}\left[s^{\frac{-1+\nu}{2}} \right] = \frac{t^{-\mu}}{\Gamma(1 - \mu)}, \qquad (8.73)$$

where $\mu = (1 + \nu)/2$.

Eventually, we obtain that the PDF $p_1(x, t)$ satisfies the following space-time fractional diffusion equation

$$_c D_{0+}^\mu p_1(x, t) = \frac{\mathcal{D}_x}{2\mathcal{D}_y^{\frac{1-\nu}{2}}} \frac{\partial^\alpha}{\partial |x|^\alpha} p_1(x, t), \qquad (8.74)$$

where $_c D_{0+}^\mu$ is the Caputo time fractional derivative of order $\mu = (1 + \nu)/2$, where $1/2 < \mu < 1$. The solution of Eq. (8.74) can be found in the form of the Fox H-function $H_{p,q}^{m,n}(z)$, where Eq. (3.112) coincides with Eq. (8.74), and arriving at the solution (3.121), we describe in great details this procedure of the solution. Therefore, the marginal PDF $p_1(x, t)$ reads

$$p_1(x, t) = \mathcal{F}^{-1}\left[E_\mu \left(-\mathcal{D}_{\mu,\alpha} t^\mu |\kappa_x|^\alpha \right) \right]$$

$$= \frac{1}{\alpha |x|} H_{3,3}^{2,1}\left[\frac{|x|}{(\mathcal{D}_{\mu,\alpha} t^\mu)^{\frac{1}{\alpha}}} \, \middle| \, \begin{matrix} (1, \frac{1}{\alpha}), (1, \frac{1+\nu}{2\alpha}), (1, \frac{1}{2}) \\ (1, 1), (1, \frac{1}{\alpha}), (1, \frac{1}{2}) \end{matrix} \right], \qquad (8.75)$$

where $\mathcal{D}_{\mu,\alpha} = \mathcal{D}_x / 2\mathcal{D}_y^{1-\mu}$ is the generalized diffusion coefficient.

Since the second moment does not exist, we can calculate the fractional moment

$$\langle |x|^q \rangle = \int_{-\infty}^\infty |x|^q p_1(x, t) \, dx, \quad 0 < q < \alpha < 2,$$

and then as the MSD we calculate the fractional moment $\langle |x|^q \rangle^{2/q}$. Taking into account that this calculation corresponds to the Mellin transform, as it is shown in Eq. (2.32), one obtains

$$\langle |x|^q \rangle^{\frac{2}{q}} = \left[\frac{2}{\alpha} (\mathcal{D}_{\mu,\alpha} t^\mu)^{\frac{q}{\alpha}} \frac{\Gamma(1 + q) \sin\left(\frac{q\pi}{2}\right)}{\Gamma\left(1 + \frac{\mu q}{\alpha}\right) \sin\left(\frac{q\pi}{\alpha}\right)} \right]^{\frac{2}{q}} \sim t^{\frac{2\mu}{\alpha}}, \qquad (8.76)$$

where $\frac{2\mu}{\alpha} = \frac{2(1+\nu)}{4-(1-\nu)(1-\bar{\nu})}$. Therefore, the fractal distribution of the backbones changes the transport exponent. When $\bar{\nu} = 1$, which corresponds to the continuous distribution of fingers and $\alpha = 2$, one arrives at the result for the fractal grid comb with MSD $\langle x^2(t) \rangle \sim t^{\frac{1+\nu}{2}}$. For $\nu = 0$ (one backbone), one finds $\langle |x|^q \rangle^{2/q} \sim t^{\frac{2}{3+\bar{\nu}}}$. We conclude here that the fractal structures of both backbones and fingers increase the transport exponent of subdiffusion along the fractal backbone structure.

8.4 Superdiffusion due to compensation kernel

In this section, we consider the fractal mesh model with a compensation memory kernel $\eta(t)$. The compensation kernel may be responsible for superdiffusion in the system with the fractal structure of fingers. In this case, the kernel compensates the trapping of particle in the fingers, for example as a result of a complex environment which accelerates the contaminant [8]. In particular, the random walks in the comb structure can be used in modeling of the RNA polymerase transcription, where the x-axis of backbones and the y-axis of fingers correspond to the active transcription and backtracking, respectively [21].

Therefore, we introduce the memory kernel inside the backbone dynamics of the fractal mesh model (8.42)

$$\frac{\partial}{\partial t}P(x,y,t) = \mathcal{D}_x \sum_{l_j \in \mathcal{S}_\nu} \delta(y - l_j) \int_0^t \eta(t-t') \frac{\partial^2}{\partial x^2}P(x,y,t')\,dt'$$

$$+ \mathcal{D}_y \sum_{r_k \in \mathcal{S}_\nu} \delta(x - r_k) \frac{\partial^2}{\partial y^2}P(x,y,t). \tag{8.77}$$

Employing the Laplace transform of Eq. (8.77), and using the separation ansatz (8.53), and integration of the equation with respect to y, we find that the marginal PDF $p_1(x,t)$ is governed by the following equation in Laplace space

$$s|x|^{-\bar{\nu}(1-\nu)/2}\hat{f}(x,s) - \frac{\mathcal{D}_x}{2\left(\mathcal{D}_y C_{\bar{\nu}}\right)^{(1-\nu)/2}}\hat{\eta}(s)s^{(1-\nu)/2}\frac{\partial^2}{\partial x^2}\hat{f}(x,s) = \delta(x),$$

$$\tag{8.78}$$

where $\hat{f}(x,s) = |x|^{\bar{\nu}(1-\nu)/2}\hat{p}_1(x,s)$. Let us use the compensation kernel in the form

$$\eta(t) = t^{-\mu}/\Gamma(1-\mu), \quad \mu = (1+\nu)/2, \tag{8.79}$$

which also accounts the fractal structure of the backbones reflected in the transport exponent μ. In order to solve Eq. (8.78) we first consider the homogeneous part of the equation, i.e.,

$$s|x|^{-\bar{\nu}(1-\nu)/2}\hat{G}(x,s) - \frac{\mathcal{D}_x}{2\,(\mathcal{D}_y C_{\bar{\nu}})^{(1-\nu)/2}}\hat{\eta}(s)s^{(1-\nu)/2}\frac{\partial^2}{\partial x^2}\hat{G}(x,s) = 0 \quad (8.80)$$

that repeats the analysis of Eqs. (8.59) and (8.60). Therefore, this is the Lommel-type equation with the solution

$$\hat{G}(x,s) = \sqrt{x}\,K_{1/\alpha}\left(\frac{2s^{1/2}x^{\alpha/2}}{\alpha}\sqrt{\frac{(2\mathcal{D}_y C_{\bar{\nu}})^{1-\mu}}{\mathcal{D}_x}}\right)$$

$$= \frac{\sqrt{x}}{2}H_{0,2}^{2,0}\left[\frac{sx^\alpha}{\alpha^2}\frac{(2\mathcal{D}_y C_{\bar{\nu}})^{1-\mu}}{\mathcal{D}_x}\;\middle|\;\begin{matrix}-\\(\frac{1}{2\alpha},1),(-\frac{1}{2\alpha},1)\end{matrix}\right], \quad (8.81)$$

where $\alpha = \frac{4-\bar{\nu}(1-\nu)}{2}$. Let us now use the function $\hat{f}(|x|,s) = C(s)\hat{G}(|x|,s) = C(s)\hat{G}(y,s)$, where $y = |x|$ and $C(s)$ is a function which depends on s. By substituting $\hat{f}(|x|,s)$ in Eq. (8.78) for $\hat{\eta}(s) = s^{\mu-1}$, we obtain the following equation

$$-2\left[\frac{\mathcal{D}_x}{(2\mathcal{D}_y C_{\bar{\nu}})^{1-\mu}}\right]\frac{\partial}{\partial y}\hat{f}(y=0,s) = 1, \quad (8.82)$$

from where we can find $C(s)$. Therefore, the solution for the marginal PDF reads

$$p_1(x,t) = \frac{\left(\frac{2(\mathcal{D}_y C_{\bar{\nu}})^{1-\mu}}{\mathcal{D}_x}\right)^{1-\frac{1}{2\alpha}}}{2\alpha^{1-1/\alpha}\Gamma\left(1-1/\alpha\right)}\frac{|x|^{\alpha-3/2}}{t^{1-\frac{1}{2\alpha}}}H_{1,0}^{0,1}\left[\frac{2\,(\mathcal{D}_y C_{\bar{\nu}})^{1-\mu}}{\alpha^2\mathcal{D}_x}\frac{|x|^\alpha}{t}\;\middle|\;\begin{matrix}-\\(-\frac{1}{2\alpha},1)\end{matrix}\right]$$

$$= \frac{\left(\frac{2(\mathcal{D}_y C_{\bar{\nu}})^{1-\mu}}{\mathcal{D}_x}\right)^{1-\frac{1}{2\alpha}}}{2\alpha^{2-1/\alpha}\Gamma\left(1-1/\alpha\right)}\frac{|x|^{\alpha-3/2}}{t^{1-\frac{1}{2\alpha}}}H_{1,0}^{0,1}\left[\left(\frac{2\,(\mathcal{D}_y C_{\bar{\nu}})^{1-\mu}}{\alpha^2\mathcal{D}_x}\right)^{1/\alpha}\frac{|x|^\alpha}{t}\;\middle|\;\begin{matrix}-\\(-\frac{1}{2\alpha},\frac{1}{\alpha})\end{matrix}\right]$$

$$= \frac{\left(\frac{2(\mathcal{D}_y C_{\bar{\nu}})^{1-\mu}}{\mathcal{D}_x}\right)^{1-\frac{1}{\alpha}}}{2\alpha^{1-2/\alpha}\Gamma\left(1-1/\alpha\right)}\frac{|x|^{\alpha-2}}{t^{1-\frac{1}{\alpha}}}H_{1,0}^{0,1}\left[\frac{2\,(\mathcal{D}_y C_{\bar{\nu}})^{1-\mu}}{\alpha^2\mathcal{D}_x}\frac{|x|^\alpha}{t}\;\middle|\;\begin{matrix}-\\(0,1)\end{matrix}\right]$$

$$= \frac{\left(\frac{2(\mathcal{D}_y C_{\bar{\nu}})^{1-\mu}}{\mathcal{D}_x}\right)^{1-\frac{1}{\alpha}}}{2\alpha^{1-2/\alpha}\Gamma\left(1-1/\alpha\right)}\frac{|x|^{\alpha-2}}{t^{1-\frac{1}{\alpha}}}\exp\left(-\frac{2\,(\mathcal{D}_y C_{\bar{\nu}})^{1-\mu}}{\alpha^2\mathcal{D}_x}\frac{|x|^\alpha}{t}\right), \quad (8.83)$$

where we applied the properties (2.25), (2.27) and (2.42), respectively. Graphical representation of the PDF (8.83) with the heavy tailed behavior is given in Fig. 8.8. This stretched exponential behavior ensures the existence of the finite MSD for superdiffusion.

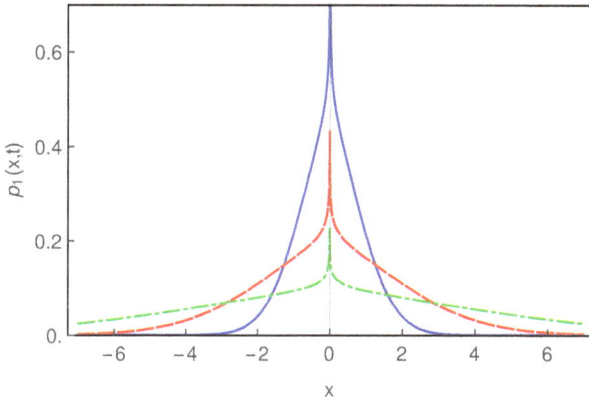

Fig. 8.8 PDF (8.83) for $\nu = \bar{\nu} = 1/2$, $\mathcal{D}_x = 1$, $\mathcal{D}_y = 1$, and $t = 1$ (solid blue line), $t = 5$ (red dashed line), $t = 20$ (green dot-dashed line). Reprinted figure with permission from T. Sandev, A. Iomin and H. Kantz, Phys. Rev. E, 95, 052107 (2017). Copyright (2017) by the American Physical Society.

From the marginal PDF (8.83), we calculate the MSD, which results to superdiffusion

$$\left\langle x^2(t) \right\rangle \simeq t^{\frac{4}{4-\bar{\nu}(1-\nu)}} \tag{8.84}$$

with the transport exponent $1 < \frac{4}{4-\bar{\nu}(1-\nu)}$ for any $0 < \nu < 1$ and $0 < \bar{\nu} < 1$. For $\bar{\nu} = 0$ (continuous distribution of fingers), normal diffusion takes place for any ν. This effect is due to the memory kernel which compensates the trapping effect of fingers. Obviously, that as the result of this compensation, superdiffusion takes place when $\bar{\nu} \neq 0$. The case with $\nu = 0$, which corresponds to the finite number of backbones, yields $\left\langle x^2 \right\rangle \sim t^{\frac{4}{4-\bar{\nu}}}$. Thus, the compensation kernel, and the fractal structure of both backbones and fingers are responsible for the appearance of superdiffusion.

In a similar way, the compensation kernel $\eta(t)$ can be considered for the fractal grid comb (8.65), which now reads

$$\frac{\partial}{\partial t} P(x,y,t) = \mathcal{D}_x \sum_{l_j \in \mathcal{S}_\nu} \delta(y - l_j) \int_0^t \eta(t-t') \frac{\partial^2}{\partial x^2} P(x,y,t') \, dt'$$

$$+ \mathcal{D}_y \mathcal{F}_{\kappa_x}^{-1} \left[|\kappa_x|^{1-\bar{\nu}} \frac{\partial^2}{\partial y^2} P(\kappa_x, y, t) \right]. \tag{8.85}$$

Let us consider $\eta(t) = t^{-(1+\nu)/2}/\Gamma\left(1 - \frac{1}{2} - \frac{\nu}{2}\right)$. Following the same procedure of solving Eq. (8.65), we find from Eq. (8.85) that the PDF $p_1(x,t)$ is governed by the space fractional diffusion equation

$$\frac{\partial}{\partial t}p_1(x,t) = \frac{\mathcal{D}_x}{2\mathcal{D}_y^{(1-\nu)/2}}\frac{\partial^\alpha}{\partial|x|^\alpha}p_1(x,t).$$ (8.86)

This equation is a particular case of Eq. (8.72) with $\mu = 1$, and its solution is given in the form of the Fox H-function (8.75)

$$p_1(x,t) = \mathcal{F}^{-1}\left[\exp\left(-\frac{\mathcal{D}_x}{2\mathcal{D}_y^{(1-\nu)/2}}t|\kappa_x|^\alpha\right)\right]$$

$$= \frac{1}{\alpha|x|}H_{3,3}^{2,1}\left[\frac{|x|}{\left(\frac{\mathcal{D}_x}{2\mathcal{D}_y^{(1-\nu)/2}}t\right)^{1/\alpha}}\left|\begin{array}{c}(1,\frac{1}{\alpha}),(1,\frac{1}{\alpha}),(1,\frac{1}{2})\\(1,1),(1,\frac{1}{\alpha}),(1,\frac{1}{2})\end{array}\right.\right].$$ (8.87)

In this case the q-th moment (8.76) reads

$$\langle|x|^q\rangle^{\frac{2}{q}} = \left[\frac{2}{\alpha}\left(\frac{\mathcal{D}_x}{2\mathcal{D}_y^{(1-\nu)/2}}\right)^{q/\alpha}\frac{t^{q/\alpha}}{\Gamma(1+q/\alpha)}\right]^{\frac{2}{q}} \sim t^{\frac{4}{4-(1-\nu)(1-\bar{\nu})}}.$$ (8.88)

Therefore, superdiffusion takes place for any $0 < \nu < 1$ and $0 < \bar{\nu} < 1$. This is a typical result for the Lévy distribution. For $\bar{\nu} = 1$ (continuous distribution of fingers), the normal diffusive behavior is obtained for any ν, which results from the compensation kernel effect. For a finite number of backbones, one obtains $\langle|x|^q\rangle^{2/q} \sim t^{\frac{4}{3+\bar{\nu}}}$. Therefore, the compensation kernel, and the fractal structure of both backbones and fingers are responsible for superdiffusion.

References

[1] T. Sandev, A. Iomin, and H. Kantz. "Fractional diffusion on a fractal grid comb". In: *Physical Review E* 91.3 (2015), p. 032108 (cit. on pp. 185, 195).

[2] A. Iomin. "Subdiffusion on a fractal comb". In: *Physical Review E* 83.5 (2011), p. 052106 (cit. on pp. 185, 202).

[3] T. Sandev, A. Iomin, and H. Kantz. "Anomalous diffusion on a fractal mesh". In: *Physical Review E* 95 (5 2017), p. 052107 (cit. on pp. 185, 195).

[4] H. Fang et al. "Linear Ultrasonic Array Development Incorporating Cantor Set Fractal Geometry". In: *2018 IEEE International Ultrasonics Symposium (IUS)* (2018), pp. 1–4 (cit. on p. 185).

[5] A. Zhokh, A. Trypolskyi, and P. Strizhak. "Relationship between the anomalous diffusion and the fractal dimension of the environment". In: *Chemical Physics* 503 (2018), pp. 71–76 (cit. on p. 185).

[6] A. Zhokh and P. Strizhak. "Advection–diffusion in a porous medium with fractal geometry: fractional transport and crossovers on time scales". In: *Meccanica* (2021), pp. 1–11 (cit. on p. 185).

[7] T. Sandev et al. "Comb model with slow and ultraslow diffusion". In: *Mathematical Modelling of Natural Phenomena* 11.3 (2016), pp. 18–33 (cit. on p. 185).

[8] T. Sandev, A. Iomin, and V. Méndez. "Lévy processes on a generalized fractal comb". In: *Journal of Physics A: Mathematical and Theoretical* 49.35 (2016), p. 355001 (cit. on pp. 185, 200, 205).

[9] A. Iomin, V. Mèndez, and W. Horsthemke. *Fractional Dynamics in Comblike Structures*. Singapore: World Scientific, 2018 (cit. on p. 185).

[10] D. Shamiryan et al. "Diffusion of solvents in thin porous films". In: *Colloids and Surfaces A: Physicochemical and Engineering Aspects* 300.1-2 (2007), pp. 111–116 (cit. on p. 186).

[11] R. Metzler and J. Klafter. "The random walk's guide to anomalous diffusion: a fractional dynamics approach". In: *Physics Reports* 339.1 (2000), pp. 1–77 (cit. on p. 190).

[12] A. A. Kilbas, O. I. Marichev, and S. G. Samko. *Fractional integrals and derivatives (theory and applications)*. 1993 (cit. on p. 192).

[13] V. E. Tarasov. "Fractional generalization of Liouville equations". In: *Chaos* 14.1 (2004), pp. 123–127 (cit. on p. 192).

[14] M. Spanner et al. "Anomalous transport of a tracer on percolating clusters". In: *Journal of Physics: Condensed Matter* 23.23 (2011), p. 234120 (cit. on p. 194).

[15] R. Metzler et al. "Anomalous diffusion models and their properties: nonstationarity, non-ergodicity, and ageing at the centenary of single particle tracking". In: *Physical Chemistry Chemical Physics* 16.44 (2014), pp. 24128–24164 (cit. on p. 194).

[16] K. Falconer. *Fractal geometry: mathematical foundations and applications*. John Wiley & Sons, 2004 (cit. on p. 195).

[17]   A. K. Golmankhaneh and A. Fernandez. "Fractal calculus of functions on Cantor tartan spaces". In: *Fractal and Fractional* 2.4 (2018), p. 30 (cit. on p. 195).

[18]   A. V. Milovanov. "Topological proof for the Alexander-Orbach conjecture". In: *Physical Review E* 56.3 (1997), p. 2437 (cit. on p. 196).

[19]   M. V. Berry, Z. V. Lewis, and J. F. Nye. "On the Weierstrass-Mandelbrot fractal function". In: *Proceedings of the Royal Society of London A. Mathematical and Physical Sciences* 370.1743 (1980), pp. 459–484 (cit. on pp. 197, 198).

[20]   B. J. West. "Sensing scaled scintillations". In: *Journal of the Optical Society of America A* 7.6 (1990), pp. 1074–1100 (cit. on pp. 197, 198).

[21]   J. Shin. Private Communication. 2016 (cit. on p. 205).

Chapter 9

Finite-velocity diffusion

It is well known that any compact initial condition, evolving due to a parabolic/diffusion equation, "spreads" instantly to infinity. This situation relates for example to the Fickian law of diffusion, or Fourier's law of heat conduction, according to equation $J = -D\nabla P$, where D is a diffusion coefficient, $P = P(x, t)$ is the PDF, while $J = J(x, t)$ is the probability flux. To overcome this unrealistic singularity with the infinite velocity of propagation, a so called telegrapher's or Cattaneo equation has been introduced [1]. The telegrapher's equation was proposed by Kelvin and Heaviside in electrodynamics theory and it was essentially employed in the heat transfer theory [2] and a persistent random walk [1, 2]. In this chapter, our main concern is the Cattaneo equation in the comb geometry, and we follow our result of Ref. [3]. This consideration is motivated by ionic transport inside neuron dendrite structure. Recent experiments, together with numerical simulations, have investigated the calcium transport inside spiny dendrites to understand the role of calcium in signal transmission and neural plasticity. This issue is well reviewed in Refs. [4, 5]. Based on these experimental finding, different theoretical approaches have been developed to explore the transport properties of spiny dendrites. It is an active field of study [6, 7, 8, 9], and new experimental findings on calcium transport and reaction transport in neuroscience [9, 10] pose new questions to understand the impact of the geometry on calcium transport and reactions in spiny dendrites and the extension of various reaction-transport models to the case of subdiffusion. Recent experiments established a relation between the geometry of the dendrite spines and the subdiffusion observed in Refs. [11, 12].

We start our consideration by discussions of the properties of the telegrapher's equation with its solution. Then concerning with the Cattaneo equation in the 2D comb geometry, we mostly pay attention on the role of the Fox H-function. Some basic consideration of the persistent random walk is borrowed from Refs. [13, 14]. Note also that the standard Cattaneo equation has been solved analytically [14, 15], and further generalization of fractional order equations has been done by several authors [13, 16, 17, 18, 19, 20, 21, 22, 23, 24].

9.1 Cattaneo equation

Prior concerning with the comb model, we discuss the properties of the telegrapher's equation

$$\tau \frac{\partial^2}{\partial t^2} u(x,t) + \frac{\partial}{\partial t} u(x,t) = \mathcal{D} \frac{\partial^2}{\partial x^2} u(x,t), \tag{9.1}$$

where a new parameter τ is a characteristic time constant, which relates to a finite propagation velocity $v = \sqrt{\frac{\mathcal{D}}{\tau}}$. This equation has been considered for various realizations of initial and boundary conditions. We consider it with the initial conditions of Eq. (9.2)

$$P(x,t=0) = \delta(x), \quad \frac{\partial}{\partial t} P(x,t=0) = 0 \tag{9.2}$$

and natural (zero) boundary conditions at infinity. In the diffusion limit $(\tau \to 0)$ of the infinite-velocity propagation, one recovers the standard diffusion equation. In the opposite limit $\tau \to \infty$, it becomes the wave equation with $\mathcal{D}/\tau = v^2$ being the finite squared speed of the wave. This equation was considered in Ref. [15], where an exact solution was obtained in the framework of Bessel functions. Here we present the solution in terms of the Fox H-function obtained in Ref. [3].

After the Fourier-Laplace transform, one obtains the expression of Eq. (9.1) in the (k,s)-space,

$$\hat{\tilde{P}}(k,s) = \frac{1+s\tau}{s(1+s\tau) + \mathcal{D}k^2}. \tag{9.3}$$

Its inverse Fourier transform yields

$$\hat{P}(x,s) = \frac{1}{2v} \left[1 + (s\tau)^{-1}\right]^{1/2} \exp\left(-\frac{s\left[1+(s\tau)^{-1}\right]^{1/2}}{v}|x|\right), \tag{9.4}$$

where $v = \sqrt{\mathcal{D}/\tau}$. Using the variable change $z = 1 + (s\tau)^{-1}$ and $\rho = \frac{|x|}{v}s$, we present Eq. (9.4) in the compact form

$$\hat{P}(x,s) = \frac{1}{2v}\sqrt{z}e^{-\rho\sqrt{z}}. \tag{9.5}$$

Now we apply Mellin's transform to Eq. (9.5). Using the identity, we have

$$\hat{P}(x,s) = \mathcal{M}^{-1}\left[\mathcal{M}[\hat{P}(x,s)]\right]$$

$$= \frac{1}{2v}\mathcal{M}^{-1}\left[\int_0^\infty z^{1/2+\xi-1}e^{-\rho\sqrt{z}}\,dz\right]$$

$$= \mathcal{M}^{-1}\left[\frac{\rho^{-2\xi-1}}{v}\Gamma(2\xi+2)\right]. \tag{9.6}$$

This eventually yields the PDF in Laplace space in the form of the Fox H-function,

$$\hat{P}(x,s) = \frac{1}{2\pi i}\frac{1}{\rho v}\int_\Omega \Gamma(2\xi+2)(\rho^2)^{-\xi}\,d\xi = \frac{1}{v\rho}H_{0,1}^{1,0}\left[\rho^2 z \left|\begin{matrix} - \\ (1,2)\end{matrix}\right.\right]. \tag{9.7}$$

Here we used the definition of the Fox H-function in Eq. (2.17) by means of the inverse Mellin transform with $\theta(\xi) = \Gamma(2\xi+2)$. Expanding function $\hat{P}(x,s)$ in the Taylor series about $z = 1$ according to Eq. (2.31), we have

$$\hat{P}(x,s) = \frac{1}{v\rho}\sum_{k=0}^\infty \frac{(z-1)^k}{k!}\frac{d^k}{dz^k}\left\{H_{0,1}^{1,0}\left[\rho^2 z \left|\begin{matrix} - \\ (1,2)\end{matrix}\right.\right]\right\}_{z=1}$$

$$= \frac{1}{|x|}\sum_{k=0}^\infty \frac{1}{k!}\tau^{-k}s^{-k-1}H_{1,2}^{1,1}\left[\frac{x^2}{v^2}s^2 \left|\begin{matrix} (0,1) \\ (1,2),(k,1)\end{matrix}\right.\right]. \tag{9.8}$$

Application of the inverse Laplace transformation in Eq. (9.8) yields

$$P(x,t) = \frac{1}{2|x|}\sum_{k=0}^\infty \frac{(-1)^k}{k!}\left(\frac{t}{\tau}\right)^k H_{2,2}^{2,0}\left[\frac{|x|}{vt} \left|\begin{matrix} (k+1,1),(0,1/2) \\ (k,1/2),(1,1)\end{matrix}\right.\right]. \tag{9.9}$$

Here the property (2.30) is also used. The asymptotic behavior of the solution in the case $\tau \to 0$ becomes Gaussian

$$P(x,t) \sim \frac{1}{\sqrt{4\pi v^2\tau t}}\exp\left(-\frac{x^2}{4v^2\tau t}\right). \tag{9.10}$$

It follows from the fact that Eq. (9.9) is the exact solution and coincides exactly with the Laplace-Fourier inversion of the PDF (9.3) for $\tau = 0$. In the opposite case of $t/\tau \to 0$, we have to account for $k = 0$ in the expansion in Eq. (9.9). In this case the Fox H-function $H_{2,2}^{2,0}\left(\frac{|x|}{vt}\right)$ reduces to the integral

$$\int_\Omega \left(\frac{|x|}{vt}\right)^{-\xi}\,d\xi = \int_{c-i\infty}^{c+i\infty} e^{Y\xi}d\xi = \delta(Y),$$

where $Y = \ln\left(\frac{|x|}{vt}\right)$. Eventually, we obtain

$$P(x,t) \sim \frac{1}{|x|} H_{0,0}^{0,0}\left[\frac{|x|^2}{(vt)^2} \left|\begin{array}{c} - \\ - \end{array}\right.\right] = \frac{\delta\left(1 - \frac{|x|^2}{(vt)^2}\right)}{|x|}$$

$$= \frac{1}{2}[\delta(x + vt) + \delta(x - vt)]. \tag{9.11}$$

Note in passing that a result of Ref. [25] and properties of the Dirac delta function are used.

The MSD can be expressed in the form of the two parameter M-L function, as well. The MSD reads (cf. Eq. (3.9))

$$\langle x^2(t)\rangle = \mathcal{L}^{-1}\left[-\frac{\partial^2}{\partial k^2}\tilde{\hat{P}}(k,s)\right]\Bigg|_{k=0},$$

from where we find

$$\langle x^2(t)\rangle = \frac{2\mathcal{D}}{\tau}\mathcal{L}^{-1}\left[\frac{s^{-2}}{s + \tau^{-1}}\right] = 2\mathcal{D}\tau\left(\frac{t}{\tau}\right)^2 E_{1,3}\left(-\frac{t}{\tau}\right)$$

$$= 2\mathcal{D}\left[t + \tau\left(e^{-t/\tau} - 1\right)\right]. \tag{9.12}$$

For the fixed τ one obtains that for the short time limit $t/\tau \to 0$ the MSD corresponds to ballistic motion

$$\langle x^2(t)\rangle \sim \mathcal{D}\tau\left(\frac{t}{\tau}\right)^2.$$

Then it changes to normal diffusion

$$\langle x^2(t)\rangle \sim 2\mathcal{D}t$$

in the long time limit $t/\tau \to \infty$. The same result one can obtain by means of the exact solution (9.9), that is

$$\langle x^2(t)\rangle = \int_{-\infty}^{\infty} x^2 P(x,t)\,dx = 2(vt)^2 \sum_{k=0}^{\infty} \frac{(-t/\tau)^k}{\Gamma(k+3)}$$

$$= 2\mathcal{D}\tau\left(\frac{t}{\tau}\right)^2 E_{1,3}\left(-\frac{t}{\tau}\right), \tag{9.13}$$

where the integration with respect to x is the Mellin transform (2.32) of the Fox H-function.

9.2 Cattaneo equation on a comb structure

Now we consider finite-velocity diffusion on a comb, described by a comb model. In this case, The Cattaneo equation on the comb reads

$$\frac{\partial}{\partial t}P(x,y,t) + \tau\frac{\partial^2}{\partial t^2}P(x,y,t) = \mathcal{D}_x\delta(y)\frac{\partial^2}{\partial x^2}P(x,y,t)$$
$$+ \mathcal{D}_y\frac{\partial^2}{\partial y^2}P(x,y,t). \tag{9.14}$$

As we already learned above, the δ-function means that diffusion along the x-direction is allowed only at $y = 0$ (the backbone). Particle moving in the backbone can be trapped in the fingers as well. The initial conditions are

$$P(x,y,t=0) = \delta(x)\delta(y), \quad \frac{\partial}{\partial t}P(x,y,t=0) = 0, \tag{9.15}$$

and the boundary conditions for the PDF $P(x,y,t)$ and for $\frac{\partial}{\partial q}P(x,y,t)$, $q = \{x,y\}$ are set to zero at infinity, $x = \pm\infty$, $y = \pm\infty$. The diffusion coefficient along the x-direction is $\mathcal{D}_x\delta(y)$, with the dimension $[\mathcal{D}_x\delta(y)] = \text{m}^2/\text{s}$, i.e., $[\mathcal{D}_x] = \text{m}^3/\text{s}$ ($[\delta(y)] = \text{m}^{-1}$). The diffusion coefficient along the fingers is $\mathcal{D}_y$ with the dimension $[\mathcal{D}_y] = \text{m}^2/\text{s}$. Correspondingly, the finite propagation velocities are $v_x = \mathcal{D}_x/[\tau\mathcal{D}_y] = \mathcal{D}_x/[v_y^2\tau^2]$ and $v_y = \sqrt{\mathcal{D}_y/\tau}$, where the x component of the velocity accounts also for the comb geometry, when the transport along the backbone depends also on the transport in fingers.

9.2.1 *Infinite domain solution*

Let us now analyze the comb Cattaneo equation (9.14). By the Fourier-Laplace transform, we obtain

$$\tilde{\tilde{\hat{P}}}(k_x, k_y, s) = \frac{1 + s\tau - \mathcal{D}_x k_x^2 \tilde{\hat{P}}(k_x, y = 0, s)}{s(1 + s\tau) + \mathcal{D}_y k_y^2}. \tag{9.16}$$

The inverse Fourier transform with respect to k_y yields

$$\tilde{\hat{P}}(k_x, y, s) = \frac{1}{2\sqrt{\mathcal{D}_y}}\frac{\left[1 + s\tau - \mathcal{D}_x k_x^2 \tilde{\hat{P}}(k_x, y = 0, s)\right]}{\sqrt{s(1 + s\tau)}}$$
$$\times \exp\left(-\frac{\sqrt{s(1 + s\tau)}}{\sqrt{\mathcal{D}_y}}|y|\right). \tag{9.17}$$

At this first step of looking for a closed form of the solution, we need to find the backbone PDF $\hat{P}(x, y = 0, s)$. From Eq. (9.17), we have

$$\tilde{\hat{P}}(k_x, y = 0, s) = \frac{1}{2\sqrt{D_y}} \frac{1 + s\tau}{\sqrt{s(1 + s\tau)} + \frac{D_x}{2\sqrt{D_y}} k_x^2}, \tag{9.18}$$

which eventually yields the closed form of the PDF in Fourier-Laplace space

$$\tilde{\hat{P}}(k_x, k_y, s) = \frac{1 + s\tau}{s(1 + s\tau) + D_y k_y^2} \times \frac{\sqrt{s(1 + s\tau)}}{\sqrt{s(1 + s\tau)} + \frac{D_x}{2\sqrt{D_y}} k_x^2}. \tag{9.19}$$

Integrating Eq. (9.14) with respect to y we obtain the marginal PDF

$$p_1(x, t) = \int_{-\infty}^{\infty} P(x, y, t)\, dy.$$

Its Fourier-Laplace image obtained from Eq. (9.19) is the Montroll-Weiss equation, which reads

$$\tilde{\hat{p}}_1(k_x, s) = \tilde{\hat{P}}(k_x, k_y = 0, s) = \frac{1}{s} \frac{\sqrt{s(1 + s\tau)}}{\sqrt{s(1 + s\tau)} + \frac{D_x}{2\sqrt{D_y}} k_x^2}, \tag{9.20}$$

and correspondingly the Laplace image is

$$\hat{p}_1(x, s) = \frac{1}{2} \sqrt{\frac{2\sqrt{D_y \tau}}{D_x}} s^{-1/2} [1 + (s\tau)^{-1}]^{1/4}$$

$$\times \exp\left(-\sqrt{\frac{2\sqrt{D_y \tau}}{D_x}} s^{1/2} [1 + (s\tau)^{-1}]^{1/4} |x|\right). \tag{9.21}$$

The inverse Laplace transform can be found in the same way as it was done for the Cattaneo equation (cf. Eqs. (9.5)–(9.8)). Rewriting Eq. (9.21) in the compact form by means of the variable change $z = 1 + (s\tau)^{-1}$ and $\rho = \sqrt{\frac{2\sqrt{D_y \tau}}{D_x}} |x| s^{1/2}$, we obtain $\hat{p}_1(x, s)$ in the form of the Fox H-function

$$\hat{p}_1(x, s) = \frac{\rho\, z^{1/4}}{2s|x|} e^{-\rho\, z^{1/4}} = \frac{1}{2s|x|} H_{0,1}^{1,0}\left[\rho z^{1/4} \,\middle|\, \begin{matrix} - \\ (1, 1) \end{matrix}\right]$$

$$= \frac{2}{s|x|} H_{0,1}^{1,0}\left[\rho^4 z \,\middle|\, \begin{matrix} - \\ (1, 4) \end{matrix}\right]. \tag{9.22}$$

Its Taylor expansion about $z = 1$ reads

$$\hat{p}_1(x, s) = \frac{2}{|x|} \sum_{k=0}^{\infty} \frac{\tau^{-k}}{k!} s^{-k-1} H_{1,2}^{1,1}\left[\left(\frac{2\sqrt{D_y \tau}}{D_x}\right)^2 |x|^4 s^2 \,\middle|\, \begin{matrix} (0, 1) \\ (1, 4), (k, 1) \end{matrix}\right]. \tag{9.23}$$

The inverse Laplace transform yields the solution for the marginal PDF,

$$p_1(x,t) = \frac{1}{|x|} \sum_{k=0}^{\infty} \frac{(-1)^k}{k!} \left(\frac{t}{\tau}\right)^k H_{2,2}^{2,0} \left[\frac{2\sqrt{\mathcal{D}_y\tau}\, x^2}{\mathcal{D}_x} \frac{x^2}{t} \middle| \begin{array}{c} (k+1,1),(0,1/2) \\ (k,1/2),(1,2) \end{array} \right].$$
(9.24)

The asymptotic behavior of the marginal PDF for $t/\tau \to \infty$ ($s\tau \to 0$) we obtain from the Laplace inversion of Eq. (9.22) in the limit $\tau \to 0$. Then, we obtain

$$p_1(x,t) \sim \frac{1}{2|x|} H_{1,1}^{1,0} \left[\frac{1}{\sqrt{\frac{\mathcal{D}_x}{2\sqrt{\mathcal{D}_y}}}} \frac{|x|}{t^{1/4}} \middle| \begin{array}{c} (1,1/4) \\ (1,1) \end{array} \right],$$
(9.25)

which is exactly the marginal PDF for a diffusion equation on the comb, see Sec. 4.3.1.

The short time solution for $t/\tau \to 0$ ($s\tau \to \infty$) corresponds to the limit $\tau \to \infty$, which is immediately obtained from the solution (9.24), by keeping only the term with $k = 0$ in the expansion. This simplification yields

$$p_1(x,t) \sim \frac{1}{|x|} H_{1,1}^{1,0} \left[\frac{2\sqrt{\mathcal{D}_y\tau}\, x^2}{\mathcal{D}_x} \frac{x^2}{t} \middle| \begin{array}{c} (1,1) \\ (1,2) \end{array} \right].$$
(9.26)

The comb geometry affects strongly the wave-diffusion transport. Indeed, the wave dynamics is attenuated and fractional dynamics becomes dominant.

9.2.2 *Fractional Cattaneo equation*

To study a fractional Cattaneo equation, let us rewrite the Montroll-Weiss equation (9.20) as follows

$$[1 + (s\tau)^{-1}]^{1/2} \left[s\tilde{p}_1(k_x, s) - 1 \right] = -\frac{\mathcal{D}_x}{2\sqrt{\mathcal{D}_y}} \sqrt{\tau} k_x^2 \tilde{\tilde{p}}_1(k_x, s),$$
(9.27)

from where by the inverse Fourier-Laplace transform we find the following generalized diffusion equation

$$\int_0^t \gamma(t - t') \frac{\partial}{\partial t'} p_1(x, t')\, dt' = \frac{\mathcal{D}_x}{2\sqrt{\mathcal{D}_y}} \sqrt{\tau} \frac{\partial^2}{\partial x^2} p_1(x, t),$$
(9.28)

where

$$\gamma(t) = \mathcal{L}^{-1} \left[\frac{s^{-1/2}}{(s + \tau^{-1})^{-1/2}} \right] = \left(\frac{t}{\tau}\right)^{-1} E_{1,0}^{-1/2} \left(-\frac{t}{\tau}\right),$$
(9.29)

while $E_{\alpha,\beta}^{\delta}(z)$ is the three parameter M-L function, see Sec. 1.3.3.3 [26].

Equation (9.28) can be rewritten as follows

$$_{\mathrm{C}}\mathcal{D}^{1/2,1}_{1,-\tau^{-1},0+}p_1(x,t) = \frac{\mathcal{D}_x}{2\sqrt{\mathcal{D}_y\tau}}\frac{\partial^2}{\partial x^2}p_1(x,t), \tag{9.30}$$

where $_{\mathrm{C}}\mathcal{D}^{\delta,\mu}_{\rho,-\nu,0+}f(t)$ is the regularized Prabhakar fractional derivative, see Eq. (1.226).

This also results in the equation for normal diffusion in the short time dynamics for $t/\tau \to 0$,

$$\frac{\partial}{\partial t}p_1(x,t) = \frac{\mathcal{D}_x}{2\sqrt{\mathcal{D}_y\tau}}\frac{\partial^2}{\partial x^2}p_1(x,t), \tag{9.31}$$

and it corresponds to the time fractional diffusion equation for $t/\tau \to \infty$,

$$_{\mathrm{C}}\mathcal{D}^{1/2}_{0+}p_1(x,t) = \frac{\mathcal{D}_x}{2\sqrt{\mathcal{D}_y}}\frac{\partial^2}{\partial x^2}p_1(x,t), \tag{9.32}$$

where $_{\mathrm{C}}\mathcal{D}^{\mu}_{0+}$ is the Caputo fractional derivative (1.196).

The MSD can be also estimated rigorously. From Eq. (9.27), we have

$$\langle x^2(t)\rangle = \mathcal{L}^{-1}\left[-\frac{\partial^2}{\partial k^2}\tilde{p}_1(k,s)\right]\Bigg|_{k=0}$$

$$= 2\left(\frac{\mathcal{D}_x}{2\sqrt{\mathcal{D}_y}}\right)\mathcal{L}^{-1}\left[\frac{s^{-3/2}}{(s+\tau^{-1})^{1/2}}\right]$$

$$= 2\left(\frac{\mathcal{D}_x}{2\sqrt{\mathcal{D}_y}}\sqrt{\tau}\right)\left(\frac{t}{\tau}\right)E^{1/2}_{1,2}\left(-\frac{t}{\tau}\right). \tag{9.33}$$

Taking into account the series representation of the three parameter M-L function, we find that for the short time limit $t/\tau \to 0$ the MSD in Eq. (9.33) results in the MSD for normal diffusion

$$\langle x^2(t)\rangle \sim 2\left(\frac{\mathcal{D}_x}{2\sqrt{\mathcal{D}_y}}\sqrt{\tau}\right)\left(\frac{t}{\tau}\right).$$

In the long time limit $t/\tau \to \infty$, subdiffusion take place with the MSD

$$\langle x^2(t)\rangle \sim 2\left(\frac{\mathcal{D}_x}{2\sqrt{\mathcal{D}_y}}\sqrt{\tau}\right)\frac{(t/\tau)^{1/2}}{\Gamma(3/2)},$$

where we apply the asymptotic formula for the associated three parameter M-L function (1.175).

It is worth noting that the MSD corresponds to the Mellin transform of the exact solution (9.24), which relates to Eq. (2.32) and yields

$$\langle x^2(t)\rangle = 2\frac{\mathcal{D}_x}{2\sqrt{\mathcal{D}_y\tau}}t\sum_{k=0}^{\infty}\frac{(-t/\tau)^k}{k!}\frac{\Gamma(k+1/2)}{\Gamma(1/2)\Gamma(k+2)}$$

$$= 2\left(\frac{\mathcal{D}_x}{2\sqrt{\mathcal{D}_y}}\sqrt{\tau}\right)\left(\frac{t}{\tau}\right)E^{1/2}_{1,2}\left(-\frac{t}{\tau}\right). \tag{9.34}$$

9.2.3 *Finite domain solution*

Note that a real structures has a finite length along the backbone. In this case, the finite boundary conditions affect strongly the transient diffusion-subdiffusion process described by Eq. (9.30). One should also bear in mind that finger's structure is finite as well. However, we suppose reasonably that the contaminant transport in this ramified structure is essentially slower than in the backbone, and the boundary conditions for the fingers can be taken at infinity, at $y = \pm\infty$. These boundary conditions are also supported by the geometrical picture (of the comb) of fractional diffusion

Therefore, the marginal PDF is described now by Eq. (9.30) with the initial condition $p_1(x, t = 0) = \delta(x)$ in the range $-L < x < L$ with the absorbing boundary conditions $p_1(x = \pm L, t) = 0$. It means that once a transporting particle reaches a boundary, it is instantly removed from the boundary.

We use the method of separation of variables $p_1(x, t) = X(x)T(t)$ that yields

$$\frac{{}_C\mathcal{D}_{1,-\tau^{-1},0+}^{1/2,1} T(t)}{T(t)} = \frac{2\sqrt{\mathcal{D}_y\tau}}{\mathcal{D}_x} \frac{X''(x)}{X(x)} = -\lambda. \tag{9.35}$$

Here λ is the separation constant, which specifies the system of two equations

$$_C\mathcal{D}_{1,-\tau^{-1},0+}^{1/2,1} T(t) + \lambda T(t) = 0, \tag{9.36}$$

$$X''(x) + \frac{2\sqrt{\mathcal{D}_y\tau}}{\mathcal{D}_x} \lambda X(x) = 0. \tag{9.37}$$

Equation (9.37) is the eigenvalue problem with the boundary condition $X(x = \pm L) = 0$, which yields a set of eigenfunctions X_n with corresponding eigenvalues λ_n. Thus, the solution of Eq. (9.30) in the finite domain is

$$p_1(x, t) = \sum_{n=0}^{\infty} T_n(t) X_n(x),$$

where $T_n(t = 0) = 1$, $\forall n$. Eventually, after accounting for the initial condition $p_1(x, 0) = \delta(x)$, the solution reads

$$p_1(x, t) = \frac{1}{L} \sum_{n=-\infty}^{\infty} e^{i \frac{(2n+1)\pi x}{2L}} T_n(t). \tag{9.38}$$

The solution of Eq. (9.36) can be found by the Laplace transform method. Thus, from relation (1.229), we have

$$\left[1 + (s\tau)^{-1}\right]^{1/2} \left[s\hat{T}_n(s) - 1\right] + \lambda_n \hat{T}_n(s) = 0, \tag{9.39}$$

i.e.,

$$\hat{T}_n(s) = \frac{\left[1 + (s\tau)^{-1}\right]^{1/2}}{s\left[1 + (s\tau)^{-1}\right]^{1/2} + \lambda_n}. \tag{9.40}$$

The solution then becomes

$$T_n(t) = \sum_{j=0}^{\infty} (-\lambda_n)^j \mathcal{L}^{-1}\left[\frac{s^{-j/2-1}}{(s + \tau^{-1})^{j/2}}\right]$$

$$= \sum_{j=0}^{\infty} (-\lambda_n)^j \, t^j E_{1,j+1}^{j/2}\left(-\frac{t}{\tau}\right), \tag{9.41}$$

with $\lambda_n = \frac{\mathcal{D}_x}{2\sqrt{\mathcal{D}_y\tau}}\left[\frac{(2n+1)\pi}{2L}\right]^2$. We note that $T_n(t = 0) = 1$, since for $t = 0$ only the first term with $j = 0$ in the sum (9.41) survives. Thus, the finite domain solution reads

$$p_1(x,t) = \frac{1}{L} \sum_{n=-\infty}^{\infty} e^{i\frac{(2n+1)\pi x}{2L}} \sum_{j=0}^{\infty} (-\lambda_n)^j \, t^j E_{1,j+1}^{j/2}\left(-\frac{t}{\tau}\right). \tag{9.42}$$

Accounting relaxation in the finite boundaries, we also find the survival probability

$$S(t) = \int_{-L}^{L} p_1(x,t)\, dx$$

which reads

$$S(t) = \frac{4}{\pi} \sum_{n=0}^{\infty} \frac{(-1)^n}{2n+1} \sum_{j=0}^{\infty} (-\lambda_n)^j \, t^j E_{1,j+1}^{j/2}\left(-\frac{t}{\tau}\right), \tag{9.43}$$

from where the first passage time PDF can be found. Recall that in random search problems (of Chapter 7) it is named by FATD $\wp(t)$, which reads

$$\wp(t) = -\frac{d}{dt}S(t) = \frac{4}{\pi} \sum_{n=0}^{\infty} \frac{(-1)^{n+1}}{2n+1} \sum_{j=0}^{\infty} (-\lambda_n)^j \, \frac{d}{dt} t^j E_{1,j+1}^{j/2}\left(-\frac{t}{\tau}\right). \tag{9.44}$$

The long time limit can be obtained by asymptotic expansion of the associated M-L function (1.175), which results in the following chain of transformations

$$\wp(t) \sim \frac{4}{\pi} \sum_{n=0}^{\infty} \frac{(-1)^{n+1}}{2n+1} \sum_{j=0}^{\infty} \left(-\lambda_n \sqrt{\tau}\right)^j \frac{t^{j/2-1}}{\Gamma(j/2)}$$

$$= \frac{4}{\pi} \sum_{n=0}^{\infty} \frac{(-1)^{n+1}}{2n+1} t^{-1} E_{1/2,0}\left(-\lambda_n \sqrt{\tau} t^{1/2}\right)$$

$$= \frac{\pi}{L^2} \frac{\mathcal{D}_x}{2\sqrt{\mathcal{D}_y}} \sum_{n=0}^{\infty} (-1)^n (2n+1) \, t^{-1/2}$$

$$\times E_{\frac{1}{2},\frac{1}{2}} \left(-\frac{\mathcal{D}_x}{2\sqrt{\mathcal{D}_y}} \frac{(2n+1)^2 \pi^2}{4L^2} t^{1/2}\right), \tag{9.45}$$

where we use the relation

$$E_{\alpha,\beta}(z) = z E_{\alpha,\alpha+\beta}(z) + \frac{1}{\Gamma(\beta)},$$

which immediately follows from the definition of the two parameter M-L function in Eq. (1.142). From here, we find a power-law decay of the form $\wp(t) \sim t^{-3/2}$, i.e.,

$$\wp(t) \sim \frac{8L^2}{\pi^{7/2}} \frac{2\sqrt{\mathcal{D}_y}}{\mathcal{D}_x} \sum_{n=0}^{\infty} \frac{(-1)^n}{(2n+1)^3} t^{-3/2}$$

$$= \frac{L^2}{8\pi^{7/2}} \frac{2\sqrt{\mathcal{D}_y}}{\mathcal{D}_x} \left[\zeta\left(3, \frac{1}{4}\right) - \zeta\left(3, \frac{3}{4}\right)\right] t^{-3/2}, \tag{9.46}$$

where $\zeta(s,a) = \sum_{k=0}^{\infty} \frac{1}{(k+a)^s}$ is the Hurwitz zeta function [27].

The obtained result has a correct limit in the infinite domain $L \to \infty$. The solution (9.42) in the Laplace space is given by

$$\hat{p}_1(x,s) = \frac{1}{L} \sum_{n=-\infty}^{\infty} e^{i\left[\frac{(2n+1)\pi}{2L}\right]x} \frac{s^{1/2-1}(1+s\tau)^{1/2}}{s^{1/2}(1+s\tau)^{1/2} + \frac{\mathcal{D}_x}{\sqrt{2\mathcal{D}_y}}\left[\frac{(2n+1)\pi}{2L}\right]^2}. \tag{9.47}$$

Taking the limit $L \to \infty$, i.e., $1/L \to 0$, the summation leads to the integration, which corresponds to the inverse Fourier transform with $k_x = \frac{(2n+1)\pi}{2L}$. This leads to an equivalent equation to Eq. (9.20) for the reduced PDF for the infinite domain case.

9.3 Persistent random walk

As admitted above, finite velocity diffusion relates to the persistent random walk (PRW). In this section, we discuss the PRW and its relation to the telegrapher's equation (or Cattaneo equation). The theory of the PRW relies on a generalization of the CTRW [28, 29], and we discuss this issue in the framework of the extended CTRW picture. The bulk of this section is a review of familiar material.[1] To demonstrate the PRW mechanism, let us consider it in the continuous time framework in the one dimensional space, $x \in \mathbb{R}$. As discussed in Chapter 3, the basic CTRW quantity is the PDF $P(x,t)$, which in its turn relates to the jump length PDF $w(x)$ and the waiting time PDF $\psi(t)$ through the Montroll-Weiss equation (3.77) with the composite jump PDF $\Upsilon(x,t) = w(x)\psi(t)$. In the renewal theory, the waiting time intervals, also called sojourns, are separated by regeneration points, in which the walk continues with the new started time and the coordinate. While the waiting time PDF $\psi(t)$ reflexes the regeneration points as the fail times [30], the survival probability $\Psi(t) = \int_t^\infty \psi(t')dt'$ describes the cumulative probability that a given time interval between consecutive regeneration points is greater than t (cf. Eq. (3.63)).

Further generalisation of the renewal process allows the walker move inside a single sojourn. In this case, the waiting time PDF (sojourn PDF) $\psi(t)$ is kept unchanged, while the jump length PDF for the complete sojourn (corresponding to regeneration points) reads now

$$w(x) = \int_0^\infty w(x,t)\psi(t)\,dt, \tag{9.48}$$

where $w(x,t)$ is the jump length PDF at time t inside a given sojourn. The composite jump PDF becomes

$$\Upsilon_\psi(x,t) = w(x,t)\psi(t)\,, \tag{9.49}$$

which is the joint PDF of the displacement being x during a complete sojourn at the waiting time being t. Another microscopic CTRW function necessary for the generalization of the renewal process is

$$\Upsilon_\Psi(x,t) = w(x,t)\Psi(t), \tag{9.50}$$

---

[1] We borrow this discussion from Refs. [13, 14], where the PRW theory is demonstrated. In some extent this familiar issue should belong to Chapter 3 as an extension of the CTRW theory. However, it is more reasonable to present this theory here as an explanation of the random walk picture of finite velocity diffusion, described in the framework of the telegrapher's (or Cattaneo) equation.

which is determined by the survival probability $\Psi(t)$ that a given time interval between consecutive regeneration points is greater than t. Another important property is that during the waiting time between any two nearest regeneration points the walker does not change the direction. Taking into account that there are two directions — "forward" and "back", we define two states of the random walk. This eventually defines the PRW as a two-state process, and for this "simple" motion on a line, the states corresponding to the directions, are forward-right denoting by $(+)$ and back-left denoting by $(-)$. Persistence assumes that the walker is in one of two states during a single sojourn. In this case, the waiting time PDFs in each state are $\psi_{\pm}(t)$, as well, the survival probabilities are $\Psi_{\pm}(t)$. The same condition is assumed for the PDF for the displacement of the random walker at time t, which in $(\pm)$ state is $w_{\pm}(x,t)$ that also provides that the random walker remains in the same state during a single sojourn. Correspondingly the composite functions defined in Eqs. (9.49) and (9.50) become

$$\Upsilon_{\psi}^{\pm}(x,t) = w_{\pm}(x,t)\psi_{\pm}(t) \quad \text{and} \quad \Upsilon_{\Psi}^{\pm}(x,t) = w(x,t)_{\pm}\Psi_{\pm}(t). \qquad (9.51)$$

The isotropic random walk in space and time is assumed, that is $\psi_{+}(t) = \psi_{-}(t)$. We also admit that the composite function $\Upsilon_{\psi}^{\pm}(x,t)(x,t)$ is a joint density for the displacement in a single sojourn in either $(\pm)$ state to be equal to x for the sojourn time t, while $\Upsilon_{\Psi}^{\pm}(x,t)$ is the PDF for the displacement to be equal to x in a single partial sojourn in the $(\pm)$ state at time t, when the total sojourn time in that state is greater than t.

Transitions between the states take place at the regeneration points, only. These transitions are determined by a transition matrix[2] with matrix elements $Q_{1,1} = Q_{2,2} = 0$ and $Q_{1,2} = Q_{2,1} = 1$. The latter describes transitions between left and right states/directions $\mathbf{Q}\binom{+}{-} = \binom{-}{+}$.

In the next step of the analysis of the PRW, we concern with the description of the random walk in each state and we establish the relation between them at the regeneration points. To this end we decompose the PDF $P(x,t)$ into the two probabilities of every state, $P(x,t) = P_{+}(x,t) + P_{-}(x,t)$, where $P_{\pm}(x,t)$ is the PDF to find the walker in the position x at time t in the $(\pm)$ state. Let us calculate the PDF $P_{+}(x,t)$ of a random walk in the state $(+)$ of any single sojourn. To this end, we need to introduce an auxiliary

---

[2]In general case of a multi-state random walk, it is assumed that transitions between different states at regeneration points are specified by a transition Markov matrix with matrix elements $Q_{j,k} \geq 0$ being the probability of the transition $j \to k$ with the property of the stochastic matrix $\sum_{k} Q_{j,k} = 1$, while $Q_{j,k}$ is the probability for a sojourn in state j to be followed by a sojourn in state k [13].

functions in each state $\rho_+(x,t)$ and $\rho_-(x,t)$, which describe the renewal process. In particular, $\rho_+(x,t)$ is the joint probability density in the $(+)$ state for the walker to be at a regeneration point at time t and position x. The same statement is valid for the PDF $\rho_-(x,t)$. Therefore, these functions obey the renewal equations, which are

$$\rho_+(x,t) = \frac{\Upsilon_\psi^+(x,t)}{2} + \int_0^t dt' \int_{-\infty}^\infty dx' \Upsilon_\psi^+(x-x',t-t')\rho_-(x',t'), \quad (9.52a)$$

$$\rho_-(x,t) = \frac{\Upsilon_\psi^-(x,t)}{2} + \int_0^t dt' \int_{-\infty}^\infty dx' \Upsilon_\psi^-(x-x',t-t')\rho_+(x',t'). \quad (9.52b)$$

The interpretation of the equations is as follows. The first terms on the rhs of the equations correspond to the first sojourn, which start at $t = 0$ in either $(\pm)$ state with probability $1/2$. These terms describes the first possibility that a regeneration point (in either state) occurs at time t, while the second terms on the rhs collect all possibilities of earlier regeneration points in either opposite $(\mp)$ state occurred at times $t' < t$ and at positions x', and no further regeneration points occurred during the time interval $t - t'$.

By means of the auxiliary functions $\rho_\pm(x,t)$ integral equations for the PDFs $P_+(x,t)$ and $P_-(x,t)$ can be obtain as well. Taking into account that $\Upsilon_\Psi(x,t)$ is the PDF for the displacement of the walker during a sojourn in either state, when the sojourn time is longer than t, we obtain

$$P_+(x,t) = \frac{\Upsilon_\Psi^+(x,t)}{2} + \int_0^t dt' \int_{-\infty}^\infty dx' \Upsilon_\Psi^+(x-x',t-t')\rho_-(x',t'), \quad (9.53a)$$

$$P_-(x,t) = \frac{\Upsilon_\Psi^-(x,t)}{2} + \int_0^t dt' \int_{-\infty}^\infty dx' \Upsilon_\Psi^-(x-x',t-t')\rho_+(x',t'). \quad (9.53b)$$

By complete analogy with Eqs. (9.52), the rhs of Eqs. (9.53) describe the PDF of the walker at the position x at time t in either state $(\pm)$, described by the first term, or a transition $\mathbf{Q}\binom{-}{+} = \binom{+}{-}$ at a regeneration point of a previous sojourn occurred at time $t' < t$ and position x', while the time to the next regeneration point is large then $t - t'$.

For the convenience of the notation in what follows analysis, let us denote

$$\alpha_+(x,t) \equiv \Upsilon_\psi^+(x,t), \quad \alpha_-(x,t) \equiv \Upsilon_\psi^-(x,t), \quad (9.54a)$$

$$\hat{\alpha}_+(k,s) \equiv \hat{\tilde{\Upsilon}}_\psi^+(k,s), \quad \hat{\alpha}_-(k,s) \equiv \hat{\tilde{\Upsilon}}_\psi^-(k,s), \quad (9.54b)$$

$$\beta_+(x,t) \equiv \Upsilon_\Psi^+(x,t), \quad \beta_-(x,t) \equiv \Upsilon_\Psi^-(x,t), \quad (9.54c)$$

$$\hat{\beta}_+(k,s) \equiv \hat{\tilde{\Upsilon}}_\Psi^+(k,s), \quad \hat{\beta}_-(k,s) \equiv \hat{\tilde{\Upsilon}}_\Psi^-(k,s). \quad (9.54d)$$

Following the standard procedures, we perform the Fourier-Laplace transform of both Eqs. (9.52) and (9.53) that yields

$$\hat{\tilde{P}}_+(k,s) = \frac{\hat{\beta}_+(k,s)}{2} + \hat{\beta}_+(k,s)\hat{\tilde{\rho}}_-(k,s), \tag{9.55a}$$

$$\hat{\tilde{P}}_-(k,s) = \frac{\hat{\beta}_-(k,s)}{2} + \hat{\beta}_-(k,s)\hat{\tilde{\rho}}_+(k,s) \tag{9.55b}$$

and

$$\hat{\tilde{\rho}}_+(k,s) = \frac{\hat{\alpha}_+(k,s)}{2} + \hat{\alpha}_+(k,s)\hat{\tilde{\rho}}_-(k,s), \tag{9.56a}$$

$$\hat{\tilde{\rho}}_-(k,s) = \frac{\hat{\alpha}_-(k,s)}{2} + \hat{\alpha}_-(k,s)\hat{\tilde{\rho}}_+(k,s). \tag{9.56b}$$

From Eqs. (9.56), we obtain

$$\hat{\tilde{\rho}}_+(k,s) = \frac{1}{2}\frac{\hat{\alpha}_+\left[1+\hat{\alpha}_-\right]}{1-\hat{\alpha}_+\hat{\alpha}_-}$$

and

$$\hat{\tilde{\rho}}_-(k,s) = \frac{1}{2}\frac{\hat{\alpha}_-\left[1+\hat{\alpha}_+\right]}{1-\hat{\alpha}_+\hat{\alpha}_-},$$

and substituting these expressions in Eqs. (9.55), we obtain that the PDF images read as follows

$$\hat{\tilde{P}}_+(k,s) = \frac{1}{2}\frac{1+\hat{\alpha}_-}{1-\hat{\alpha}_+\hat{\alpha}_-}\hat{\beta}_+, \quad \hat{\tilde{P}}_-(k,s) = \frac{1}{2}\frac{1+\hat{\alpha}_+}{1-\hat{\alpha}_+\hat{\alpha}_-}\hat{\beta}_-. \tag{9.57}$$

Eventually, the result is expressed in the form of microscopic characteristics of the continuous time PRW, namely $\psi_\pm(t)$ and $w_\pm(x,t)$. In general, these functions have a variety of realizations, and for the consideration of finite velocity diffusion these functions are

$$\psi_+(t) = \psi_-(t) = \frac{1}{\tau}e^{-t/\tau}, \quad w_\pm(x,t) = \delta(x \mp vt). \tag{9.58}$$

It is worth noting that this microscopic description according to Eq. (9.58) is a dichotomy, namely the waiting time density $\psi_\pm(t)$ is responsible for the random walk, while the jump length density $w_\pm(x,t)$ reflects the deterministic wave dynamics, where the walker is on the wave cone with the wave velocity v.

Accounting for the expressions in Eq. (9.58) and the notation in Eqs. (9.54), we obtain the composite PDFs as follows

$$\alpha_\pm(x,t) \equiv \Upsilon_\psi^\pm(x,t) = \delta(x \mp vt)e^{-t/\tau}/\tau \tag{9.59a}$$

$$\beta_\pm(x,t) \equiv \Upsilon_\Psi^\pm(x,t) = \delta(x \mp vt)e^{-t/\tau}. \tag{9.59b}$$

In Fourier-Laplace space, the system of Eqs. (9.59) reads

$$\hat{\alpha}_{\pm} = \frac{1}{\tau} \int_0^\infty e^{-st} e^{-t/\tau} \int_{-\infty}^\infty \delta(x \mp vt) e^{-\imath kx} \, dx dt = \frac{1}{\tau s + 1 \pm \imath k v \tau} \quad (9.60a)$$

$$\hat{\beta}_{\pm} = \int_0^\infty e^{-st} e^{-t/\tau} \int_{-\infty}^\infty \delta(x \mp vt) e^{-\imath kx} \, dx dt = \frac{\tau}{\tau s + 1 \pm \imath k v \tau} \quad (9.60b)$$

Substituting these results in Eqs. (9.55) and taking into account that the Fourier-Laplace image is the superposition, $\hat{\tilde{P}}(k, s) = \hat{\tilde{P}}_+(k, s) + \hat{\tilde{P}}_-(k, s)$, we obtain

$$\hat{\tilde{P}}(k, s) = \frac{2 + s\tau}{s^2 \tau + 2 + k^2 v^2 \tau}. \quad (9.61)$$

Replacing $\tau/2 \to \tau$ and taking into account that $v^2 = \mathcal{D}/\tau$, we arrive at Eq. (9.3) for the Cattaneo equation in Fourier-Laplace space.

We note that one can further extend the PRW model to derive different forms of time fractional telegrapher's (or Cattaneo) equations [31, 32]. Two and three dimensional telegrapher's equations has been introduced and analyzed in Refs. [33, 34, 35], as well.

References

[1] C. Cattaneo. "Sulla conduzione del calore". In: *Atti del Seminario Matematico e Fisico dell'Università di Modena* 3 (1948), p. 83 (cit. on p. 211).

[2] D. D. Joseph and L. Preziosi. "Heat waves". In: *Reviews of Modern Physics* 61.1 (1989), p. 41 (cit. on p. 211).

[3] T. Sandev and A. Iomin. "Finite-velocity diffusion on a comb". In: *Europhysics Letters (EPL)* 124.2 (2018), p. 20005 (cit. on pp. 211, 212).

[4] E. Korkotian and M. Segal. "Spatially confined diffusion of calcium in dendrites of hippocampal neurons revealed by flash photolysis of caged calcium". In: *Cell Calcium* 40.5-6 (2006), pp. 441–449 (cit. on p. 211).

[5] M. Segal. "Dendritic spines and long-term plasticity". In: *Nature Reviews Neuroscience* 6.4 (2005), pp. 277–284 (cit. on p. 211).

[6] V. Méndez and A. Iomin. "Comb-like models for transport along spiny dendrites". In: *Chaos, Solitons & Fractals* 53 (2013), pp. 46–51 (cit. on p. 211).

[7] A. Iomin and V. Méndez. "Reaction-subdiffusion front propagation in a comblike model of spiny dendrites". In: *Physical Review E* 88.1 (2013), p. 012706 (cit. on p. 211).

[8] S. B. Yuste, E. Abad, and A. Baumgaertner. "Anomalous diffusion and dynamics of fluorescence recovery after photobleaching in the random-comb model". In: *Physical Review E* 94.1 (2016), p. 012118 (cit. on p. 211).

[9] J. Rose, S. X. Jin, and A. M. Craig. "Heterosynaptic molecular dynamics: locally induced propagating synaptic accumulation of CaM kinase II". In: *Neuron* 61.3 (2009), pp. 351–358 (cit. on p. 211).

[10] B. A. Earnshaw and P. C. Bressloff. "A diffusion-activation model of CaMKII translocation waves in dendrites". In: *Journal of Computational Neuroscience* 28.1 (2010), pp. 77–89 (cit. on p. 211).

[11] F. Santamaria et al. "Anomalous diffusion in Purkinje cell dendrites caused by spines". In: *Neuron* 52.4 (2006), pp. 635–648 (cit. on p. 211).

[12] F. Santamaria et al. "The diffusional properties of dendrites depend on the density of dendritic spines". In: *European Journal of Neuroscience* 34.4 (2011), pp. 561–568 (cit. on p. 211).

[13] J. Masoliver and K. Lindenberg. "Continuous time persistent random walk: a review and some generalizations". In: *European Physical Journal B* 90.6 (2017), pp. 1–13 (cit. on pp. 212, 222, 223).

[14] G. H. Weiss. "Some applications of persistent random walks and the telegrapher's equation". In: *Physica A* 311.3-4 (2002), pp. 381–410 (cit. on pp. 212, 222).

[15] J. Masoliver and G. H. Weiss. "Finite-velocity diffusion". In: *European Journal of Physics* 17.4 (1996), p. 190 (cit. on p. 212).

[16] A. Compte and R. Metzler. "The generalized Cattaneo equation for the description of anomalous transport processes". In: *Journal of Physics A: Mathematical and General* 30.21 (1997), p. 7277 (cit. on p. 212).

[17] M. A. Olivares-Robles and L. S. Garcia-Colin. "On different derivations of Telegrapher's type kinetic equations". In: *Journal of Non-Equilibrium Thermodynamics* 21 (1996), p. 361 (cit. on p. 212).

[18] H. Qi and X. Jiang. "Solutions of the space-time fractional Cattaneo diffusion equation". In: *Physica A* 390.11 (2011), pp. 1876–1883 (cit. on p. 212).

[19] G. Fernandez-Anaya, F. J. Valdes-Parada, and J. Alvarez-Ramirez. "On generalized fractional Cattaneo's equations". In: *Physica A* 390.23-24 (2011), pp. 4198–4202 (cit. on p. 212).

[20] H. Qi, H. Y. Xu, and X. W. Guo. "The Cattaneo-type time fractional heat conduction equation for laser heating". In: *Computers and Mathematics with Applications* 66.5 (2013), pp. 824–831 (cit. on p. 212).

[21] L. Liu, L. Zheng, and X. Zhang. "Fractional anomalous diffusion with Cattaneo–Christov flux effects in a comb-like structure". In: *Applied Mathematical Modelling* 40.13-14 (2016), pp. 6663–6675 (cit. on p. 212).

[22] S. M Cvetićanin, D. Zorica, and M. R. Rapaić. "Generalized time-fractional telegrapher's equation in transmission line modeling". In: *Nonlinear Dynamics* 88.2 (2017), pp. 1453–1472 (cit. on p. 212).

[23] L. Liu et al. "Fractional anomalous convection diffusion in comb structure with a non-Fick constitutive model". In: *Journal of Statistical Mechanics: Theory and Experiment* 2018.1 (2018), p. 013208 (cit. on p. 212).

[24] F. Ferrillo, R. Spigler, and M. Concezzi. "Comparing Cattaneo and fractional derivative models for heat transfer processes". In: *SIAM Journal on Applied Mathematics* 78.3 (2018), pp. 1450–1469 (cit. on p. 212).

[25] N. Südland and G. Baumann. "On the Mellin Transforms of Dirac'S Delta Function, The Hausdorff Dimension Function, and The Theorem by Mellin". In: *Fractional Calculus and Applied Analysis* 7.4 (2004), pp. 409–420 (cit. on p. 214).

[26] T. R. Prabhakar. "A singular integral equation with a generalized Mittag-Leffler function in the kernel". In: *Yokohama Mathematical Journal* 19 (1971), pp. 7–15 (cit. on p. 217).

[27] A. M. Mathai, R. K. Saxena, and H. J. Haubold. *The H-function: theory and applications*. Springer Science & Business Media, 2009 (cit. on p. 221).

[28] E. W. Montroll. "Markoff Chains and Excluded Volume Effect in Polymer Chains". In: *Journal of Chemical Physics* 18 (1950), pp. 734–743 (cit. on p. 222).

[29] G. H. Weiss. "The Two-State Random Walk". In: *Journal of Statistical Physics* 15.2 (1976), pp. 157–165 (cit. on p. 222).

[30] D. R. Cox. *Renewal Theory*. London: Methuen, 1967 (cit. on p. 222).

[31] J. Masoliver. "Fractional telegrapher's equation from fractional persistent random walks". In: *Physical Review E* 93.5 (2016), p. 052107 (cit. on p. 226).

[32] K. Górska et al. "Generalized Cattaneo (telegrapher's) equations in modeling anomalous diffusion phenomena". In: *Physical Review E* 102.2 (2020), p. 022128 (cit. on p. 226).

[33] J. Masoliver and K. Lindenberg. "Two-dimensional telegraphic processes and their fractional generalizations". In: *Physical Review E* 101.1 (2020), p. 012137 (cit. on p. 226).

[34] J. Masoliver. "Three-dimensional telegrapher's equation and its fractional generalization". In: *Physical Review E* 96.2 (2017), p. 022101 (cit. on p. 226).

[35] J. Masoliver. "Telegraphic transport processes and their fractional generalization: a review and some extensions". In: *Entropy* 23.3 (2021), p. 364 (cit. on p. 226).

Chapter 10

Appendices

Appendix A

Functional calculus

In general case, functionals are defined as variable quantities (numbers) whose values are determined by the choice of one or several functions, see e.g., [1]. For example, the length l of a curve between two points (x_1, y_1) and (x_2, y_2) on a plane is a functional. If the equation of the curve $y = y(x)$ is given, then

$$l[y(x)] = \int_{x_1}^{x_2} \sqrt{1 + (y')^2}\, dx, \quad y' \equiv \frac{dy}{dx}.$$

In the same way one determines the area S of a surface as a functional, since the latter is determines by the equation of the surface $z = z(x, y)$ as follows

$$S = \iint_{p_{x,y}} \sqrt{1 + \left(\frac{\partial}{\partial x} z\right)^2 + \left(\frac{\partial}{\partial y} z\right)^2}\, dx dy,$$

where $p_{x,y}$ is a projection of the surface on the xy-plane.

A.1 Functional derivative

The calculus of variations is explored to investigate extremum values of functionals, like the geodesic problem for surfaces in three or more dimensional spaces. Another well known task is the minimum of the principal Hamiltonian function in Lagrangian or Hamiltonian mechanics. In all these cases the functionals depend on a set of functions $\{y(x)\}$ and their derivatives,

$$F[y(x)] = \int_{x_1}^{x_2} f\left(\{y(x)\}, \{y'(x)\}\dots\{y^{(n)}(x)\}\right) dx, \qquad \text{(A.1)}$$

where $f(z)$ is a known function.

We consider functionals which are formal solutions of stochastic equations, and which are our prime concern in Sec. 3.6. These are the functionals [2, 3], which are determined by functions $\xi(t)$ as follows

$$F_1[\xi(t)] = \int_{t_1}^{t_2} A(\tau)\xi(\tau)\,d\tau, \tag{A.2a}$$

$$F_2[\xi(t)] = \iint_{t_1}^{t_2} B(\tau_1, \tau_2)\xi(\tau_1)\xi(\tau_2)\,d\tau_1 d\tau_2, \tag{A.2b}$$

$$F_3[\xi(t)] = f\left(\Phi[\xi(t)]\right). \tag{A.2c}$$

Here $A(t), B(t_1, t_2)$ are known fixed functions, while $f(z)$ is a known function and $\Phi[\xi(t)]$ is a functional by itself.

Many properties of functionals are analogous to the properties of functions [1]. For example, by this analogy, Eq. (A.2a) defines the linear functional. By this analogy, the increment of the functional can be defined as well,

$$\delta F[\xi(\tau)] = F[\xi(\tau) + \delta\xi(\tau)] - F[\xi(\tau)],$$

where $\delta\xi(\tau)$ is the increment of $\xi(\tau)$ on the same time interval $t \in (t - \frac{1}{2}\Delta T, t + \frac{1}{2}\Delta T)$. Thus the variation of a functional is the principal (main linear) part of the increment of the functional with respect to $\delta\xi(t)$. Therefore, the functional of the variational (functional) derivative is [2]

$$\frac{\delta F[\xi(t)]}{\delta\xi(t)} = \lim_{\Delta T \to 0} \frac{\delta F[\xi(t)]}{\int_{\Delta T} \xi(t)\,dt}. \tag{A.3}$$

This also suppose that the increment $\delta\xi(\tau)$ is not zero for $t - \Delta T < \tau < t + \Delta T$, that is $\delta\xi(\tau) \equiv \delta\xi(\tau)\Theta(\Delta T - |\tau - t|)$. Let us employ this definition for calculation of the functional derivatives of the functionals defined in Eqs. (A.1) and (A.2).

First, we consider the derivative of Eq. (A.2a). The variation of the functional reads

$$\delta F_1[\xi(\tau)] = F_1[\xi + \delta\xi] - F_1[\xi] = \int_{t-\Delta T/2}^{t+\Delta T/2} A(\tau)\,\delta\xi(\tau)\,d\tau.$$

In the limit $\Delta T \to 0$, the midpoint evaluation of the integral yields

$$\delta F_1[\xi(\tau)] = A(t) \int_{t-\Delta T/2}^{t+\Delta T/2} \delta\xi(\tau)\,d\tau$$

that eventually yields

$$\frac{\delta F_1[\xi(\tau)]}{\delta\xi(t)} = A(t). \tag{A.4}$$

Accounting for the linear part with respect to the increment $\delta\xi(t)$ in Eq. (A.2b), one obtains

$$\frac{\delta F_2[\xi(\tau)]}{\delta\xi(t)} = \int_{t_1}^{t_2} [B(t,\tau) + B(\tau,t)]\,\xi(\tau)\,d\tau. \qquad (A.5)$$

From these two examples, we have formally the following expression

$$\delta F[\xi(\tau)] = \int_{\Omega} \frac{\delta F[\xi(\tau)]}{\delta\xi(\tau)}\,\delta\xi(\tau)\,d\tau. \qquad (A.6)$$

We also account that

$$\frac{\delta\xi(\tau)}{\delta\xi(t)} = \delta(\tau - t), \qquad (A.7)$$

which immediately follows from the choice of $F[\xi(\tau)] = \xi(\tau)$.

Considering functional $F_3[\xi(\tau)]$ in Eq. (A.2c), we explore some other properties of the functional derivatives. For example, if $f\left(\Phi[\xi(t)]\right) = \frac{\delta\Phi[\xi]}{\delta\xi(\tau)}$. Then

$$\frac{\delta}{\delta\xi(t)}\left[\frac{\delta\Phi[\xi]}{\delta\xi(\tau)}\right] = \frac{\delta^2\Phi[\xi]}{\delta\xi(t)\,\delta\xi(\tau)}.$$

Now let us consider Leibniz formula for the product $f\left(\Phi[\xi(t)]\right) = F_1[\xi]F_2[\xi]$. Then, we have

$$\frac{\delta}{\delta\xi(t)}\,[F_1[\xi]F_2[\xi]] = \frac{\delta F_1[\xi]}{\delta\xi(t)}F_2[\xi] + F_1[\xi]\frac{\delta F_2[\xi]}{\delta\xi(t)}.$$

Another way of presenting the variation of the functional $\delta F[\xi(t)]$ can be considered for the functional (A.1). To that end it depends now on a parameter ε as follows [1]

$$F[\xi(t),\varepsilon] = \int_{t_1}^{t_2} f\left(t,\xi(t,\varepsilon),\dot{\xi}(t,\varepsilon)\right) dt$$

where $\xi(t,\varepsilon) = \xi + \varepsilon\,\delta\xi$ and $\dot{\xi}(t,\varepsilon) = \dot{\xi}(t) + \varepsilon\,\delta\dot{\xi}(t)$. Then, as shown in [1], the functional variation is

$$\delta F[\xi(t)] = \left.\frac{\partial F[\xi(t),\varepsilon]}{\partial\varepsilon}\right|_{\varepsilon=0}$$

$$= \int_{t_1}^{t_2} dt \left[\frac{\partial f\left(t,\xi(t,\varepsilon),\dot{\xi}(t,\varepsilon)\right)}{\partial\xi}\frac{\partial\xi(t,\varepsilon)}{\partial\varepsilon} + \frac{\partial f\left(t,\xi(t,\varepsilon),\dot{\xi}(t,\varepsilon)\right)}{\partial\dot{\xi}}\frac{\partial\dot{\xi}(t,\varepsilon)}{\partial\varepsilon}\right]_{\varepsilon=0}$$

$$= \int_{t_1}^{t_2} \left[\frac{\partial f\left(t,\xi,\dot{\xi}\right)}{\partial\xi}\delta\xi + \frac{\partial f\left(t,\xi,\dot{\xi}\right)}{\partial\dot{\xi}}\delta\dot{\xi}\right] dt$$

$$= \int_{t_1}^{t_2} \left[\frac{\partial f\left(t,\xi,\dot{\xi}\right)}{\partial\xi} - \frac{d}{dt}\frac{\partial f\left(t,\xi,\dot{\xi}\right)}{\partial\dot{\xi}}\right]\delta\xi\,dt. \qquad (A.8)$$

Then, the functional derivative is

$$\frac{\delta F[\xi(t)]}{\delta \xi(\tau)} = \int_{t_1}^{t_2} \left[\frac{\partial f\left(t, \xi, \dot{\xi}\right)}{\partial \xi} - \frac{d}{dt} \frac{\partial f\left(t, \xi, \dot{\xi}\right)}{\partial \dot{\xi}} \right] \frac{\delta \xi(t)}{\delta \xi(\tau)} dt$$

$$= \frac{\partial f\left(\tau, \xi, \dot{\xi}\right)}{\partial \xi} - \frac{d}{d\tau} \frac{\partial f\left(\tau, \xi, \dot{\xi}\right)}{\partial \dot{\xi}}. \tag{A.9}$$

A.2 Functional integral

Let us perform averaging of functional $F[\xi]$ over the Gaussian distribution. This procedure is important for consideration of correlation functions or characteristic functions of random fields or functions, say $\xi = \xi(t)$. The Gaussian PDF $\rho(\xi)$ is the functional by itself,

$$[d\rho(\xi)] = \exp\left[-\frac{1}{4D} \int \xi^2(t) dt \right] [d\xi(t)] = \rho[\xi][d\xi(t)], \tag{A.10}$$

where $[d\xi(t)] = \prod_{\tau=0}^{t} d\xi(\tau)/\sqrt{4\pi D/d\tau}$. Note that the notation: $D[\xi(t)] \equiv [d\xi(t)]$ is used as well [4, 5]. The normalization condition is the functional integral as well,

$$\int [d\rho(\xi)] = \prod_{\tau=0}^{t} \left\{ \int_{-\infty}^{\infty} \frac{d\xi(\tau)}{\sqrt{4\pi D/d\tau}} \exp\left[-\frac{1}{4D} \xi^2(\tau) \, d\tau \right] \right\} = \prod_{\tau=0}^{t} 1 = 1. \tag{A.11}$$

Let us consider the integral $I(t) = \int [d\rho(\xi)] \, F[\xi]$. Obviously if $F[\xi] = F_1[\xi(t)]$ from Eq. (A.2a) then $I(t) = 0$. Let $F[\xi] = \xi(t)\xi(t')$, then

$$\langle \xi(t)\xi(t') \rangle = \int [d\rho(\xi)] \, \xi(t)\xi(t'),$$

and taking into account that

$$\xi(t') \exp\left[-\frac{1}{4D} \xi^2(\tau) \, d\tau \right] = -2D \frac{\delta}{\delta \xi(t')} \exp\left[-\frac{1}{4D} \xi^2(\tau) \, d\tau \right], \tag{A.12}$$

and performing integration by parts, we have

$$\langle \xi(t)\xi(t') \rangle = -2D \int \xi(t) \frac{\delta}{\delta \xi(t')} [d\rho(\xi)] = 2D \int \frac{\delta \xi(t)}{\delta \xi(t')} [d\rho(\xi)] = 2D \, \delta(t - t'). \tag{A.13}$$

Therefore, $\xi(t)$ is a delta correlated (white) Gaussian noise. Moreover, the averaging of the product $\xi(t)F[\xi]$ is

$$\left\langle \xi(t)F[\xi] \right\rangle = 2D \left\langle \frac{\delta F[\xi]}{\delta \xi(t)} \right\rangle. \tag{A.14}$$

Let $F[\xi] = F_2[\xi(t)]$ in Eq. (A.2b), then we have a chain of transformations

$$I(t) = \int \left[\iint_{t_1}^{t_2} B(\tau_1, \tau_2)\xi(\tau_1)\xi(\tau_2)\, d\tau_1 d\tau_2 \right] [d\rho(\xi)]$$

$$= -2D \int [d\rho(\xi)] \iint_{t_1}^{t_2} B(\tau_1, \tau_2)\xi(\tau_1) \frac{\delta}{\delta\xi(\tau_2)}\, d\tau_1 d\tau_2$$

$$= 2D \int [d\rho(\xi)] \iint_{t_1}^{t_2} B(\tau_1, \tau_2) \frac{\delta\xi(\tau_1)}{\delta\xi(\tau_2)}$$

$$= 2D \int [d\rho(\xi)] \iint_{t_1}^{t_2} B(\tau_1, \tau_2)\delta(\tau_2 - \tau_1)\, d\tau_1 d\tau_2$$

$$= 2D \int_{t_1}^{t_2} B(\tau, \tau)\, d\tau. \tag{A.15}$$

Let us estimate the characteristic functional, which reads

$$\Phi[\kappa(\tau)] = \left\langle \exp\left[i \int \kappa(\tau)\xi(\tau)\, d\tau \right] \right\rangle = \int \exp\left[i \int \kappa(\tau)\xi(\tau)\, d\tau \right] [d\rho(\xi)]$$

$$\int [d\xi(t)] \exp\left\{ -\frac{1}{4D} \int \left[\xi^2(\tau) - i4D\kappa(\tau)\xi(\tau) \right] d\tau \right\}$$

$$= \exp\left[-D \int \kappa^2(\tau)\, d\tau \right]. \tag{A.16}$$

From here, the correlation function (A.13) reads

$$\langle \xi(t)\xi(t') \rangle = \frac{\delta}{i\delta\kappa(t)} \frac{\delta}{i\delta\kappa(t')} \Phi[\kappa(\tau)] \bigg|_{\kappa(\tau)=0} = 2D\,\delta(t - t').$$

Appendix B

Stochastic differential equations

B.1 Itô stochastic calculus

The Langevin equation (3.126) has been considered as an illustrative example of a simplified one dimensional random dynamics in the presence of the additive Gaussian white noise $\xi(t)$,

$$dX(t) = f[X(t)]\,dt + dB(t). \tag{B.1}$$

Equation (B.1) describes Brownian motion, and in the absence of the drift term, $f = 0$, it is just a Wiener process $W(t)$, which is

$$X(t) = \int_0^t dB(\tau)\,d\tau = B(t) \equiv W(t). \tag{B.2}$$

It is supposed that $X(t)$ is continuous and while nowhere differentiable, we have $W(t + dt) - W(t) = \xi(t)\,dt$. It is important to admit that the integral in Eq. (B.2) is well/uniquely defined as a limit of partial sums $S_N = \sum_{i=1}^{N}[W(t_i) - W(t_{i-1})] = W(t_N) - W(t_0)$, which is independent of the choice of intermediate points at the partition of the interval t on subintervals Δt_i.

The situation changes dramatically for a multiplicative noise, when Eq. (B.1) becomes

$$dX(t) = f[X(t)]\,dt + g(t)\,dW(t), \tag{B.3}$$

where $g(t)$ is an arbitrary function. In general, case $g(t)$ can be an implicit function of time, $g[X(t)]$, as well. In this case the limit of partial sums $S_N = \sum_{i=1}^{N} g(\tau_i)[W(t_i) - W(t_{i-1})]$ does depend on the choice of the intermediate points $\tau_i \in \Delta t_i = t_i - t_{i-1}$. This issue is well explained and discussed with great details in Refs. [6, 7]. To explain this deficiency, we first list some properties of the Wiener process $W(t)$, taking the initial time $t = 0$

and $W(t = 0) = 0$. Then we have [7], irregularity upon continuity, non-differentiability and independence of increments. The latter follows from the Markov nature of the Wiener process and means that for all time sub-intervals Δt_i the increments $\Delta W_i = W(t_i) - W(t_{i-1})$ are independent of each other. Another important properties are relates to averages,

$$\langle W(t) \rangle = 0, \tag{B.4a}$$

$$\langle W_j(t) W_k(t) \rangle = t \, \delta_{j,k}, \tag{B.4b}$$

$$\langle W(t) W(t') \rangle = \min(t, t'). \tag{B.4c}$$

Let us obtain Eq. (B.4c) for the correlation function, which is correct for both $t > t'$ and $t' > t$. For example for $t > t'$, one obtains from Eq. (B.4c) that

$$\langle W(t) W(t') \rangle = \langle [W(t) - W(t')] W(t') \rangle + \langle W^2(t') \rangle$$
$$= \langle [W(t) - W(t')][W(t') - W(0)] \rangle + \langle W^2(t') \rangle = t'. \tag{B.5}$$

Using these properties, let us estimate the averaged value of the partial sum, for example for $g(t) = W(t)$. Then, we have

$$\langle S_N \rangle = \sum_{i=1}^{N} \langle W(\tau_i)[W(t_i) - W(t_{i-1})] \rangle$$
$$= \sum_{i=1}^{N} [\min(\tau_i, t_i) - \min(\tau_i, t_{i-1})] = \sum_{i=1}^{N} (\tau_i - t_{i-1}). \tag{B.6}$$

If for example we set $\tau_i = \alpha t_i + (1 - \alpha) t_{i-1}$, $\forall i$, where $0 \leq \alpha \leq 1$, then

$$\langle S_N \rangle = \sum_{i=1}^{N} (t_i - t_{i-1}) = t\alpha. \tag{B.7}$$

In this case the chaotic integral $\int_0^t g(\tau) \, dW(\tau)$ is not uniquely defined and depends on choice of the intermediate points. Moreover, the partial sum S_N converges to the stochastic integral in the limit $N \to \infty$ as the mean squared value. For example, for $g(t) = W(t)$

$$\int_0^t W(\tau) \, dW(\tau) = \text{ms} \cdot \lim_{N \to \infty} S_N. \tag{B.8}$$

That is the averaged value of the partial sum is first estimated and then the limit is taken.

Let us first consider the Itô stochastic integral, when $W(\tau_i) = W(t_{i-1})$ and we denote $W(t_i) \equiv W_i$ [7]. Then, we have from Eqs. (B.8) and (B.6)

$$S_N = \sum_{i=1}^{N} W_{i-1}[W_i - W_{i-1}] \equiv \sum_{i=1}^{N} W_{i-1}\Delta W_i$$

$$= \frac{1}{2}\sum_{i=1}^{N}[(W_{i-1} + \Delta W_i)^2 - W_i^2 - \Delta W_i^2]$$

$$= \frac{1}{2}[W^2(t) - W^2(t_0)] - \frac{1}{2}\sum_{i=1}^{N}\Delta W_i^2. \quad \text{(B.9)}$$

Note that $t_0 = 0$ and $W(0) = 0$. In the mean squared limit the last term is equal to t. Therefore, the Itô stochastic integral is

$$\text{Itô:} \quad \int_{t_0}^{t} W(\tau)\,dW(\tau) = \frac{1}{2}[W^2(t) - W^2(t_0) - (t - t_0)]. \quad \text{(B.10)}$$

An alternative approach has been suggested by Stratonovich, when the middle point term in the partial sum is $[W(t_i) + W(t_{i-1})]/2$. Then the Stratonovich stochastic integral in the mean squared limit reads [7]

$$\text{St:} \quad \int_{t_0}^{t} W(\tau)\,dW(\tau) = \lim_{N\to\infty} \sum_{j=1}^{N} \frac{1}{2}[W(t_i) + W(t_{i-1})][W_j - W_{j-1}]$$

$$= \frac{W^2(t) - W^2(t_0)}{2}. \quad \text{(B.11)}$$

We do not discuss all mathematical details of stochastic calculus, which can be found in Refs. [6, 7]. However, for the ensuing discussion we shall need the Itô rule/formula for an arbitrary function of $X(t)$ and t, say $F[X(t)] = F[X(t), t]$. Then, $F[X(t), t]$ obeys the stochastic differential equation according to Eq. (B.3) as follows [6]

$$dF[X(t)] = \left\{ \frac{\partial}{\partial t}F[X(t)] + f[X(t)]\frac{\partial}{\partial x}F[X(t)] + \frac{1}{2}g^2[X(t)]\frac{\partial^2}{\partial x^2}F[X(t)] \right\} dt$$

$$+ g[X(t)]\frac{\partial}{\partial x}F[X(t)]dW(t), \quad \text{(B.12)}$$

where $[dW(t)]^2 = dt$ is used in this expansion.

B.2 Fokker-Planck equation

Let us obtain the Fokker-Planck equations in the Itô and Stratonovich interpretations of the stochastic Langevin equations.

B.2.1 *Fokker-Planck equation – Itô interpretation*

Let $F(X, t)$ be an arbitrary infinitely often differentiable function with respect to X and t, which vanishes identically outside a bounded interval of the form $[x_1, x_2] \times [t_1, t_2]$. Then, taking integration with respect to time and averaging [6], we obtain from the Itô formula (B.12) that

$$\langle \{F[X(T), T] - F(X(0), 0]\} \rangle = \int_0^T \left\langle \left\{ \frac{\partial}{\partial t} F[X(t)] + f[X(t)] \frac{\partial}{\partial x} F[X(t)] \right. \right.$$
$$\left. \left. + \frac{1}{2} \frac{\partial^2}{\partial x^2} F[X(t)] g^2[X(t)] \right\} dt \right\rangle,$$
(B.13)

where it is accounted that the last term in the Itô formula does not contribute, namely

$$\int_0^T \left\langle \frac{\partial}{\partial x} F[X(t)] g[X(t)] \, dW(t) \right\rangle = 0.$$
(B.14)

Then, setting $T = \infty$, we obtain that $F[X(\infty), \infty] = F(X(0), 0] = 0$. The averaging procedure corresponds to the integration with the transition PDF $P(x, t|y, 0)$ with zero boundary conditions. Eventually, integrating Eq. (B.13) by parts and taking into account that $F[X(t), t]$ is an arbitrary functions, we obtain the Fokker-Planck equation for the transition PDF $P(x, t|y, 0)$ as follows

$$\frac{\partial}{\partial t} P(x, t|y, 0) = -\frac{\partial}{\partial x} f(x) P(x, t|y, 0) + \frac{1}{2} \frac{\partial^2}{\partial x^2} \left[g^2(x) P(x, t|y, 0) \right].$$
(B.15)

B.2.2 *Fokker-Planck equation – Stratonovich interpretation*

In the Stratonovich approach, the stochastic integral in Eq. (B.14) is not zero anymore. Let us estimate it as the partial sum [7],

$$\int_0^t \frac{\partial}{\partial x} F[X(\tau)] g[X(\tau)] \, dW(\tau) \equiv \int_0^t G[X(\tau)] \, dW(\tau)$$

$$= \text{ms} \cdot \lim_{N \to \infty} \sum_{i=1}^N G\left[\frac{X(t_i) + X(t_{i-1})}{2} \right] [W(t_i) - W(t_{i-1})].$$
(B.16)

Let us present $X(t_i)$ as the shift $X(t_i) = X(t_{i-1}) + dX(t_{i-1})$, and then we use the Itô stochastic equation for the increment

$$dX(t_i) = f[X(t_{i-1})](t_i - t_{i-1}) + g[X(t_{i-1})][W(t_i) - W(t_{i-1})].$$
(B.17)

Then performing the expansion according to the Itô formula, we have

$$G\left[\frac{X(t_i)+X(t_{i-1})}{2}\right] = G\left[X(t_i) + \tfrac{1}{2}dX(t_{i-1})\right] = G[X(t_i)]$$

$$+ \frac{1}{2}\left\{f[X(t_{i-1})]\frac{\partial}{\partial x}G[X(t_{i-1})] + \frac{1}{4}g^2[X(t_{i-1})]\frac{\partial^2}{\partial x^2}G[X(t_{i-1})]\right\}(t_i - t_{i-1})$$

$$+ \frac{1}{2}g[X(t_{i-1})]\frac{\partial}{\partial x}G[X(t_{i-1})][W(t_i) - W(t_{i-1})]. \qquad (B.18)$$

Eventually, the Stratonovich interpretation of the stochastic equation leads to the additional term in the drift in its Itô analog. That is, for the Stratonovich stochastic equation

$$\text{St:} \quad dX(t) = F[X(t)]\,dt + G[X(t)]\,dW(t), \qquad (B.19)$$

its Itô analog reads

$$\text{Itô:} \quad dX(t) = \left\{F[X(t)] + \frac{1}{4}\frac{\partial}{\partial x}G^2[X(t)]\right\}dt + G[X(t)]\,dW(t), \qquad (B.20)$$

which eventually appears in the drift term in the FPE (B.15),

$$f(x) \rightarrow f(x) + \frac{1}{4}\frac{d}{dx}G^2(x). \qquad (B.21)$$

Note that Eq. (B.21) coincides exactly with the drift term obtained in the framework of correlation functions approach [8].

We also note that analogous analysis can be performed for $\tau_i = t_i$, which corresponds to the Klimontovich-Hänggi interpretation, see Chapter 5. Disregarding the drift term, $f(x) = 0$ and setting $G(x) = \sqrt{2\mathcal{D}(x)}$, one arrives at the Langevin equation (5.1) with the position dependent diffusion coefficient $\mathcal{D}(x)$ [9],

$$\dot{x}(t) = \sqrt{2\,\mathcal{D}(x)}\,\xi(t). \qquad (B.22)$$

Therefore, according to the Itô stochastic calculus, considered above for various interpretations of the stochastic integrals, the Langevin equation (B.22) corresponds to the Fokker-Planck equation for the transition PDF

$$\frac{\partial}{\partial t}P(x,t|y,0) = \frac{\partial}{\partial x}\left\{\mathcal{D}(x)^{1-A/2}\frac{\partial}{\partial x}\left[\mathcal{D}(x)^{A/2}P(x,t|y,0)\right]\right\}, \qquad (B.23)$$

which coincides exactly with Eq. (5.2) for the PDF $P(x,t)$. Here, the cases with $A = 2, 1, 0$ correspond to the Itô, Stratonovich, and Klimontovich-Hänggi interpretations, respectively.

Appendix C

Large deviation principle

C.1 Large deviation function

We present a brief insight into the large deviation principle, considered as the Legendre transform according to the Gärtner-Ellis theorem. We borrow this material from the explanatory review [10]. Let us consider a real random variable X_t parametrized by the positive t with the probability $P(X_t \in [a, a+da])$ to be in the interval da, or with the PDF $p_0(X_t)$. Then, the scaled cumulant generating function of X_t is defined by the limit [10]

$$\lambda(\kappa) = \lim_{t \to \infty} \frac{1}{t} \ln \left[\langle e^{t\kappa X_t} \rangle \right], \tag{C.1}$$

where $\kappa \in \mathbb{R}$ and

$$\langle e^{t\kappa X_t} \rangle = \int e^{t\kappa a} P(X_t \in [a, a+da]) \int e^{t\kappa a} p_0(X_t = a)\, da. \tag{C.2}$$

If $\lambda(\kappa)$ is a differentiable function, then according to the Gärtner-Ellis theorem, X_t satisfies a large deviation principle, that is

$$P(X_t \in [a, a+da]) = p_0(X_t)\, da \asymp e^{tI(a)}\, da, \tag{C.3}$$

where a rate function $I(a)$ is given by the Legendre transform as follows

$$I(a) = \max_{\kappa \in \mathbb{R}} \{\kappa a - \lambda(\kappa)\}. \tag{C.4}$$

Equation (C.4) supposes that $\lambda(\kappa)$ is a convex function.

Let us apply the large deviation principle employed in Sec. 6.2.3. To this end one considers the integration as the averaging procedure, which determines the scaled cumulant generating function

$$\lambda\tau = \lim_{t \to \infty} \frac{1}{t} \ln \left[\langle e^{-t\kappa\tau} \rangle \right] = \int_0^1 f(t')\, e^{-t\kappa t'}\, e^{-\Phi(t,t')}\, dt'. \tag{C.5}$$

From the Gärtner-Ellis theorem, we have that the asymptotics of the PDF and the rate function are

$$p_0(\tau, t) \asymp e^{-tI(\tau_0)}, \tag{C.6a}$$

$$I(\tau_0) = \min_{\kappa}\{\kappa\tau - \lambda(\kappa)\}. \tag{C.6b}$$

Here we use $\min_{\kappa}$ instead of $\max_{\kappa}$, since $\lambda(\kappa)$ is concave.

C.2 Legendre transform

We borrow this brief insight into the Legendre transform from lectures [11].

Definition C.1. A function $f(x)$ defined over some interval $x \in [a, b]$ is called *convex* over $[a, b]$ if for every choice of numbers $\{x_1, x_2\} \in [a, b]$ and every $\epsilon \in [0, 1]$, we have

$$f(\epsilon x_1 + (1 - \epsilon)x_2) \leq \epsilon f(x_1) + (1 - \epsilon)f(x_2). \tag{C.7}$$

Geometrically, this means that the graph of the function lies below the line segment joining any two points of the graph.

If a function $-f(x)$ is convex, then its negative $f(x)$ is called *concave*.

Some properties of convex/concave functions are in turn. Convex and concave functions are necessarily continuous. These are differentiable except zero measure points. If a convex function $f(x)$ is everywhere differentiable, then it lies above any of its tangents. If a convex function $f(x)$ is everywhere differentiable twice, then its second derivative is non-negative.

Let us consider a transformation from $f(x)$ to a new function $\check{f}(p)$, where the new variables is

$$p = \frac{df(x)}{dx} \equiv f'(x).$$

This transformation is reversible iff $f(x)$ is ether convex or concave, and it is known as the Legendre transformation, defined as follows

Definition C.2.

$$\check{f}(p) := \begin{cases} \max_x\{xp - f(x)\} & \text{if } f(x) \text{ is convex,} \\ \min_x\{xp - f(x)\} & \text{if } f(x) \text{ is concave.} \end{cases} \tag{C.8}$$

This definition determines the extremum condition for the function $xp - f(x)$, which leads to the definition of the new variable p. Therefore, the extremum solution of $x = x(p)$ determines the function $\check{f}(p) = x(p)p - f(x(p))$. Then, we have

Proposition C.1. *The condition*

$$\breve{\tilde{f}} = f \tag{C.9}$$

is valid for both convex and concave functions f.

Proof. Twice Legendre transforming $f(x)$, we have

$$\breve{\tilde{f}} = \max_q \{pq - \tilde{f}(p)\} = p(q)q - \tilde{f}(p(q)). \tag{C.10}$$

Taking into account that $\tilde{f}(p) = x(p)p - f(x(p))$ and inserting this expression in Eq. (C.10), and accounting that $x(p(q)) = q$, we obtain

$$\breve{\tilde{f}} = p(q)q - \tilde{f}(p(q)) = p(q)q - [x(p(q))p(q) - f(x(p(q)))] = f(q).$$

$\square$

Appendix D

Fractals and fractal dimension

The concept of a fractal was introduced by mathematicians, such as Cantor and Peano, at the end of the 19th century. The relation of fractals to natural phenomena was emphasized by Mandelbrot [12], who also coined the term "fractal" in 1975. This idea was first expressed by Kolmogorov in [13] for fractional Brownian motion, and later rediscovered by Mandelbrot and Van Ness, who explored this phenomenon in greater detail in [14]. A fractal is a set which has a fractal dimension. To make this definition precise, it is necessary to define a dimension applicable to a large class of sets. Mathematically, a fractal can be defined as a set of points in a d dimensional metric space for which it is impossible to determine any conventional measure with integer dimension, i.e., length, area, or volume. The rigorous mathematical concept of the fractal dimension, the so-called Hausdorff-Besicovich dimension $\dim_H S$, relates to the fractional Hausdorff measure [15] $\mu_\nu(S)$ of a set S such that

$$\dim_H S = \inf\{\nu : \ \mu_\nu(S) = 0\} = \sup\{\nu : \mu_\nu(S) = \infty\}, \qquad \text{(D.1)}$$

which yields the definition

$$\mu_\nu(S) = \begin{cases} \infty & \text{if } \nu > \dim_H S, \\ 0 & \text{if } \nu < \dim_H S. \end{cases} \qquad \text{(D.2)}$$

The mathematical concept of the Hausdorff-Besicovich dimension is not really useful for most physical applications. In physics, fractals can be characterized not only by the Hausdorff-Besicovich dimension but by a number of other dimensions that are easy to find in experiments, real or numerical, and that provide a versatile description of the object's properties. One possible simplified dimension, proposed by Kolmogorov, is the *capacity* or *fractal dimension* related to box counting. There is vast literature on the subject. We cite only few references where this subject is well discussed for random walks and chaotic dynamics [12, 16, 17, 18]. Let $\mathcal{S} \subset R^d$ be a compact set. For every $\epsilon > 0$ denote by $N(\epsilon)$ the minimum number of d dimensional hypercubes, cubes of linear size ϵ, needed to cover $\mathcal{S}$. The fractal dimension of $\mathcal{S}$ is defined by the formula

$$\nu_f = \dim \mathcal{S} = \lim_{\epsilon \to 0} \frac{\log N(\epsilon)}{\log(1/\epsilon)}, \qquad (D.3)$$

if the limit exists. This coincides with the *scaling dimension*, which is defined when $\mathcal{S}$ can be subdivided into P replicas of itself, each of which is similar to $\mathcal{S}$ and with linear magnification by a factor Q becomes congruent to $\mathcal{S}$. The scaling dimension is given by

$$\nu_s = \dim \mathcal{S} = \frac{\log P}{\log Q}. \qquad (D.4)$$

Another important definition of the fractal set dimension, which is important for random walks in the large scale asymptotics, is the mass dimension. Assigning a unit density to $\mathcal{S}$, we define the mass of the set of radius l as $m(l) \sim l^\nu$. We can also define it as a measure or volume: $\mu(l) \sim l^\nu$. Note that $\nu = \nu_f = \nu_s$, and we call it the fractal dimension of a fractal set $\mathcal{S}_\nu$.

D.1   Fractal diffusion and dynamical dimensions

The fractal dimension is an extensive characteristic of a fractal structure, which depends on a decimation procedure. Intensive dynamical characteristics, which determine the random walk properties on the fractal, are spectral and walk dimensions. To describe these dynamical characteristics, let us consider a D dimensional lattice as a basis for a $d_f = \nu$ dimensional fractal structure $\mathcal{S}_\nu \equiv \mathcal{S}_{d_f}$ embedded inside the lattice. The diffusion equation, which describes the time evolution of the probability $P_i(t)$ to find a walking particle at site i at time t is [19, 20, 21]

$$\frac{\partial P_i(t)}{\partial t} = \sum_j T_{i,j}[P_j(t) - P_i(t)] \equiv \bar{\Delta} P_i(t), \qquad (D.5)$$

where $\bar{\Delta}$ is a quasi-Laplacian

$$-\bar{\Delta}_{i,j} = \delta_{i,j} \sum_k T_{i,k} - T_{i,j} \qquad (D.6)$$

and the initial condition is $P_i(t = 0) = \delta_{i,i_0}$. Here j corresponds to the nearest neighbours of site i in the fractal structure. Note that it is not a stochastic matrix, since $\sum_j \bar{\Delta}_{ij} = 0$.

To solve Eq. (D.5), we expand $P_i(t)$ over the eigenfunctions $u_n(i) \equiv u_{E_n}(i)$ of the quasi-Laplacian with the eigenvalues/eigenspectrum E_n. In the continuous limit approximation,[1] $u_n(i) \to u_n(r) \equiv u_{E_n}(r)$, the eigenvalue problem is

$$-\bar{\Delta}u_n(r) = E_n u_n(r). \tag{D.7}$$

Therefore, the solution to the diffusion Eq. (D.5) is

$$P(r,t) = \sum_n G_n(t)u_n(r), \tag{D.8}$$

where taking into account the initial conditions, for example as $i_0 \to r = 0$, one obtains $G_n(t) \equiv G(E_n, t) = u_n^*(0)e^{-E_n t}$. Then from Eq. (D.8), we obtain

$$P(r,t) = \sum_n u_n^* u_n(r)e^{-E_n t} = \sum_{E_n} \int \delta(E - E_n)u_E^* u_E(r)e^{-Et}\, dE$$

$$= \int \rho(E)u_E^* u_E(r)e^{-Et}\, dE, \tag{D.9}$$

where $\rho(E) = \sum_{E_n} \delta(E - E_n)$ is the spectral density.

It is interesting to admit that the eigenvalue problem (D.7) relates also to the vibration spectrum ω of the fractal structure $\mathcal{S}_{d_f}$, defined by a wave equation with the same Laplacian $\bar{\Delta}$ and correspondingly with the same eigenfunctions $u_E(r)$, such that $E = \omega^2$. These modes are named *fractons* [23]. It was postulated [24] that the fractional number of modes $N(\omega)$, with frequencies less than ω, varies as

$$N(\omega) = \int_0^\omega \rho(\omega')\, d\omega' \underset{\omega \to 0}{\sim} \omega^{d_s}. \tag{D.10}$$

Correspondingly, we have for the spectral density

$$\rho(\omega) \sim \omega^{d_s - 1}, \quad \omega \to 0, \tag{D.11a}$$

$$\rho(E) \sim E^{-1 + d_s/2}, \quad E \to 0, \tag{D.11b}$$

where d_s is the spectral, or fracton dimension [23].

Note that the solution (D.9) establishes the relation between the spectral density and the return PDF as follows

$$P(0,t) \approx \int \rho(E)e^{-Et}dE. \tag{D.12}$$

---

[1] Note that the transition to the continuous r-space is more sophisticate, see Eqs. (3.27) and (3.28) in Ref. [22].

Then substituting Eq. (D.11b) in Eq. (D.12), we obtain the asymptotic behavior of the return PDF as follows

$$P(0,t) \approx t^{-d_s/2}\Gamma(d_s/2), \quad t \to \infty. \tag{D.13}$$

Let us establish a relation between the spectral and fractal dimensions. We recall that diffusion of a particle on a fractal structure can be described in terms of the effective dimension of the random walk, d_w, which relates to the transport exponent $\alpha = 2/d_w$ and the MSD reads [25]

$$\langle r^2(t) \rangle \sim t^{2/d_w}. \tag{D.14}$$

The dimensions d_s and d_w are independent values, which define the PDF $P(r,t)$ for fractal diffusion in term of the two-parameter scaling solution as follows. Presenting the PDF in the normalized form [19], which according to Eq. (D.14) reads

$$P(r,t) = r^{-d_f} g(r^{d_w}/t) \sim t^{-d_f/d_w} g\left(r^{d_w}/t\right), \tag{D.15}$$

and compare the obtained result in Eq. (D.15) with the return PDF in Eq. (D.13), we obtain the relation between the spectral dimension and the fractal dimension as follows

$$d_s = \frac{2d_f}{d_w}. \tag{D.16}$$

Note also that the normalization condition, which is the integration over the fractal measure (D.2), reads

$$\int P(r,t)\, d\mu_\nu(r) = \int r^{-d_f} g(r^{d_w}/t)\, d\mu_{d_f} = 1.$$

Appendix E

Implementation of Wolfram Mathematica

In the appendix we present some functions and transforms, which are implemented in Wolfram Mathematica and used in the main text of the book. We begin with the Fourier and the Laplace transforms, which are the main treatments in the fractional calculus.

- Fourier transform: `FourierTransform[f, x, k]`.
  In Wolfram Mathematica the Fourier transform is defined by
  $$\mathcal{F}[f(x)](k) = F(k) = \frac{1}{\sqrt{2\pi}} \int_{-\infty}^{\infty} f(z) \, e^{-\imath kx} \, dx. \qquad (E.1)$$
  For the first time it was introduced in Mathematica 4.0 in 1999.
- Inverse Fourier transform: `InverseFourierTransform[F, k, x]`.
  In Wolfram Mathematica the inverse Fourier transform is defined by
  $$\mathcal{F}^{-1}[F(k)](x) = f(x) = \frac{1}{\sqrt{2\pi}} \int_{-\infty}^{\infty} F(k) \, e^{\imath kx} \, dk. \qquad (E.2)$$
  For the first time it was introduced in Mathematica 4.0 in 1999.
- Fourier cosine transform: `FourierCosTransform[f, x, k]`.
  In Wolfram Mathematica the Fourier cosine transform is defined by
  $$\mathcal{F}[f(x)](k) = F(k) = \sqrt{\frac{2}{\pi}} \int_{-\infty}^{\infty} f(z) \, \cos(kx) \, dx. \qquad (E.3)$$
  For the first time it was introduced in Mathematica 4.0 in 1999.
- Inverse Fourier cosine transform:
  `InverseFourierCosTransform[F, k, x]`.
  In Wolfram Mathematica the inverse Fourier cosine transform is defined by
  $$\mathcal{F}^{-1}[F(k)](x) = f(x) = \sqrt{\frac{2}{\pi}} \int_{-\infty}^{\infty} F(k) \, \cos(kx) \, dk. \qquad (E.4)$$
  For the first time it was introduced in Mathematica 4.0 in 1999.

- Fourier sine transform: `FourierSinTransform[f, x, k]`.
  In Wolfram Mathematica the Fourier sine transform is defined by

$$\mathcal{F}[f(x)](k) = F(k) = \sqrt{\frac{2}{\pi}} \int_{-\infty}^{\infty} f(z)\,\sin(kx)\,dx. \qquad (E.5)$$

  For the first time it was introduced in Mathematica 4.0 in 1999.
- Inverse Fourier sine transform:
  `InverseFourierSinTransform[F, k, x]`.
  In Wolfram Mathematica the inverse Fourier sine transform is defined by

$$\mathcal{F}^{-1}[F(k)](x) = f(x) = \sqrt{\frac{2}{\pi}} \int_{-\infty}^{\infty} F(k)\,\sin(kx)\,dk. \qquad (E.6)$$

  For the first time it was introduced in Mathematica 4.0 in 1999.
- Laplace transform: `LaplaceTransform[f, t, s]`.
  For the first time it was introduced in Mathematica 4.0 in 1999.
- Inverse Laplace transform: `InverseLaplaceTransform[F, s, t]`.
  For the first time it was introduced in Mathematica 4.0 in 1999.
- Mellin transform: `MellinTransform[f, x, q]`.
  For the first time it was introduced in Mathematica 11.0 in 2016.
- Inverse Mellin transform: `InverseMellinTransform[F, q, x]`.
- Exponential function: `Exp[z]`.
- Error function: `Erf[z]`.
  For the first time it was introduced in Mathematica 1.0 in 1988.
- Complementary error function: `Erfc[z]`.
  For the first time it was introduced in Mathematica 2.0 in 1991.
- Generalized exponential integral function: `ExpIntegralE[n, z]`.
  For the first time it was introduced in Mathematica 1.0 in 1988.
- Gamma function: `Gamma[z]`.
  For the first time it was introduced in Mathematica 1.0 in 1988.
- Beta function: `Beta[a, b]`.
  For the first time it was introduced in Mathematica 1.0 in 1988.
- One parameter M-L function: `MittagLefflerE[`α`, z]`.
  For the first time it was introduced in Mathematica 9.0 in 2012.
- Two parameter M-L function: `MittagLefflerE[`α, β`, z]`.
  For the first time it was introduced in Mathematica 9.0 in 2012.
- Pochhammer symbol: `Pochhammer[`γ`, k]`.
  For the first time it was introduced in Mathematica 1.0 in 1988.

- Fox H-function:

  FoxH[$\{\{\{a_1, A_1\}, \ldots, \{a_n, A_n\}\}, \{\{a_{n+1}, A_{n+1}\}, \ldots, \{a_p, A_p\}\}\}$,

  $\{\{\{b_1, B_1\}, \ldots, \{b_m, B_m\}\}, \{\{b_{m+1}, B_{m+1}\}, \ldots, \{b_q, B_q\}\}\}, z]$.

  For the first time it was introduced in Mathematica 12.3 in 2021.
- Normal distribution: NormalDistribution[μ, σ].

  For the first time it was introduced in Mathematica 6.0 in 2007.
- Log-normal distribution: LogNormalDistribution[μ, σ].

  For the first time it was introduced in Mathematica 6.0 in 2007.

E.1 Numerical inversion of Laplace transform

In the next examples, we illustrate NumericalInversion.m package from
the Mathematica [26], where the inverse Laplace transform techniques by
Durbin [27], Stehfest [28, 29], Weeks [30], Piessens [31], and Crump [32]
can be used.

Example E.1. We use the Durbin technique of the inverse Laplace transform of

$$F(s) = \frac{s^{\alpha-1}}{s^\alpha + \lambda}, \quad \lambda > 0,$$

which is the Laplace image of the associated one parameter M-L function,

$$f(t) = E_\alpha\left(-\lambda t^\alpha\right),$$

see relations (1.137) and (1.138). Both plots, obtained by the numerical
Laplace inversion and by using MittagLefflerE[α, z] are exactly the same,
and are presented in Fig. E.1. For $\alpha = 1$, the inverse Laplace transform
yields exponential function.

```
Clear["Global`*"]
<< NumericalInversion.m
F[α_,λ_,s_]:=s^(α-1)/(s^α+λ); f[α_,λ_,t_]:=Durbin[F[α,λ,s]s,t];
Plot[{f[1/4,1,t],f[1,1,t],f[7/4,1,t]},{t,0,12},PlotRange → All,
Frame → True,FrameLabel → {"t","f(t)"},
FrameStyle → (FontFamily → "Helvetica"),LabelStyle → (FontSize → 12),
PlotStyle → {{Black,Dashing[{Large,Medium}]},Black,
{Black,Dashing[{0,Small,Large,Small}]}}]
```

```
Plot [{MittagLefflerE [1/4,-t^(1/4)] ,Exp[-t],MittagLefflerE [7/4,-t^(7/4)]},
{t,0,12},PlotRange → All,Frame → True,FrameLabel → {"t","f(t)"},
FrameStyle → (FontFamily → "Helvetica"),LabelStyle → (FontSize → 12),
PlotStyle → {{Black,Dashing[{Large,Medium}]},Black,
{Black,Dashing[{0,Small,Large,Small}]}}]
```

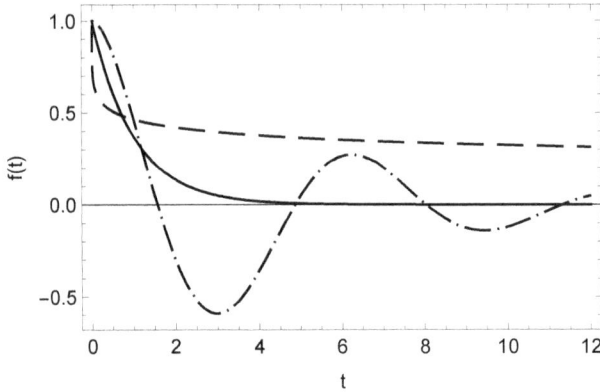

Fig. E.1 Graphical representation of the one parameter M-L function for $\lambda = 1$ and $\alpha = 1/4$ (dashed line), $\alpha = 1$ (exponential function – solid line), and $\alpha = 7/4$ (dot-dashed line).

Example E.2. The numerical inverse Laplace transform is applied for

$$F(s) = \frac{s^{\alpha-\beta}}{s^{\alpha} + \lambda}, \quad \lambda > 0,$$

where the Weeks technique is used. The obtained result is compared with the exact expression for the associated two parameter M-L function,

$$f(t) = t^{\beta-1} E_{\alpha,\beta}\left(-\lambda t^{\alpha}\right),$$

according to Eqs. (1.154) and (1.155), and the Fox H-function,

$$f(t) = t^{\beta-1} H_{1,2}^{1,1}\left[\lambda t^{\alpha} \left| \begin{array}{c} (0,1) \\ (0,1), (1-\beta,\alpha) \end{array} \right.\right],$$

see Eq. (2.43). The plots, obtained by the numerical Laplace inversion, and by MittagLefflerE[α, β, z] and FoxH[{...}, z] are exactly the same and shown in Fig. E.2.

```
Clear["Global`*"]
<< NumericalInversion.m
F[α_, β_, λ_, s_]:=s^(α-β)/(s^α+λ); f[α_, β_, λ_, t_]:=Weeks[F[α, β, λ, s], s, t];

Plot[{f[3/2, 3/2, 1, t], f[3/4, 1, 1, t], f[7/4, 1/2, 1, t]}, {t, 0, 12},
PlotRange → {-0.8, 1.5}, Frame → True, FrameLabel → {"t", "f(t)"},
FrameStyle → (FontFamily → "Helvetica"), LabelStyle → (FontSize → 12),
PlotStyle → {{Black, Dashing[{Large, Medium}]}, Black,
{Black, Dashing[{0, Small, Large, Small}]}}]
```

Plot $\left[\left\{t^{1/2} \, \text{MittagLefflerE} \left[3/2, 3/2, -t^{3/2}\right], \text{MittagLefflerE} \left[3/4, 1, -t^{3/4}\right],\right.\right.$
$t^{-1/2} \, \text{MittagLefflerE} \left[7/4, 1/2, -t^{7/4}\right]\}, \{t, 0, 12\}, \text{PlotRange} \to \{-0.8, 1.5\},$
Frame $\to$ True, FrameLabel $\to$ {"t", "f(t)"}, FrameStyle $\to$
(FontFamily $\to$ "Helvetica"), LabelStyle $\to$ (FontSize $\to$ 12),
PlotStyle $\to$ {{Black, Dashing[{Large, Medium}]}, Black,
{Black, Dashing[{0, Small, Large, Small}]}}]

Plot[{t$^{1/2}$ FoxH[{{{0, 1}}, {}}, {{{0, 1}}, {{1 − 3/2, 3/2}}}, t$^{3/2}$],
FoxH[{{{0, 1}}, {}}, {{{0, 1}}, {{1 − 1, 3/4}}}, t$^{3/4}$],
t$^{-1/2}$ FoxH[{{{0, 1}}, {}}, {{{0, 1}}, {{1 − 1/2, 7/4}}}, t$^{7/4}$]},
{t, 0, 12}, PlotRange $\to$ {−0.8, 1.5}, Frame $\to$ True, FrameLabel $\to$ {"t", "f(t)"},
FrameStyle $\to$ (FontFamily $\to$ "Helvetica"), LabelStyle $\to$ (FontSize $\to$ 12),
PlotStyle $\to$ {{Black, Dashing[{Large, Medium}]}, Black,
{Black, Dashing[{0, Small, Large, Small}]}}]

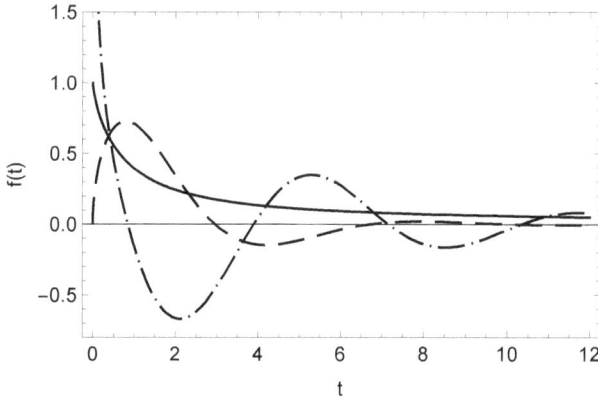

Fig. E.2 Graphical representation of the associated two parameter M-L function for $\lambda = 1$, and $\alpha = 3/2$, $\beta = 3/2$ (dashed line), $\alpha = 3/4$, $\beta = 1$ (solid line), and $\alpha = 7/4$, $\beta = 1/2$ (dot-dashed line).

Example E.3. The Weeks technique of the inverse Laplace transform is demonstrated for the image

$$F(s) = \frac{s^{\alpha\gamma - \beta}}{(s^\alpha + \lambda)^\gamma}, \quad \lambda > 0.$$

The obtained result is compared with the exact expression for the Fox H-function,

$$f(t) = t^{\beta - 1} E_{\alpha,\beta}^\gamma \left(-\lambda t^\alpha\right) = \frac{1}{\Gamma(\gamma)} t^{\beta - 1} H_{1,2}^{1,1} \left[\lambda t^\alpha \, \middle| \, \begin{array}{l} (1 - \gamma, 1) \\ (0, 1), (1 - \beta, \alpha) \end{array}\right],$$

see Eq. (2.45). Note that the three parameter M-L function is not implemented in Wolfram Mathematica. Both plots, obtained by the numerical Laplace inversion and by FoxH[{...}, z] are exactly the same and shown in Fig. E.3.

```
Clear["Global`*"]
<< NumericalInversion.m
F[α_, β_, γ_, λ_, s_]:=s^(αγ−β)/(s^α+λ)^γ; f[α_, β_, γ_, λ_, t_]:=Weeks[F[α, β, γ, λ, s], s, t];

Plot[{f[1/2, 1, 3/2, 1, t], f[3/2, 3/2, 3/2, 1, t], f[1, 3/4, 5/4, 1, t]}, {t, 0, 12},
  PlotRange → {−0.5, 1.3}, Frame → True, FrameLabel → {"t", "f(t)"},
  FrameStyle → (FontFamily → "Helvetica"), LabelStyle → (FontSize → 12),
  PlotStyle → {{Black, Dashing[{Large, Medium}]}, Black,
  {Black, Dashing[{0, Small, Large, Small}]}}]


Plot[{1/Gamma[3/2] FoxH[{{{1 − 3/2, 1}}, {}, {{{0, 1}}, {{1 − 1, 1/2}}}, t^(1/2)],
  t^(1/2)/Gamma[3/2] FoxH[{{{1 − 3/2, 1}}, {}, {{{0, 1}}, {{1 − 3/2, 3/2}}}},
  t^(3/2)], t^(−1/4)/Gamma[5/4] FoxH[{{{1 − 5/4, 1}}, {}, {{{0, 1}}, {{1 − 3/4, 1}}}, t]},
  {t, 0, 12}, PlotRange → {−0.5, 1.3}, Frame → True, FrameLabel → {"t", "f(t)"},
  FrameStyle → (FontFamily → "Helvetica"), LabelStyle → (FontSize → 12),
  PlotStyle → {{Black, Dashing[{Large, Medium}]}, Black,
  {Black, Dashing[{0, Small, Large, Small}]}}]
```

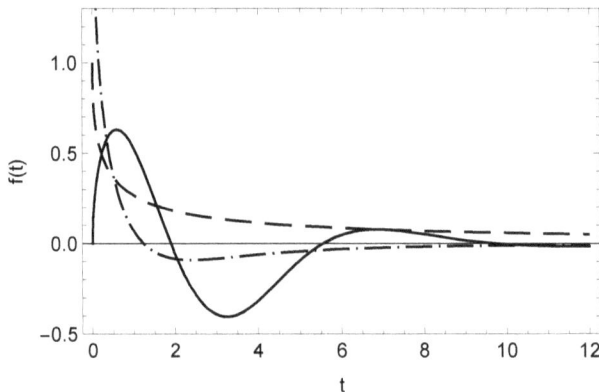

Fig. E.3 Graphical representation of the associated three parameter M-L function for $\lambda = 1$, and $\alpha = 1/2$, $\beta = 1$, $\gamma = 3/2$ (dashed line), $\alpha = 3/2$, $\beta = 3/2$, $\gamma = 3/2$ (solid line), and $\alpha = 1$, $\beta = 3/4$, $\gamma = 5/4$ (dot-dashed line).

Example E.4. The numerical inverse Laplace transform is applied for

$$F(s) = \frac{s^{-\beta}}{1 \pm \sum_{j=1}^{n} \lambda_j s^{-\alpha_j}}, \qquad \alpha_j, \lambda_j > 0, \tag{E.7}$$

where the Crump technique is used. It is known that the inverse Laplace transform of Eq. (E.7) results in the associated multinomial M-L function, see Refs. [33, 34, 35],

$$f(t) = \mathcal{L}^{-1} \left[\frac{s^{-\beta}}{1 \pm \sum_{j=1}^{n} \lambda_j s^{-\alpha_j}} \right] = \mathcal{E}_{(\alpha_1, \alpha_2, \ldots, \alpha_n), \beta} \left(t; \pm \lambda_1, \pm \lambda_2, \ldots, \pm \lambda_n \right). \tag{E.8}$$

The associated multinomial M-L function is defined as follows

$$\mathcal{E}_{(\alpha_1, \alpha_2, \ldots, \alpha_n), \beta} \left(t; \pm \lambda_1, \pm \lambda_2, \ldots, \pm \lambda_n \right)$$
$$= t^{\beta-1} E_{(\alpha_1, \alpha_2, \ldots, \alpha_n), \beta} \left(\mp \lambda_1 t^{\alpha_1}, \mp \lambda_2 t^{\alpha_2}, \ldots, \mp \lambda_n t^{\alpha_n} \right), \tag{E.9}$$

where

$$E_{(\alpha_1, \alpha_2, \ldots, \alpha_n), \beta} \left(z_1, z_2, \ldots, z_n \right)$$
$$= \sum_{k=0}^{\infty} \sum_{\substack{l_1 + l_2 + \cdots + l_n = k \\ l_1 \geq 0, l_2 \geq 0, \ldots, l_n \geq 0}} \binom{k}{l_1, \ldots, l_n} \frac{\prod_{i=1}^{n} z_i^{l_i}}{\Gamma \left(\beta + \sum_{i=1}^{n} \alpha_i l_i \right)} \tag{E.10}$$

is the multinomial M-L function, and

$$\binom{k}{l_1, \ldots, l_n} = \frac{k!}{l_1! l_2! \ldots l_n!}$$

are the multinomial coefficients.

From the definition of the associated multinomial M-L function (E.9), one finds that for $n = 1$ (i.e., $\lambda_1 = \lambda$, $\alpha_1 = \alpha$) it corresponds to the associated two parameter M-L function (1.154),

$$\mathcal{E}_{(\alpha), \beta} \left(t; \pm \lambda \right) = \mathcal{L}^{-1} \left[\frac{s^{-\beta}}{1 \pm \lambda s^{-\alpha}} \right]$$
$$= \mathcal{L}^{-1} \left[\frac{s^{\alpha-\beta}}{s^{\alpha} \pm \lambda} \right]$$
$$= t^{\beta-1} E_{\alpha, \beta} \left(\mp \lambda_1 t^{\alpha_1} \right) = \mathcal{E}_{\alpha, \beta} \left(t; \pm \lambda \right). \tag{E.11}$$

Moreover, for $n = 2$, applying the series expansion approach (see Ref. [36]), we have

$$\mathcal{E}_{(\alpha_1,\alpha_2),\beta}\left(t;\pm\lambda_1,\pm\lambda_2\right) = \mathcal{L}^{-1}\left[\frac{s^{-\beta}}{1 \pm \lambda_1 s^{-\alpha_1} \pm \lambda_2 s^{-\alpha_2}}\right]$$

$$= \mathcal{L}^{-1}\left[\frac{s^{-\beta}}{1 \pm \lambda_1 s^{-\alpha_1}}\frac{1}{1 \pm \lambda_2 \frac{s^{-\alpha_2}}{1\pm\lambda_1 s^{-\alpha_1}}}\right]$$

$$= \sum_{k=0}^{\infty}(\mp\lambda_2)^k \, \mathcal{L}^{-1}\left[\frac{s^{-(\alpha_2-\alpha_1)k+\alpha_1-\beta}}{\left(s^{\alpha_1} \pm \lambda_1\right)^{k+1}}\right]$$

$$= \sum_{k=0}^{\infty}(\mp\lambda_2)^k t^{\alpha_2 k+\beta-1} E^{k+1}_{\alpha_1,\alpha_2 k+\beta}\left(\mp\lambda_1 t^{\alpha_1}\right)$$

$$= \sum_{k=0}^{\infty}(\mp\lambda_2)^k \mathcal{E}^{k+1}_{\alpha_1,\alpha_2 k+\beta}(t;\pm\lambda_1), \qquad (E.12)$$

where we also use the Laplace transform (1.174) of the associated three parameter M-L function. Thus, the associated multinomial M-L function (E.9) reduces to infinite series of the associated three parameter M-L functions (1.173), which is shown to be convergent (see Appendix C in Ref. [37], as well as Ref. [38]).

In Fig. E.4 we give graphical representation of the associated multinomial M-L function for $n = 2$ by using the numerical inverse Laplace transform. We note that for the graphical representation one can also use the result (E.12) by truncation of the infinite series.

The (associated) multinomial M-L function has many applications in the theory of multi term fractional differential equations, fractional diffusion and generalized Langevin equations, see Refs. [33, 35, 39, 40, 41, 42].

```
Clear["Global`*"]
<< NumericalInversion.m
F[α1_,α2_,β_,λ1_,λ2_,s_]:=(s^-β)/(1+λ1 s^-α1+λ2 s^-α2);
f[α1_,α2_,β_,λ1_,λ2_,t_]:=Crump[F[α1,α2,β,λ1,λ2,s],s,t];
```

```
Plot[{f[1,2,1,1,1,t],f[1/2,3/4,7/8,1,1,t],f[5/4,7/4,7/4,1,1,t]},
{t,0,12},PlotRange → {-0.4,1.2},Frame → True,FrameLabel → {"t","f(t)"},
FrameStyle → (FontFamily → "Helvetica"),LabelStyle → (FontSize → 12),
PlotStyle → {{Black,Dashing[{Large,Medium}]},Black,
{Black,Dashing[{0,Small,Large,Small}]}}]
```

Example E.5. We apply the Stehfest technique for the numerical inverse

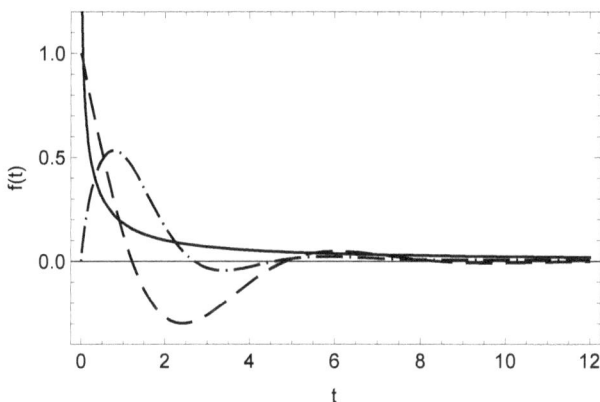

Fig. E.4 Graphical representation of the associated multinomial M-L function for $\lambda_1 = \lambda_2 = 1$, and $\alpha_1 = 1$, $\alpha_2 = 2$, $\beta = 1$ (dashed line), $\alpha_1 = 1/2$, $\alpha_2 = 3/4$, $\beta = 7/8$ (solid line), and $\alpha_1 = 5/4$, $\alpha_2 = 7/4$, $\beta = 7/4$ (dot-dashed line).

Laplace transform of

$$\hat{P}(x,s) = \frac{s^{\alpha/2-1}}{2\sqrt{\mathcal{D}}} e^{-\sqrt{\frac{s^\alpha}{\mathcal{D}}}|x|}.$$

The result is presented in Fig. E.5. The exact result, which is the solution to the time fractional diffusion equation, is given by the Fox H-function, see Eq. (3.122),

$$P(x,t) = \frac{1}{2|x|} H_{1,1}^{1,0}\left[\frac{|x|}{\sqrt{\mathcal{D}t^{\alpha/2}}} \, \middle| \, \begin{matrix}(1,\alpha/2)\\(1,1)\end{matrix}\right]. \tag{E.13}$$

The cusp at the origin is clearly seen in the figure.

```
Clear["Global`*"]
<< NumericalInversion.m
F[α_,d_,x_,s_]:=s^{α/2-1}/(2√d) Exp[-√s^α/dAbs[x]];
f[α_,d_,x_,t_]:=Stehfest[F[α,d,x,s],s,t];

Plot[{f[1/2,1,x,0.1], f[1/2,1,x,1], f[1/2,1,x,10]}, {x,-5,5},
PlotRange → All, Frame → True, FrameLabel → {"x","P(x,t)"},
FrameStyle → (FontFamily → "Helvetica"), LabelStyle → (FontSize → 12),
PlotStyle → {Black, {Black, Dashing[{Large, Medium}]},
{Black, Dashing[{0, Small, Large, Small}]}}]]
```

Example E.6. In Wolfram Mathematica, the numerical inverse Laplace transform can be implemented together with the numerical inverse Fourier

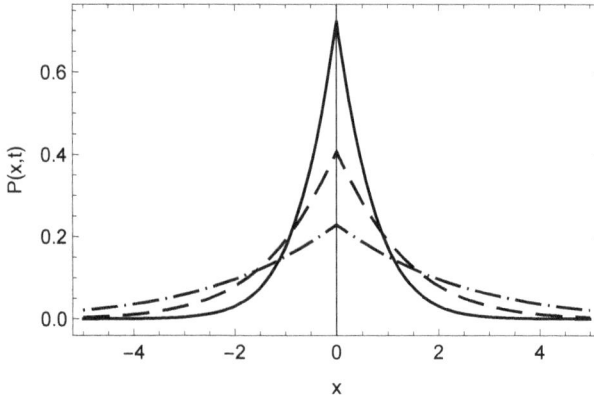

Fig. E.5 Graphical representation of (E.13) for $\alpha = 1/2$, $\mathcal{D} = 1$, and $t = 0.1$ (solid line), $t = 1$ (dashed line), $t = 10$ (dot-dashed line).

transform. We demonstrate the Stehfest technique of the inverse Fourier-Laplace transform of

$$\hat{\tilde{P}}(k, s) = \frac{s^{\alpha - 1}}{s^{\alpha} + \mathcal{D}|k|^{\beta}}, \tag{E.14}$$

which is the solution of the space-time fractional diffusion equation in Fourier-Laplace space, see Eq. (3.114). The exact result is as follows, see Eq. (3.121)

$$P(x, t) = \frac{1}{\beta|x|} H_{3,3}^{2,1} \left[\frac{|x|}{\left(\mathcal{D}t^{\alpha}\right)^{\frac{1}{\beta}}} \left| \begin{array}{c} (1, \frac{1}{\beta}), (1, \frac{\alpha}{\beta}), (1, \frac{1}{2}) \\ (1, 1), (1, \frac{1}{\beta}), (1, \frac{1}{2}) \end{array} \right. \right]. \tag{E.15}$$

The result of the numerical inverse Fourier-Laplace transform is shown in Fig. E.6. Note that, when we perform the numerical inverse Fourier transform of Eq. (E.14), we multiply the expression by a factor $\frac{1}{\sqrt{2\pi}}$ since the inverse Fourier transform in Mathematica is defined by Eq. (E.2).

```
Clear["Global`*"]
<< NumericalInversion.m
F[α_, β_, d_, s_, k_]:=s^(α-1)/(s^α + d Abs[k]^β);
f[α_, β_, d_, t_, x_]:=1/(2 Pi)^(1/2) NFourierStehfest[F[α, β, d, s, k], {s, k}, {t, x}];

Plot[{f[0.5, 2, 1, 1, x], f[0.25, 1.75, 1, 1, x], f[1, 1.5, 1, 1, x]}, {x, -5, 5},
PlotRange → All, Frame → True, FrameLabel → {"x", "P(x,t)"},
FrameStyle → (FontFamily → "Helvetica"), LabelStyle → (FontSize → 12),
PlotStyle → {Black, {Black, Dashing[{Large, Medium}]},
{Black, Dashing[{0, Small, Large, Small}]}}]
```

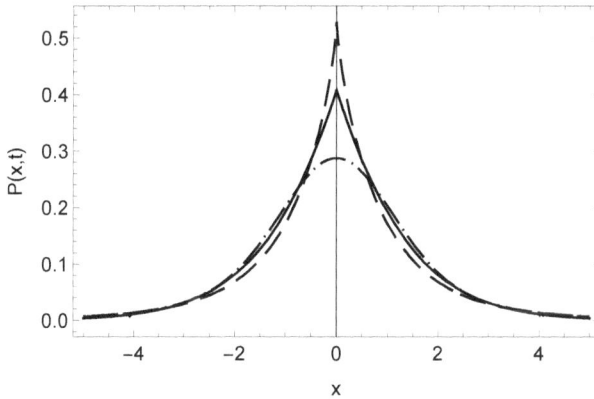

Fig. E.6 Graphical representation of (E.15) for $t = 1$, $\mathcal{D} = 1$ and $\alpha = 0.5$, $\beta = 2$ (solid line), $\alpha = 0.25$, $\beta = 1.75$ (dashed line), $\alpha = 1$, $\beta = 1.5$ (dot-dashed line).

Example E.7. We apply the numerical inverse Laplace transform to the expression

$$\hat{p}_1(x, s) = \frac{s^{-1/2}}{2D\sqrt{\frac{s^{1/2}}{\mathcal{D}} + \frac{V^2}{4\mathcal{D}^2}}} e^{\frac{V}{2\mathcal{D}}x - \sqrt{\frac{s^{1/2}}{\mathcal{D}} + \frac{V^2}{4\mathcal{D}^2}}|x|}, \tag{E.16}$$

where $V = \frac{v}{2\sqrt{\mathcal{D}_y}}$ and $\mathcal{D} = \frac{\mathcal{D}_x}{2\sqrt{\mathcal{D}_y}}$, which is the solution in Laplace space to the time fractional drift-diffusion equation (4.94). We compare the results obtained by the numerical inverse Fourier-Laplace transform with the solution (4.86) in the Fourier-Laplace space,

$$\hat{\hat{p}}_1(k_x, s) = \frac{s^{-1/2}}{s^{1/2} + \imath V k_x + \mathcal{D}k_x^2}. \tag{E.17}$$

Both methods give same results, and corresponding plots are shown in Fig. E.7.

```
Clear["Global`*"]
<< NumericalInversion.m
F[V_,d_,x_,s_]:=1/(2 d)s^{-1/2}/(s^{1/2}/d + V^2/(4 d^2))^{1/2}
Exp[V x/(2 d) - (s^{1/2}/d + V^2/(4 d^2))^{1/2} Abs[x]];
f[V_,d_,x_,t_]:=Stehfest[F[V,d,x,s],s,t];

Plot[{f[1,1,x,0.1],f[1,1,x,1],f[1,1,x,10]},{x,-5,7},
 PlotRange → All, Frame → True, FrameLabel → {"x", "p_1(x,t)"},
 FrameStyle → (FontFamily → "Helvetica"), LabelStyle → (FontSize → 12),
 PlotStyle → {Black,{Black,Dashing[{Large,Medium}]},
 {Black,Dashing[{0,Small,Large,Small}]}}]
```

Clear["Global`*"]
<< NumericalInversion.m
F[v_, d_, s_, w_]:=s$^{-0.5}$/(s$^{0.5}$ + I v w + d Abs[w]^2);
f[v_, d_, t_, x_]:=
1/(2 Pi)$^{1/2}$ NFourierStehfest[F[v, d, s, w], {s, w}, {t, x}];

Plot[{f[1, 1, 0.1, x], f[1, 1, 1, x], f[1, 1, 10, x]}, {x, −5, 7}, PlotRange → All,
Frame → True, FrameLabel → {"x", "p_1(x,t)"} , FrameStyle →
(FontFamily → "Helvetica"),
LabelStyle → (FontSize → 12), PlotStyle → {Black, {Black,
Dashing[{Large, Medium}]},
{Black, Dashing[{0, Small, Large, Small}]}}]

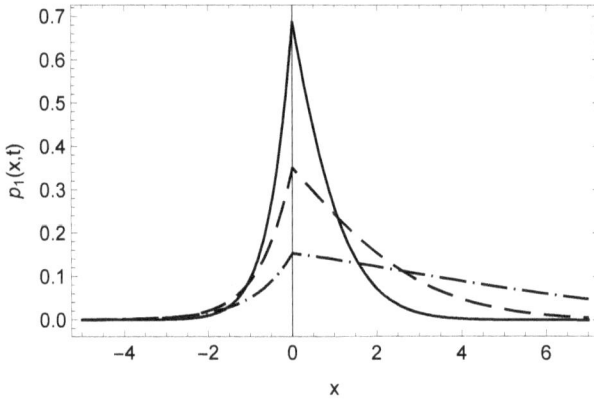

Fig. E.7 Graphical representation of the solution for $V = 1$, $\mathcal{D} = 1$, and $t = 0.1$ (solid line), $t = 1$ (dashed line), $t = 10$ (dot-dashed line).

References

[1] L. Elsgolts. *Differential Equations and the Calculus of Variations*. Moscow: Mir, 1977 (cit. on pp. 231–233).

[2] V. I. Klyatskin. *Stohasticheskie uravneniya i volny v sluchaino-neodnorodnyh sredah (Stochastic equations and waves in randomly inhomogeneous media)*. Moscow: Nauka, 1980 (cit. on p. 232).

[3] V. I. Klyatskin. *Stochastic Equations through the Eye of the Physicist Basic Concepts, Exact Results and Asymptotic Approximations*. Amsterdam: Elsevier, 2005 (cit. on p. 232).

[4] R. P. Feynman and A. R. Hibbs. *Quantum Mechanics and Path Integrals*. New York: McGraw-Hill, 1965 (cit. on p. 234).

[5] L. Schulman. *Techniques and Applications of Path Integration*. New York: Wiley, 1981 (cit. on p. 234).

[6] W. Horsthemke and R. Lefever. *Noise-Induced Transitions. Theory and Applications in Physics, Chemistry, and Biology*. Berlin: Springer-Verlag, 1984 (cit. on pp. 237, 239, 240).

[7] C. W. Gardiner. *Handbook of Stochastic Methods for Physics, Chemistry and the Natural Sciences*. 2nd. Berlin: Springer-Verlag, 1990 (cit. on pp. 237–240).

[8] S. M. Rytov, Yu. A. Kravtsov, and V. I. Tatarskii. *Principles of Statistical Radiophysics 1: Elements of Random Process Theory*. Berlin: Springer-Verlag, 1987 (cit. on p. 241).

[9] N. Leibovich and E. Barkai. "Infinite ergodic theory for heterogeneous diffusion processes". In: *Physical Review E* 99.4 (2019), p. 042138 (cit. on p. 241).

[10] H. Touchette. "The large deviation approach to statistical mechanics". In: *Physics Reports* 478.1 (2009), pp. 1–69. ISSN: 0370-1573 (cit. on p. 243).

[11] M. Desern. "Statistical Physics". In: *Lecture Course/33-765* (2012). URL: https://www.andrew.cmu.edu/course/33-765/pdf/Legendre.pdf (cit. on p. 244).

[12] B. B. Mandelbrot. *The Fractal Geometry of Nature*. New York: W.H. Freeman and Company, 1983 (cit. on pp. 247, 248).

[13] A. N. Kolmogoroff. "Wiensersche Spiralen und einige andere interessante Kurven im Hilbertschen Raum." German. In: *Doklady Akademii Nauk SSSR* 26 (1940), pp. 115–118. ISSN: 1819-0723 (cit. on p. 247).

[14] B. B. Mandelbrot and J. W. Van Ness. "Fractional Brownian Motions, Fractional Noises and Applications". In: *SIAM Review* 10.4 (1968), pp. 422–437 (cit. on p. 247).

[15] K. Falconer. *Fractal geometry: mathematical foundations and applications*. John Wiley & Sons, 2004 (cit. on p. 247).

[16] A. Lasota and M.C. Mackey. *Fractals and Noise, Stochastic Aspects of Dynamics*. New York: Springer-Verlag, 1994 (cit. on p. 248).

[17] K. T. Alligood, T. D. Sauer, and J. A. Yorke. *Chaos. An Introduction to Dynamical Systems*. New York: Springer, 2000 (cit. on p. 248).

[18] B. D. Hughes. *Random Walks and Random Environments: Volume 1: Random Walks*. Oxford: Clarendon Press, 1995 (cit. on p. 248).

[19] R. Dekeyser, A. Maritan, and A. L. Stella. "Diffusion on fractal substrates". In: *Diffusion Processes: Experiment, Theory, Simulations*. Ed. by Andrzej Pękalski. Berlin, Heidelberg: Springer, 1994, pp. 21–36. DOI: https://doi.org/10.1007/BFb0031116 (cit. on pp. 248, 250).

[20] J.-F. Gouyet. *Physics and Fractal Structures*. Paris: Paris: Masson, 1996 (cit. on p. 248).

[21] A. Barrat, M. Barthelemy, and A. Vespignani. *Dynamical processes on complex networks*. Cambridge: Cambridge University Press, 2013 (cit. on p. 248).

[22] S. Havlin and A. Bunde. "Percolation II". In: *Fractals and Disordered Systems*. Ed. by S Havlin and A. Bunde. Berlin Heidelberg: Springer-Verlag, 1996, pp. 115–177 (cit. on p. 249).

[23] S. Alexander and R Orbach. "Density of states on fractals: "fractons"". In: *J. Physique Lett.* 43.17 (1982), pp. 625–631. DOI: 10.1051/jphyslet:019820043017062500 (cit. on p. 249).

[24] D. Dhar. "Lattices of effectively nonintegral dimensionality". In: *Journal of Mathematical Physics* 18.4 (1977), pp. 577–585. DOI: 10.1063/1.523316. URL: https://doi.org/10.1063/1.523316 (cit. on p. 249).

[25] D. ben Avraham and S. Havlin. *Diffusion and Reactions in Fractals and Disordered Systems*. Cambridge, UK: Cambridge University Press, 2000 (cit. on p. 250).

[26] A. Mallet. "Numerical inversion of Laplace transform, 2000". In: *Wolfram Library Archive. http://library. wolfram. com/infocenter/MathSource/2691* (2015) (cit. on p. 253).

[27] F. Durbin. "Numerical inversion of Laplace transforms: an efficient improvement to Dubner and Abate's method". In: *Computer Journal* 17.4 (1974), pp. 371–376 (cit. on p. 253).

[28] H. Stehfest. "Algorithm 368: Numerical inversion of Laplace transforms [D5]". In: *Communications of the ACM* 13.1 (1970), pp. 47–49 (cit. on p. 253).

[29] H. Stehfest. "Remark on algorithm 368: Numerical inversion of Laplace transforms". In: *Communications of the ACM* 13.10 (1970), p. 624 (cit. on p. 253).

[30] W. T. Weeks. "Numerical inversion of Laplace transforms using Laguerre functions". In: *Journal of the ACM (JACM)* 13.3 (1966), pp. 419–429 (cit. on p. 253).

[31] R. Piessens. "Gaussian quadrature formulas for the numerical integration of Bromwich's integral and the inversion of the Laplace transform". In: *Journal of Engineering Mathematics* 5.1 (1971), pp. 1–9 (cit. on p. 253).

[32] K. S. Crump. "Numerical inversion of Laplace transforms using a Fourier series approximation". In: *Journal of the ACM (JACM)* 23.1 (1976), pp. 89–96 (cit. on p. 253).

[33] Yu. Luchko. "Operational method in fractional calculus". In: *Fractional Calculus and Applied Analysis* 2.4 (1999), pp. 463–488 (cit. on pp. 257, 258).

[34] R. Hilfer, Yu. Luchko, and Z. Tomovski. "Operational method for the solution of fractional differential equations with generalized Riemann-Liouville fractional derivatives". In: *Fract. Calc. Appl. Anal* 12.3 (2009), pp. 299–318 (cit. on p. 257).

[35] T. Sandev and Z. Tomovski. *Fractional Equations and Models*. Springer Nature, 2019 (cit. on pp. 257, 258).

[36] I. Podlubny. *Fractional differential equations: an introduction to fractional derivatives, fractional differential equations, to methods of their solution and some of their applications*. Elsevier, 1998 (cit. on p. 258).

[37] T. Sandev, Z. Tomovski, and J. L. A. Dubbeldam. "Generalized Langevin equation with a three parameter Mittag-Leffler noise". In: *Physica A: Statistical Mechanics and its Applications* 390.21-22 (2011), pp. 3627–3636 (cit. on p. 258).

[38] J. Paneva-Konovska. *From Bessel to multi-index Mittag-Leffler functions: Enumerable families, series in them and convergence*. World Scientific, 2016 (cit. on p. 258).

[39] T. Sandev et al. "Distributed-order diffusion equations and multifractality: Models and solutions". In: *Physical Review E* 92.4 (2015), p. 042117 (cit. on p. 258).

[40] T. Sandev et al. "Diffusion and Fokker-Planck-Smoluchowski equations with generalized memory kernel". In: *Fractional Calculus and Applied Analysis* 18.4 (2015), pp. 1006–1038 (cit. on p. 258).

[41] T. Sandev and Z. Tomovski. "Langevin equation for a free particle driven by power law type of noises". In: *Physics Letters A* 378.1-2 (2014), pp. 1–9 (cit. on p. 258).

[42] E. Bazhlekova. "Completely monotone multinomial Mittag-Leffler type functions and diffusion equations with multiple time-derivatives". In: *Fractional Calculus and Applied Analysis* 24.1 (2021), pp. 88–111 (cit. on p. 258).

Index

* 9 7 8 9 8 1 1 2 5 2 9 4 5 *